Mechanical Tribology and Surface Technology

Mechanical Tribology and Surface Technology

Guest Editor
Zhenpeng He

Basel • Beijing • Wuhan • Barcelona • Belgrade • Novi Sad • Cluj • Manchester

Guest Editor
Zhenpeng He
School of Aeronautical Engineering
Civil Aviation University of China
Tianjin
China

Editorial Office
MDPI AG
Grosspeteranlage 5
4052 Basel, Switzerland

This is a reprint of the Special Issue, published open access by the journal *Lubricants* (ISSN 2075-4442), freely accessible at: https://www.mdpi.com/journal/lubricants/special_issues/YYWR0OSEVJ.

For citation purposes, cite each article independently as indicated on the article page online and as indicated below:

Lastname, A.A.; Lastname, B.B. Article Title. *Journal Name* **Year**, *Volume Number*, Page Range.

ISBN 978-3-7258-3611-6 (Hbk)
ISBN 978-3-7258-3612-3 (PDF)
https://doi.org/10.3390/books978-3-7258-3612-3

© 2025 by the authors. Articles in this book are Open Access and distributed under the Creative Commons Attribution (CC BY) license. The book as a whole is distributed by MDPI under the terms and conditions of the Creative Commons Attribution-NonCommercial-NoDerivs (CC BY-NC-ND) license (https://creativecommons.org/licenses/by-nc-nd/4.0/).

Contents

Chun-Wei Yao, Ian Lian, Jiang Zhou, Paul Bernazzani, Mien Jao and Md Ashraful Hoque
Corrosion Resistance and Nano-Mechanical Properties of a Superhydrophobic Surface
Reprinted from: *Lubricants* 2025, 13, 16, https://doi.org/10.3390/lubricants13010016 1

Lijie Ma, Fengnan Li, Shijie Ba, Zunyan Ma, Xinhui Mao, Qigao Feng and Kang Yang
Investigation of Anti-Friction Properties of MoS_2 and SiO_2 Nanolubricants Based on the Friction Pairs of Inconel 718 Superalloy and YG6 Carbide
Reprinted from: *Lubricants* 2025, 13, 4, https://doi.org/10.3390/lubricants13010004 11

Robert Mašović, Suzana Jakovljević, Ivan Čular, Daniel Miler and Dragan Žeželj
Investigation of Effect of Surface Modification by Electropolishing on Tribological Behaviour of Worm Gear Pairs
Reprinted from: *Lubricants* 2024, 12, 408, https://doi.org/10.3390/lubricants12120408 28

Zhenpeng He, Lanhao Jia, Jiaxin Si, Ning Li, Hongyu Wang, Baichun Li, et al.
Research on the Sealing Performance of Segmented Annular Seals Based on Fluid–Solid–Thermal Coupling Model
Reprinted from: *Lubricants* 2024, 12, 407, https://doi.org/10.3390/lubricants12120407 46

Man-Soo Joun, Yun Heo, Nam-Hyeon Kim and Nam-Yun Kim
On Lubrication Regime Changes during Forward Extrusion, Forging, and Drawing
Reprinted from: *Lubricants* 2024, 12, 352, https://doi.org/10.3390/lubricants12100352 64

Kyungmok Kim
Development of a Machine Vision System for the Average Roughness Measurement of Shot- and Sand-Blasted Surfaces
Reprinted from: *Lubricants* 2024, 12, 339, https://doi.org/10.3390/lubricants12100339 80

Zhenpeng He, Yuhang Guo, Jiaxin Si, Ning Li, Lanhao Jia, Yuchen Zou and Hongyu Wang
Numerical Optimization Analysis of Floating Ring Seal Performance Based on Surface Texture
Reprinted from: *Lubricants* 2024, 12, 241, https://doi.org/10.3390/lubricants12070241 96

Youngjun Park, Gwanghee Hong, Sanghyun Jun, Jeongmook Choi, Taegyu Kim, Minsoo Kang and Gunhee Jang
Thermo-Fluid–Structural Coupled Analysis of a Mechanical Seal in Extended Loss of AC Power of a Reactor Coolant Pump
Reprinted from: *Lubricants* 2024, 12, 212, https://doi.org/10.3390/lubricants12060212 119

Oleksandr Stelmakh, Hongyu Fu, Serhii Kolienov, Vasyl Kanevskii, Hao Zhang, Chenxing Hu and Valerii Grygoruk
Criteria for Evaluating the Tribological Effectiveness of 3D Roughness on Friction Surfaces
Reprinted from: *Lubricants* 2024, 12, 209, https://doi.org/10.3390/lubricants12060209 136

Martin Ovsik, Klara Fucikova, Lukas Manas and Michal Stanek
Influence of Polymer Flow on Polypropylene Morphology, Micro-Mechanical, and Tribological Properties of Injected Part
Reprinted from: *Lubricants* 2024, 12, 202, https://doi.org/10.3390/lubricants12060202 154

Miroslav Tomáš, Stanislav Németh, Emil Evin, František Hollý, Vladimír Kundracik, Juliy Martyn Kulya and Marek Buber
Comparison of Friction Properties of GI Steel Plates with Various Surface Treatments
Reprinted from: *Lubricants* 2024, 12, 198, https://doi.org/10.3390/lubricants12060198 179

Jozef Jurko, Katarína Paľová, Peter Michalík and Martin Kondrát
Optimization of Sustainable Production Processes in C45 Steel Machining Using a Confocal Chromatic Sensor
Reprinted from: *Lubricants* **2024**, *12*, 99, https://doi.org/10.3390/lubricants12030099 . . . . **192**

Liang Yan, Linyi Guan, Di Wang and Dingding Xiang
Application and Prospect of Wear Simulation Based on ABAQUS: A Review
Reprinted from: *Lubricants* **2024**, *12*, 57, https://doi.org/10.3390/lubricants12020057 . . . . **212**

Article

Corrosion Resistance and Nano-Mechanical Properties of a Superhydrophobic Surface

Chun-Wei Yao [1,*], Ian Lian [2], Jiang Zhou [1], Paul Bernazzani [3], Mien Jao [4] and Md Ashraful Hoque [1]

[1] Department of Mechanical Engineering, Lamar University, Beaumont, TX 77710, USA; zhoujx@lamar.edu (J.Z.); mhoque4@lamar.edu (M.A.H.)
[2] Department of Biology, Lamar University, Beaumont, TX 77710, USA; ilian@lamar.edu
[3] Department of Chemistry and Biochemistry, Lamar University, Beaumont, TX 77710, USA; pbernazzan@lamar.edu
[4] Department of Civil and Environmental Engineering, Lamar University, Beaumont, TX 77710, USA; jaomu@lamar.edu
* Correspondence: cyao@lamar.edu; Tel.: +1-409-880-7008

Abstract: Nanoindentation has been used to characterize the mechanical and creep properties of various materials. However, research on the viscoelastic and creep properties of superhydrophobic surfaces remains limited. In this study, a superhydrophobic coating was developed and its corrosion resistance was evaluated initially. Electrochemical impedance spectroscopy (EIS) results quantitatively confirm the enhanced anti-corrosion performance of the superhydrophobic coating. Subsequently, this study investigates the creep, hardness, strain rate sensitivity, and viscoelastic behavior of the superhydrophobic surface at the nanoscale before and after accelerated corrosion exposure. Our findings reveal that during the creep tests, the logarithmic values of creep strain rate and stress exhibited a good linear relationship. Additionally, the surface retains its key viscoelastic properties (hardness, storage modulus, loss modulus, and tan δ) even after exposure to corrosion. These results highlight the surface's robustness under corrosive conditions, a crucial factor for applications requiring both mechanical integrity and environmental resilience.

Keywords: corrosion resistance; nanoscale dynamic mechanical analysis; creep; superhydrophobic coating

1. Introduction

Superhydrophobic coatings possess significant properties, including anti-biofouling [1], self-cleaning [2], and anti-icing capabilities [3], and have introduced novel approaches for corrosion mitigation [4]. A key characteristic of superhydrophobic coating is the creation of unique hierarchical micro/nanostructures on substrates, which can trap significant amounts of air in an atmospheric environment, substantially reducing the contact area between water droplets and the surface [5–8]. To develop such coatings, two major requirements must be met: high surface roughness, which can be achieved by altering the surface topography at the micro/nanoscale [9], and low surface free energy, which can be achieved by modifying the surface chemistry [10]. Various methods can be employed to fabricate superhydrophobic coating, including nanocomposite coatings [11–14], chemical vapor deposition [15], self-assembled monolayers [16], and template methods [17]. These fabrication processes ensure that both surface roughness and low surface free energy are achieved to fulfill the essential requirements for superhydrophobic coatings.

Recent studies have employed nanoindentation to characterize the mechanical or creep properties of various materials. For example, studies have utilized nanoindentation tests with different indentation loads to evaluate mechanical parameters or creep behavior, providing insights into the effects of load and loading strain rate on these properties [18–22]. Additionally, research on the nanoindentation creep behavior of nanocrystalline materials such as Ni, Ni-Fe alloys, and supercrystalline nanocomposites has revealed underlying mechanisms and apparent activation volumes [23,24]. Furthermore, nanoindentation has been extensively used to study the creep–fatigue interaction and local creep behavior of materials like P92 steel welded joints, offering valuable insights into the mechanical properties and creep deformation at the micro/nanoscale [25]. These studies underscore the importance of nanoindentation in understanding the nano-mechanical properties and creep behavior of various materials. Dynamic mechanical analysis (DMA) is crucial for measuring properties such as storage modulus, loss modulus, and damping capability (tan δ). This technique involves applying a sinusoidal stress to a material and measuring the resulting strain to determine its modulus. DMA is widely used to characterize a material's mechanical responses by tracking changes in its dynamic properties as a function of frequency, temperature, or time. Many researchers have utilized DMA as a supplementary test to evaluate the viscoelastic properties of nanomaterials used in micro/nano structured surface fabrication, alongside their tribological performance. More recently, researchers have conducted nano dynamic mechanical analysis (nano-DMA), where a sinusoidal stress is applied via a nanoindenter to characterize the viscoelastic properties of materials, such as storage modulus, loss modulus, and tangent delta [26–28]. These advanced techniques collectively contribute to a comprehensive understanding of material behavior at the nanoscale. However, reports on the nano-mechanical properties and creep behavior of superhydrophobic coatings are rarely seen. Typically, the performance of superhydrophobic coatings as corrosion inhibitors is evaluated by examining the corrosion behavior at the macroscale, often utilizing techniques like electrochemical impedance spectroscopy (EIS). Therefore, in this study, electrochemical impedance spectroscopy (EIS) was initially conducted on both steel and superhydrophobic coated substrates to understand their macroscale corrosion behavior. Subsequently, the nanohardness, strain rate sensitivity, and creep behavior of a superhydrophobic coating at the nanoscale were investigated before and after accelerated corrosion exposure. To the authors' knowledge, this study is the first to report the viscoelastic properties, including hardness, storage modulus, loss modulus, and tan δ, of a superhydrophobic surface, even after exposure to accelerated corrosion, at the nanoscale. The results provide crucial insights for the applications of the superhydrophobic coating, highlighting its potential for various industrial and engineering applications.

2. Materials and Methods

In this study, A653 steel substrates were used. The chemicals involved included acetone (Sigma Aldrich, St. Louis, MO, USA), isopropyl alcohol (Sigma Aldrich, St. Louis, MO, USA), anhydrous ethanol (Sigma Aldrich, St. Louis, MO, USA), and silane-modified hydrophobic SiO_2 nanoparticles (RX-50) with an average diameter of 55 ± 15 nm (Evonik, Piscataway, NJ, USA). Additionally, the polydimethylsiloxane (PDMS) elastomer kit (Sylgard 184) from Dow Corning was utilized.

The A653 steel substrates were degreased and cleaned ultrasonically for 20 min at room temperature. After cleaning, the substrates were rinsed thoroughly with isopropyl alcohol, ethanol, and deionized water [29]. To prepare the superhydrophobic surface solution, 1.7 g of PDMS and 11 g of toluene were combined and mixed in an ultrasonic mixer for 1 min to ensure proper dispersion. In a separate step, 2 g of silica nanoparticles was mixed with 10 g of toluene for 30 s in a planetary centrifugal mixer. The two solutions were then combined and mixed for an additional 30 s. Following this, 0.17 g of curing agent was added to the mixture, which was mixed again for 30 s in the centrifugal mixer, followed by a 30 s defoaming process. The coating was applied using a spray gun at 80 MPa and room temperature, with the nozzle positioned 10 cm from the substrate and moved horizontally across the surface. Finally, the coated substrates were cured at ambient temperature (25 °C) in a fume hood for 3 days.

A drop shape analyzer (DSA25E, Krüss, Matthews, NC, USA) was employed to measure the static contact angle using 10 µL deionized water droplets at ambient temperature, as shown in Figure 1. The surface morphology of the superhydrophobic coated steel (SCS) substrate was evaluated using an atomic force microscope (AFM, Park NX10, Park System Co., Santa Clara, CA, USA), as shown in Figure 2a, and a scanning electron microscope (SEM, JSM-7500F, JEOL, Peabody, MA, USA), as shown in Figure 2b. An accelerated corrosion test was performed with a potentiostat system (Autolab PGSTAT204, Metrohm, Riverview, FL, USA) in a 3.5 wt.% NaCl solution at 0.135 V vs. open circuit potential for 90 min. A silver/silver chloride (Ag/AgCl) electrode served as the reference electrode, while a platinum wire mesh and the test specimen functioned as the counter electrode and working electrode, respectively. EIS measurements were conducted in a 3.5 wt% NaCl solution, with frequencies spanning from 0.01 Hz to 100 kHz and a wave amplitude of 10 mV, at room temperature. Nanoindentation tests and nano dynamic mechanical analysis were performed using a TI 980 TriboIndenter (Bruker, Eden Prarie, MN, USA) equipped with a three-sided Berkovich diamond tip before and after 90 min of accelerated corrosion. All creep and nano-DMA tests were conducted at a maximum nanoindentation force of 1000 µN at an ambient temperature of 25 °C. The thickness of the superhydrophobic surface was measured to be 35 µm using a digital coating thickness meter (Elcometer). At least three tests were performed for each measurement.

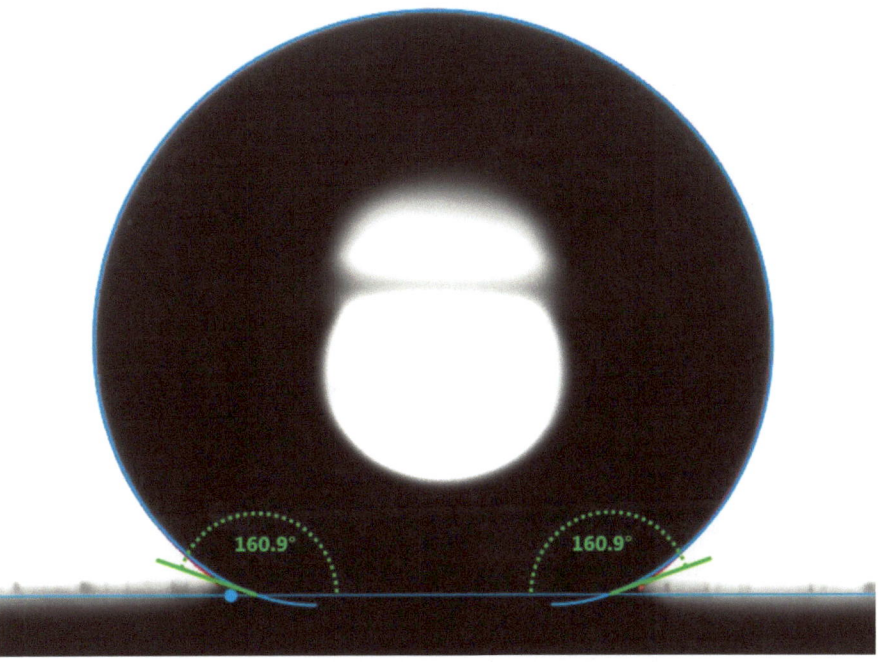

Figure 1. Static contact angle image for a water droplet of 10 µL on the superhydrophobic surface.

Figure 2. (**a**) Three-dimensional AFM image of the superhydrophobic surface; (**b**) SEM image of the superhydrophobic surface.

3. Results and Discussions

Figure 1 shows the static contact angle of the SCS sample. The incorporation of silica nanoparticles with PDMS resulted in a high static contact angle of 160.9°, indicative of the material's hydrophobic nature. The static contact angle was measured at five random locations, confirming the uniformity of the coating. Figure 2a shows the three-dimensional AFM image of the SCS sample. The microscale roughness, caused by the aggregation of nanoparticles, along with the nanoscale roughness from individual nanoparticles, created a hierarchical structure that enhanced the hydrophobicity of the SCS sample. Meanwhile, Figure 2b shows scanning electron microscopy (SEM) images of the SCS sample, where the dispersed nanoparticles within the network structure are clearly visible, further demonstrating the sample's textured surface.

The corrosion resistance of superhydrophobic-coated steel substrates was examined through electrochemical impedance spectroscopy (EIS) [13]. Figure 3a,b illustrate the equivalent circuit models for the steel and superhydrophobic coated substrates, respectively. The circuit elements in Figure 3a used to model the behavior of various interfaces and media in the system are defined as follows: Rs denotes the resistance of the electrolyte solution, and Rtop represents the resistance of the top steel surface. Qtop is the constant phase element associated with the top steel surface–electrolyte interface. Rd denotes the resistance of the diffusion layer over the top steel surface due to electrochemical reactions, while Qd represents the constant phase element for the diffusion layer at this interface. The Warburg impedance element (W) models the diffusional impedance for an infinitely thick diffusion layer. Figure 3b shows the equivalent electrical circuit used to simulate the case of the superhydrophobic coated substrate. In addition to Rs and W, the case introduces an additional resistance, Rscs, which represents the resistance attributed to the superhydrophobic coating, and Qscs, the constant phase element associated with the coating, based on its dielectric properties. Additionally, two more elements, Rd_scs and Qd_scs, represent the charge transfer resistance and constant phase element related to the diffusion layer at the interface. These elements are included to account for the distinct chemical interactions between the corrosive solution and the superhydrophobic coated substrate as compared to the steel substrate.

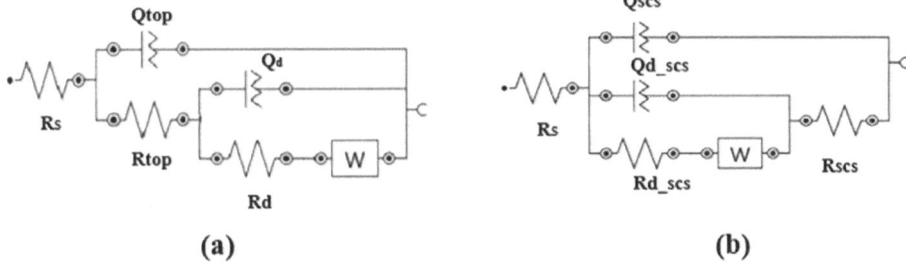

Figure 3. Electrical equivalent circuits for EIS of (**a**) steel substrate and (**b**) the superhydrophobic coated substrate.

Figure 4a presents the Nyquist plot, illustrating the relationship between the imaginary component of impedance ($-Z''$) and the real component (Z') for both substrate types. This plot shows a small semicircle for the steel substrate, corresponding to the diffusion layer resistance at the top surface–electrolyte interface, with an approximate diameter of 700 $\Omega \cdot cm^2$. For the superhydrophobic coated substrate, the semicircle diameter is notably larger, at 2.29 $k\Omega \cdot cm^2$, indicating a higher resistance of the diffusion layer and suggesting superior corrosion resistance for the coated substrate. Figure 4b provides the Bode plot of the impedance modulus ($|Z|$) as a function of log frequency for both substrates. At 1 MHz, the maximum applied frequency, the steel substrate exhibits an AC impedance modulus of 4.71 $\Omega \cdot cm^2$, while the superhydrophobic coated substrate achieves a significantly greater modulus of 700.55 $\Omega \cdot cm^2$. This elevated impedance at high frequency further reflects the resistive properties of the superhydrophobic coating, underscoring its enhanced corrosion resistance. The Bode plot of the phase angle (Φ) variation with frequency is displayed in Figure 4c. Here, the steel substrate shows a phase angle of 68.42°, whereas the superhydrophobic coating exhibits a reduced negative phase angle of 23.32°. The phase angle (Φ) can be determined by the following equation:

$$\Phi = \arctan(\frac{|Z_{Imaginary}|}{|Z_{Real}|}) \tag{1}$$

As Z_{Real} increases, the semicircle diameter also expands, signifying a high-resistance diffusion layer over the coated substrate. Hence, the decrease in phase angle further supports the improved corrosion resistance of the coated substrate. In addition, a strong agreement between experimental data and model fitting is observed as shown in Figure 4, suggesting that the equivalent circuit models fit very well with the experimental data. As one of the features of the superhydrophobic surface, air is trapped within the nanostructured morphology, which likely hinders the electrolyte solution from penetrating the surface structures and creates a high-resistance diffusion layer over the coated substrate.

Figure 5a shows the evolution of the creep displacement with a creep time of 600 s with the SCS sample before corrosion occurred and after 90 min of accelerated corrosion. The SCS sample's creep behaviors did not show any significant difference even after 90 min of accelerated corrosion. Both creep–time curves show a two-stage upward trend: an initial rapid rise within the first ~50 s, which corresponds to the instable primary creep, followed by a slow linear increase, which corresponds to a more stable secondary creep. Therefore, it indicated that SCS possesses viscoelasticity at room temperature.

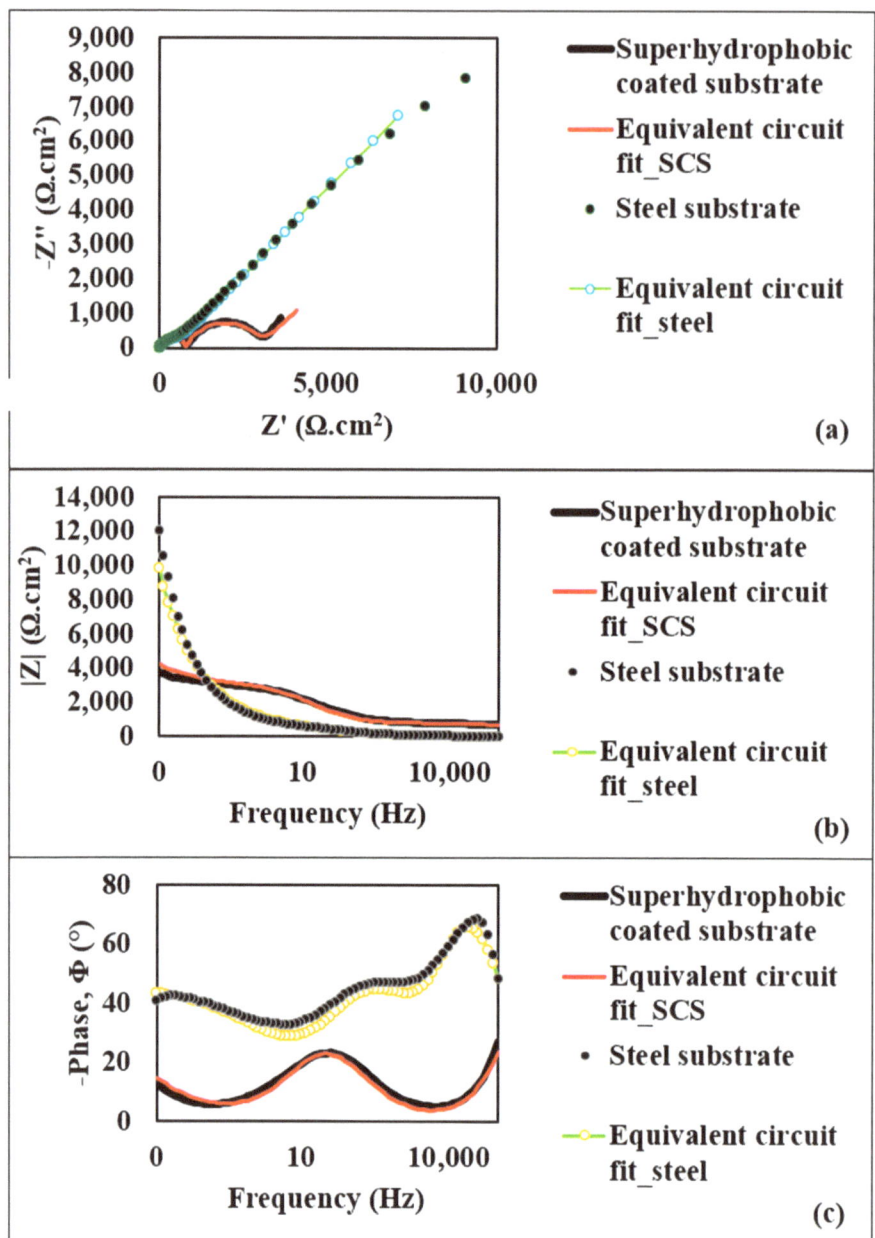

Figure 4. (**a**) Nyquist plot for steel substrate and the superhydrophobic coated substrate; (**b**) Bode modulus diagrams for steel substrate and the superhydrophobic coated substrate; (**c**) Bode phase angle diagrams for steel substrate and the superhydrophobic coated substrate.

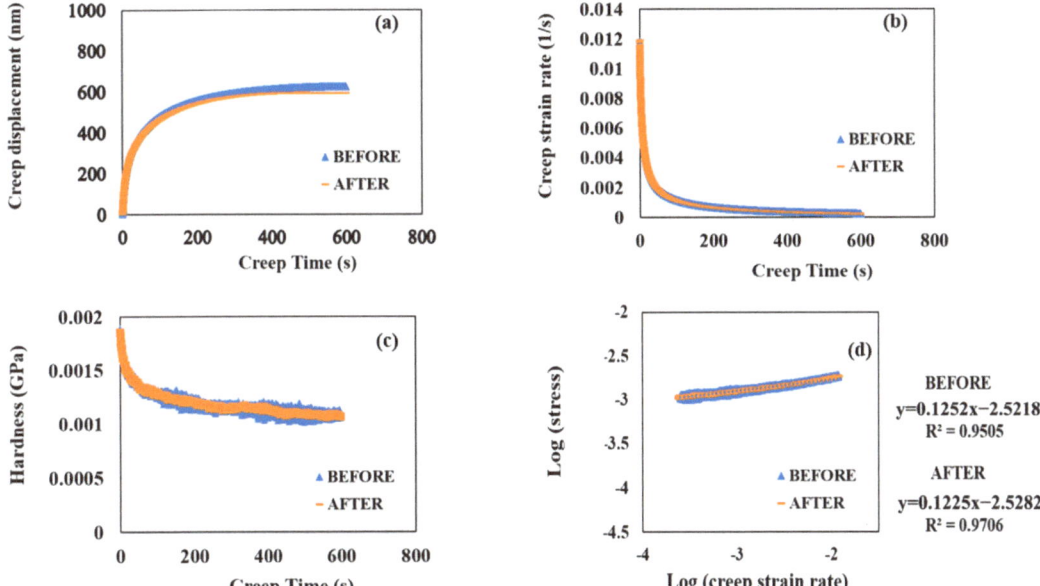

Figure 5. (**a**) Creep displacement versus creep time. (**b**) Creep strain rate versus creep time. (**c**) Hardness versus creep time. (**d**) Logarithmic stress versus logarithmic creep strain rate before and after 90 min of accelerated corrosion.

Figure 5b,c plot the values of creep strain rate and hardness against creep time, both before and after 90 min of accelerated corrosion. The creep strain rates significantly decreased from 0.01177 to 0.0019 s^{-1} during the first ~50 s. In comparison, the hardness values gradually decreased during the first ~50 s before declining linearly. As creep time increased, both hardness values continued to decrease. A logarithmic plot of the creep strain rate and stress yielded the creep strain rate sensitivity (m), a characteristic that indicates the stress response of samples to sudden changes in strain rate [24,30]:

$$\log(\text{stress}) = (1/m) \times \log(\text{strain rate}) + a1. \quad (2)$$

where a1 is the fitting coefficient

Notably, the strain rate sensitivity of SCS after 90 min of accelerated corrosion was evaluated as 0.122 from the creep tests under a load of 1000 μN (Figure 5d), which is very close to the pre-corrosion value of 0.125. This indicates that the m value of SCS is insensitive to corrosion. Additionally, during the creep tests, the logarithmic values of creep strain rate and stress exhibited a good linear relationship, increasing correspondingly from the lower left to the upper right corner (Figure 5d). Furthermore, no significant differences in creep resistance to indentation and comprehensive mechanical properties were observed after 90 min of accelerated corrosion.

To further characterize the viscoelasticity of the SCS sample, the nanoscale dynamic mechanical behavior of the material was investigated through nano-DMA. Figure 4 presents the hardness, storage modulus, loss modulus, and tan δ as functions of penetration depth.

Figure 6a,b show that both hardness and storage modulus exhibit a decreasing trend as penetration depth increases. The storage modulus, which represents the stiffness of the viscoelastic material and is proportional to the energy stored during deformation, initially decreases rapidly but then transitions to a slower, linear decrease. Importantly, there is no noticeable difference in the trends of hardness and storage modulus when comparing the SCS sample before and after 90 min of accelerated corrosion, suggesting that the material retains its stiffness and resistance to deformation even after exposure to corrosive conditions.

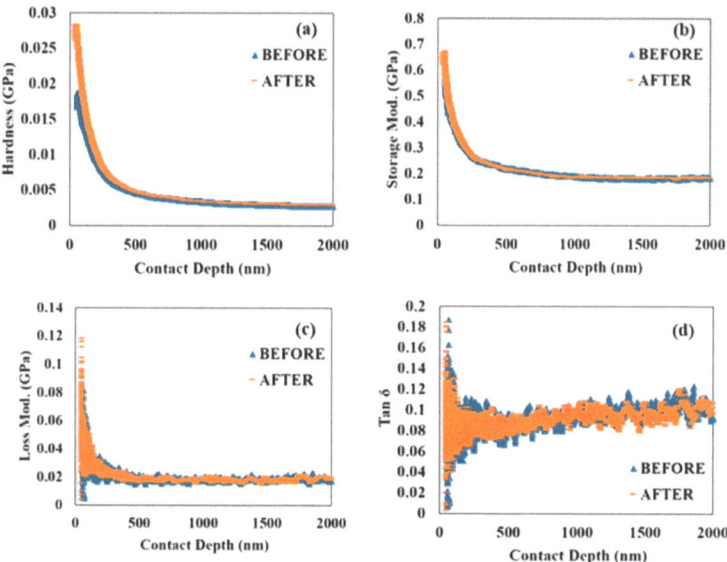

Figure 6. Nano dynamical mechanical analysis: (**a**) hardness versus contact depth; (**b**) storage modulus versus contact depth; (**c**) loss modulus versus contact depth; (**d**) tan δ versus contact depth before and after 90 min of accelerated corrosion.

Figure 6c,d provide a comparison of the loss modulus and tan δ, which are crucial indicators of the material's viscoelastic properties. The loss modulus initially shows a rapid increase. This behavior is indicative of the material's microstructural response to dynamic loading, where the initial response involves significant energy dissipation. Similarly, tan δ, the ratio of the loss modulus to the storage modulus, reflects the damping characteristics of the material. Both the loss modulus and tan δ decrease after the initial rise, eventually stabilizing at constant values.

The higher initial values of the loss modulus and tan δ suggest that the SCS sample's viscoelastic properties are significantly influenced by its nano/microstructures at the finest level. However, as penetration depth increases, these properties stabilize, indicating a consistent viscoelastic response beyond the top surface layer. Notably, the SCS sample shows no major differences in the loss modulus and tan δ before and after 90 min of accelerated corrosion. This implies that the material's ability to dissipate energy and its damping characteristics remain largely unaffected by the corrosive environment.

4. Conclusions

In this study, we characterized the corrosion resistance and mechanical and creep properties of superhydrophobic steel substrate at the nanoscale. Electrochemical impedance spectroscopy (EIS) results quantitatively confirm the enhanced anti-corrosion performance of the superhydrophobic coating. The nano-DMA results demonstrate that the superhydrophobic surface retains its viscoelastic properties, including hardness, storage modulus, loss modulus, and tan δ, even after exposure to accelerated corrosion. Based on the authors' understanding, this is the first time that these properties of a superhydrophobic surface have been reported at the nanoscale. The results also indicate the material's robustness under a corrosive condition, making it suitable for applications where both mechanical integrity and resistance to environmental degradation are critical. The superhydrophobic coating has the potential to serve as an effective top coating, significantly enhancing the corrosion resistance of steel substrates.

Author Contributions: C.-W.Y. designed and performed the experiments and analyzed the data. I.L., J.Z., P.B. and M.J. reviewed and edited the paper. M.A.H. completed the formal analysis. All authors have read and agreed to the published version of the manuscript.

Funding: This research received no external funding.

Data Availability Statement: The original contributions presented in the study are included in the article, further inquiries can be directed to the corresponding author.

Acknowledgments: The authors appreciate the Texas Manufacturing Assistance Center (TMAC) at Lamar University and the Center for Innovation, Commercialization and Entrepreneurship (CICE) at Lamar University for providing lab space.

Conflicts of Interest: The authors declare no conflicts of interest.

References

1. Yang, W.; Li, J.; Zhou, P.; Zhu, L.; Tang, H. Superhydrophobic copper coating: Switchable wettability, on-demand oil-water separation, and antifouling. *Chem. Eng. J.* **2017**, *327*, 849–854. [CrossRef]
2. Wang, Z.; Li, Q.; She, Z.; Chen, F.; Li, L. Low-cost and large-scale fabrication method for an environmentally-friendly superhydrophobic coating on magnesium alloy. *J. Mater. Chem.* **2012**, *22*, 4097–4105. [CrossRef]
3. Jung, S.; Dorrestijn, M.; Raps, D.; Das, A.; Megaridis, C.M.; Poulikakos, D. Are superhydrophobic surfaces best for icephobicity? *Langmuir* **2011**, *27*, 3059–3066. [CrossRef]
4. Li, C.; Ma, R.; Du, A.; Fan, Y.; Zhao, X.; Cao, X. One-step fabrication of bionic superhydrophobic coating on galvanized steel with excellent corrosion resistance. *J. Alloys Compd.* **2019**, *786*, 272–283. [CrossRef]
5. Zhang, S.J.; Cao, D.L.; Xu, L.K.; Lin, Z.F.; Meng, R.Q. Fabrication of a superhydrophobic polypropylene coating on magnesium alloy with improved corrosion resistance. *Int. J. Electrochem. Sci.* **2020**, *15*, 177–187. [CrossRef]
6. Gong, A.; Zheng, Y.; Yang, Z.; Guo, X.; Gao, Y.; Li, X. Spray fabrication of superhydrophobic coating on aluminum alloy for corrosion mitigation. *Mater. Today Commun.* **2021**, *26*, 101828. [CrossRef]
7. Huang, J.; Lou, C.; Xu, D.; Lu, X.; Xin, Z.; Zhou, C. Cardanol-based polybenzoxazine superhydrophobic coating with improved corrosion resistance on mild steel. *Prog. Org. Coat.* **2019**, *136*, 105191. [CrossRef]
8. Yang, H.; Dong, Y.; Li, X.; Gao, Y.; He, W.; Liu, Y.; Mu, X.; Zhao, Y.; Fu, W.; Wang, X.; et al. Development of a mechanically robust superhydrophobic anti-corrosion coating using micro-hBN/nano-Al_2O_3 with multifunctional properties. *Ceram. Int.* **2025**, *51*, 491–505. [CrossRef]
9. She, Z.; Li, Q.; Wang, Z.; Li, L.; Chen, F.; Zhou, J. Novel method for controllable fabrication of a superhydrophobic CuO surface on AZ91D magnesium alloy. *ACS Appl. Mater. Interface* **2012**, *4*, 4348–4356. [CrossRef] [PubMed]
10. Zhu, X.; Zhang, Z.; Xu, X.; Men, X.; Yang, J.; Zhou, X.; Xue, Q. Facile fabrication of a superamphiphobic surface on the copper substrate. *J. Colloid Interface Sci.* **2012**, *367*, 443–449. [CrossRef] [PubMed]
11. Schaeffer, D.A.; Polizos, G.; Smith, D.B.; Lee, D.F.; Hunter, S.R.; Datskos, P.G. Optically transparent and environmentally durable superhydrophobic coating based on functionalized SiO_2 nanoparticles. *Nanotechnology* **2015**, *26*, 055602. [CrossRef] [PubMed]
12. Zhang, Z.; Ge, B.; Men, X.; Li, Y. Mechanically durable, superhydrophobic coatings prepared by dual-layer method for anticorrosion and self-cleaning. *Colloids Surf. A* **2016**, *490*, 182–188. [CrossRef]
13. Sebastian, D.; Yao, C.; Lian, I. Multiscale corrosion analysis of superhydrophobic coating on 2024 aluminum alloy in a 3.5 wt% NaCl solution. *MRS Commun.* **2020**, *10*, 305–311. [CrossRef]
14. Qing, Y.; Yang, C.; Hu, C.; Zheng, Y.; Liu, C. A facile method to prepare superhydrophobic fluorinated polysiloxane/ZnO nanocomposite coatings with corrosion resistance. *Appl. Surf. Sci.* **2015**, *326*, 48–54. [CrossRef]
15. Yu, J.; Qin, L.; Hao, Y.; Kuang, S.; Bai, X.; Chong, Y.M.; Zhang, W.; Wang, E. Vertically aligned boron nitride nanosheets: Chemical vapor synthesis, ultraviolet light emission, and superhydrophobicity. *ACS Nano* **2010**, *4*, 414–422. [CrossRef]
16. Boinovich, L.B.; Emelyanenko, A.M. The behaviour of fluoro-and hydrocarbon surfactants used for fabrication of superhydrophobic coatings at solid/water interface. *Colloids Surf. A* **2015**, *481*, 167–175. [CrossRef]
17. Tao, C.; Yan, H.; Yuan, X.; Yao, C.; Yin, Q.; Zhu, J.; Ni, W.; Yan, L.; Zhang, L. Synthesis of shape-controlled hollow silica nanostructures with a simple soft-templating method and their application as superhydrophobic antireflective coatings with ultralow refractive indices. *Colloids Surf. A* **2016**, *501*, 17–23. [CrossRef]

18. Wang, J.; Yang, C.; Liu, Y.; Li, Y.; Xiong, Y. Using Nanoindentation to Characterize the Mechanical and Creep Properties of Shale: Load and Loading Strain Rate Effects. *ACS Omega* **2022**, *7*, 14317–14331. [CrossRef]
19. Yang, C.; Liu, Y.; Wang, J.; Wu, D.; Liu, L.; Su, Z.; Xiong, Y. Application of nanoindentation technique in mechanical characterization of organic matter in shale: Attentive issues, test protocol, and technological prospect. *Gas Sci. Eng.* **2023**, *113*, 204966. [CrossRef]
20. Zhang, P.; Zhang, D.; Zhao, J. Control of fracture toughness of kerogen on artificially-matured shale samples: An energy-based nanoindentation analysis. *Gas Sci. Eng.* **2024**, *124*, 205266. [CrossRef]
21. Huang, H.; Zhang, W.; Shi, H.; Ni, J.; Ding, L.; Yang, B.; Zheng, Y.; Li, X. Experimental investigation of microscale mechanical alterations in shale induced by fracturing fluid contact. *Gas Sci. Eng.* **2024**, *124*, 205264. [CrossRef]
22. Wang, J.; Dziadkowiec, J.; Liu, Y.; Jiang, W.; Zheng, Y.; Xiong, Y.; Peng, P.A.; Renard, F. Combining atomic force microscopy and nanoindentation helps characterizing in-situ mechanical properties of organic matter in shale. *Int. J. Coal Geol.* **2024**, *281*, 104406. [CrossRef]
23. Sun, W.; Jiang, Y.; Zhang, Z.; Ma, Z.; Sun, G.; Hu, J.; Jiang, Z.; Zhang, X.; Ren, L. Nanoindentation creep behavior of nanocrystalline Ni and Ni-20 wt% Fe alloy and underlying mechanisms revealed by apparent activation volumes. *Mater. Des.* **2023**, *225*, 111479. [CrossRef]
24. Yan, C.; Bor, B.; Plunkett, A.; Domènech, B.; Maier-Kiener, V.; Giuntini, D. Nanoindentation creep of supercrystalline nanocomposites. *Mater. Des.* **2023**, *231*, 112000. [CrossRef]
25. Song, Y.; Ma, Y.; Pan, Z.; Li, Y.; Zhang, T.; Gao, Z. Nanoindentation Characterization of Creep-fatigue Interaction on Local Creep Behavior of P92 Steel Welded Joint. *Chin. J. Mech. Eng.* **2021**, *34*, 131. [CrossRef]
26. Ma, X.; Ma, J.; Bian, X.; Tong, X.; Han, D.; Jia, Y.; Wu, S.; Zhang, N.; Geng, C.; Li, P.; et al. The role of nano-scale elastic heterogeneity in mechanical and tribological behaviors of a Cu–Zr based metallic glass thin film. *Intermetallics* **2021**, *133*, 107159. [CrossRef]
27. Rath, A.; Mathesan, S.; Ghosh, P. Nanomechanical characterization and molecular mechanism study of nanoparticle reinforced and cross-linked chitosan biopolymer. *J. Mech. Behav. Biomed. Mater.* **2016**, *55*, 42–52. [CrossRef] [PubMed]
28. Díaz-Guillén, J.; Naeem, M.; Hdz-García, H.; Acevedo-Davila, J.; Díaz-Guillén, M.; Khan, M.; Iqbal, J.; Mtz-Enriquez, A. Duplex plasma treatment of AISI D2 tool steel by combining plasma nitriding (with and without white layer) and post-oxidation. *Surf. Coat. Technol.* **2020**, *385*, 125420. [CrossRef]
29. Hoque, M.A.; Yao, C.-W.; Lian, I.; Zhou, J.; Jao, M.; Huang, Y.-C. Enhancement of corrosion resistance of a hot-dip galvanized steel by superhydrophobic top coating. *MRS Commun.* **2022**, *12*, 415–421. [CrossRef]
30. Holz, H.; Merle, B. Novel nanoindentation strain rate sweep method for continuously investigating the strain rate sensitivity of materials at the nanoscale. *Mater. Des.* **2023**, *236*, 11247. [CrossRef]

Disclaimer/Publisher's Note: The statements, opinions and data contained in all publications are solely those of the individual author(s) and contributor(s) and not of MDPI and/or the editor(s). MDPI and/or the editor(s) disclaim responsibility for any injury to people or property resulting from any ideas, methods, instructions or products referred to in the content.

Article

Investigation of Anti-Friction Properties of MoS$_2$ and SiO$_2$ Nanolubricants Based on the Friction Pairs of Inconel 718 Superalloy and YG6 Carbide

Lijie Ma [1,*], Fengnan Li [1], Shijie Ba [1], Zunyan Ma [2], Xinhui Mao [2], Qigao Feng [1,*] and Kang Yang [3]

1. School of Mechanical and Electrical Engineering, Henan Institute of Science and Technology, Xinxiang 453003, China
2. School of Mechanical and Power Engineering, Harbin University of Science and Technology, Harbin 150080, China
3. Department of Mechanical Engineering, Anyang Institute of Technology, Anyang 455000, China
* Correspondence: ma_lj@hist.edu.cn (L.M.); fqg@hist.edu.cn (Q.F.); Tel.: +86-0373-304-0394 (L.M.); +86-0373-304-0394 (Q.F.)

Abstract: In order to improve the anti-friction property of common mineral oil and develop a high-performance lubricant, MoS$_2$ and SiO$_2$ nano-additives were individually dispersed into the 350SN mineral oil at various weight percentages to prepare nanolubricants. Then, the viscosity, wettability, and tribological properties of the nanolubricants were measured and analyzed with a rotary viscometer, a contact angle measuring instrument, and a friction tester. Finally, the action mechanism of two nano-additives was explained based on the energy spectrum test results of the abrasion surface. The results show that MoS$_2$ and SiO$_2$ nano-additives could improve the viscosity of the base fluid and change its wettability, giving nanolubricants better anti-friction performance than the base fluid. Due to the difference in physical properties, SiO$_2$ and MoS$_2$ nanolubricants presented different friction reduction rules with the increase in nano-additive percentage. Under experimental conditions, SiO$_2$ nanolubricants showed better anti-friction effects than MoS$_2$ nanolubricants. When the SiO$_2$ percentage was 10 wt% and 15 wt%, the maximum friction coefficient was reduced to 0.06, which was about 1/3 of that with the base fluid. In this case, the abrasion surface quality was significantly improved, and the abrasion trace size was about half that of the base fluid. The energy spectrum test results show that the action mechanism of the MoS$_2$ nano-additive is the adsorption film effect and mending effect of nanoparticles, while the main action mechanism of the SiO$_2$ nano-additive should be the polishing effect and rolling effect of nanoparticles.

Keywords: anti-friction properties; nanolubricant; MoS$_2$; SiO$_2$; friction coefficient; abrasion surface topography; abrasion trace size; mechanism

1. Introduction

Friction is a common phenomenon in the process of mechanical operation and is also the main cause of energy loss and mechanical failure [1]. Lubrication is the main way to reduce energy waste and restrain mechanical wear. According to the different mediums, lubrication methods can be divided into fluid lubrication, solid lubrication, and semi-solid lubrication, of which fluid lubrication is the most widely used one [2]. Nanofluid lubrication is an emerging efficient fluid lubrication technology that has become one of the research highlights in the field of tribology [3–5].

During nanofluid lubrication, nanoscale particle additives are dispersed into the base liquid (including water, emulsion, oil, etc.) to prepare nanolubricants, which improves the lubrication effect of the base liquid through the adhesion, filling, embedding, and rolling of nano-additives between the friction interfaces [6,7]. At present, many nanoparticles can be used to prepare nanolubricants, including metals, metal oxides, sulfides, nanocomposites, and carbon nanotubes; most of them have a large specific surface area, a high chemical activity, and excellent friction reduction characteristics [8,9]. Therefore, nanolubricants have many advantages over common lubricants, such as high thermostability, high load capacity, high efficiency, and low pollution.

Due to the differences in physicochemical properties of nano-additives, nanolubricants with various additives have different anti-friction performance and mechanism. Jatti et al. [10] found that the CuO nano-additive could be brought into the moving friction pairs of LM6 aluminum alloy and EN31 bearing steel to transform sliding friction into rolling friction, thereby reducing the shear resistance and improving the lubricating performance of machine oil. Sia et al. [11] disclosed that the SiO_2 nano-additive could roll into the friction interface of chrome-plated carbon steel bearings to fill and repair the defective area, decreasing the friction temperature and wear rate. Sujan et al. [12] studied the influence of mixed Al_2O_3 and WS_2 additives on the anti-friction performance of the base fluid. Their results show that the Al_2O_3 nanoparticles with a higher hardness could be pressed into the friction surface, reducing the wear rate, while the flaky WS_2 microparticles were adsorbed on the friction surface to form a solid lubricating film, thereby decreasing the friction coefficient. Su et al. [13] investigated the performance of graphite nanolubricants with LB2000 vegetable oil as the base fluid and found that graphite additives form a physical deposition film between friction interfaces, thereby decreasing the friction coefficient and the wear rate, and the nano-additives with a 35 nm size exhibited better anti-friction performance than the nano-additives with an 80 nm particle size. The research by Yu also showed that MoS_2 nanoparticles could form an isolation layer between friction interfaces, thus preventing the direct contact of frictional pairs and reducing the fluctuation and amplitude of the instantaneous friction coefficient [14].

In addition to the properties of nano-additives, the additive percentage is also an important factor in the performance of nanolubricants. Mo et al. [15] developed several Sn nanolubricants with different additive percentages and applied them to the wear resistance tests between steel–brass friction pairs, with the results showing that Sn nanolubricants could repair the worn surface, and the higher the additive percentage is, the better the repair effect. Singh et al. [16] carried out lubricating performance tests of SiO_2 nanolubricants through pin-on-disc frictional experiments. Their results show that the friction coefficient and the wear rate first decreased and then increased with the rise in the SiO_2 percentage, and the lowest friction coefficient and the best abrasion trace morphology were achieved when the SiO_2 percentage was 0.6 wt%. However, when the SiO_2 percentage exceeded 1.2 wt%, the SiO_2 nanolubricant could not perform better than the base oil. Chu et al. [17] studied the anti-friction performance of diamond nanolubricants using ring-on-block friction tests on SKD11 and SKD61 steel and found that when the additive percentage was 2 vol%, the friction coefficient was the minimum, and when the additive percentage was 3 vol%, the wear rate was the lowest. Gupta et al. [18] investigated the property of graphite nanolubricants through four-ball frictional experiments, and the results indicate that the wear scar size also exhibited a changing trend of decreasing first and then increasing with the rise in the additive percentage, and the wear rate was the lowest when the additive percentage was 3 wt%. The research of Luo disclosed that when Al_2O_3 nanolubricants were applied in the four-ball and thrust-ring friction tests, the Al_2O_3 nanoparticles adhered to the friction surface to form a self-laminating protective film, and the 0.1 wt% nanolubricant displayed a better anti-friction performance than the 0.05 wt%, 0.5 wt%, and 1 wt% nanolubricants [19].

In recent years, the application of nanolubricants in the field of transportation, electronics equipment, nuclear systems, and machining operations has also achieved some good results. Peyghambarzadeh et al. [20] found that Al_2O_3/water nanolubricant can enhance heat transfer efficiency by 45% in the automobile radiator than pure water. The research of Mohamed [21] also showed that the nanolubricant dispersed with copper (Cu) and graphene (Gr) nanoparticles in a fully formulated engine oil is a promising lubricating medium for increasing the durability of frictional sliding components and fuel economy in automobile engines. Kulkarni et al. [22] applied Al_2O_3 nanolubricant as a coolant in a diesel–electric generator to reduce equipment temperature. Buongiorno [23] discovered that nanolubricants have potential applications in improving the economy and safety of nuclear reactors. Syafiq et al. [24] carried out a minimum quantity lubrication (MQL) research with SiO_2 nanolubricant during the turning of Al319 aluminum alloy and found that the cutting temperature decreased and the surface roughness improved compared with the common MQL lubricant, and the best effect was achieved at the 0.5 wt–1.5 wt% SiO_2 percentage. Ni et al. [25] conducted the broaching tests on AISI 1045 steel under the CuO nanolubricant and found that the maximum broaching force was decreased to 62.7% of that of pure sesame oil when the CuO percentage was approximately 0.54 wt%. According to the drilling test on AISI321 stainless steel, Pal et al. reported that the Al_2O_3 nanolubricant could significantly improve the machining quality and the tool life compared with the common MQL lubricant [26].

In summary, nano-additives can significantly improve the anti-friction properties of the base fluid, and the type and percentage of nano-additives are two important parameters that affect the anti-friction properties of nanolubricants. Nowadays, although some progress has been made in the application of nanolubricants in the machining field, most of the related research is still focused on easy-to-machining materials, such as aluminum alloy, copper alloy, and medium carbon steel.

As a kind of common industrial raw material, mineral oil is widely applied as a lubricant and coolant in the field of machining. However, with the industrial application of a large number of difficult-to-machining materials, common mineral oil has been difficult to meet the practical needs of this field [27,28]. In this study, eight nanolubricants were prepared by dispersing different weight percentages of MoS_2 and SiO_2 additives into 350SN mineral oil. Then, the viscosity, wettability, and tribological properties of the nanolubricants were intensively investigated. Finally, the anti-friction mechanisms of the two nano-additives were discussed. This study helps reveal the action mechanism of nano-additives, enriches the theory of particle tribology, and has an important practical value to the development and application of high-performance lubricants.

2. Material and Methodology

2.1. Preparation of Nanolubricants

To compare the anti-friction effect of different nano-additives, MoS_2 and SiO_2 particles with large property differences were selected as additives. MoS_2 is a common soft lubricating material with a small friction coefficient (0.04–0.09) and good chemical and thermal stability. It is widely applied in the aerospace, casting, and forging industries. SiO_2 is a typical hard and brittle material with excellent high-temperature performance and chemical stability and is widely used as an antiadhesion agent or flow aid in the industrial field [29–31]. The particle size of the two additives is approximately 30 nm, and their morphology and physical properties are shown in Figure 1 and Table 1, respectively. In this study, 350SN mineral oil produced by Jinan Xinhuihuang Chemical Co., Ltd. was selected as the base fluid. It is a commonly used industrial oil that can not only be used independently as a lubricant in the manufacturing field but also as a base liquid. The main properties of 350SN mineral oil are listed in Table 2.

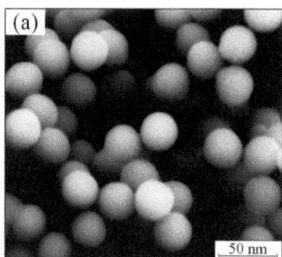 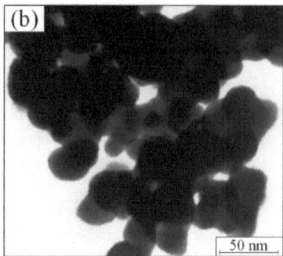

Figure 1. SEM images of nano-additives: (**a**) SiO_2 (**b**) MoS_2.

Table 1. Physical properties of SiO_2 and MoS_2 nano-additives.

Nanoparticle	Molecular Weight	Density (g/cm^3)	Melting Point (°C)	Moh's Hardness
SiO_2	60.08	2.2–2.6	1723	7
MoS_2	160.07	4.8	2375	1.0–1.5

Table 2. Properties of 350SN mineral oil.

Indexes	Density	Kinematic Viscosity (40 °C) (mm^2/s)	Flash Point (°C)	Pour Point (°C)
Value	0.87	62–68	238	−12

The nano-additives have large surface energy and specific surface area, which makes them easy to agglomerate in the base fluid [32,33]. To improve their dispersibility, according to references [34,35], KH-550 (γ-Aminopropyl triethoxysilane) and SP80 (Sorbitol fatty acid ester) were applied as the dispersants of SiO_2 and MoS_2 nano-additives, respectively, which have good dispersion effect on nano-additives and have little impact on the properties of the base fluid. The dosage of the dispersants was approximately 50% of the weight of nano-additives [34,35]. In light of our pre-machining tests on nickel-based alloy, it was difficult for the nanolubricant with a low additive percentage to have a better anti-friction effect than the base fluid. Therefore, the weight percentage of the nano-additive was determined to be 5 wt%, 10 wt%, 15 wt%, and 20 wt%. On the basis of the above, different proportions of base fluid, nano-additive, and dispersant were mixed together. Then, the mixture was stirred with a magnetic stirrer and oscillated with an ultrasonic oscillator for 30 min to obtain the nanolubricants. Figure 2 shows the eight nanolubricants prepared in this research.

Figure 2. Nanolubricants: (**a**) MoS$_2$; (**b**) SiO$_2$.

2.2. Viscosity and Wettability Test of Nanolubricants

Viscosity is an important indicator for evaluating the fluidity of nanolubricants, which is closely related to the lubricating and cooling performance [36,37]. As shown in Figure 3a, a NDJ-5s rotary viscometer (Wincom, Changsha, China) was used to measure the dynamic viscosity. Given that MoS$_2$ nanolubricants are similar to Newtonian fluids, their viscosities were measured using the traditional Newtonian fluid viscosity measurement method. SiO$_2$ nanolubricants have obvious shear thinning and thixotropic features, and their viscosities decrease gradually with the increase in shear time [38,39]. Therefore, the viscosity values at the 120 s shear time were recorded as the final result.

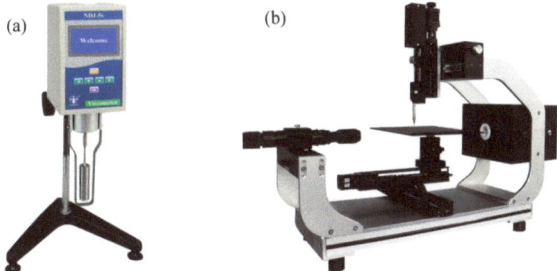

Figure 3. Viscosity and wettability test equipment: (**a**) NDJ-5s rotary viscometer; (**b**) SDC-100 contact angle measuring instrument.

The wettability was evaluated by the contact angle of the droplet spreading, which was tested using a SDC-100 contact angle measuring instrument (SINDIN, Dongguan, China) (see Figure 3b). In this measurement, a 15 mm × 15 mm × 3 mm Inconel 718 superalloy rectangular block with surface roughness Ra 150 nm was used for droplet spreading, and the same sample was also used as the lower specimen for the subsequent friction test. Before the measurement, the samples were ultrasonically cleaned in a 5 vol% acetone aqueous solution for 20 min and then wiped with ethanol absolute and dried at room temperature. For each lubricant, the mean of five tests was taken as the final result.

2.3. Tribological Tests and Conditions

As shown in Figure 4, the friction test was carried out on a MWF-500 ball-on-disk reciprocating friction tester (MWF Detector, Singapore). The lower specimen was an Inconel 718 rectangular block, and its chemical composition and mechanical properties are shown in Tables 3 and 4, respectively. Inconel 718 superalloy has excellent physical mechanics properties such as high strength, good toughness, high-temperature resistance, and corrosion resistance, and it is widely used in the field of aviation, aerospace, and nuclear industries. However, it is also a typical difficult-to-cutting material that requires high-performance cutting lubricants [40]. The upper specimen was a YG6 cemented carbide ball with a diameter of ϕ6.5 mm and a surface roughness of Ra 25.4 nm. YG6 is a common tool material, and its chemical composition and mechanical properties are shown in Table 5. In this research, the anti-friction properties of lubricants are evaluated by friction coefficient, surface topography, and abrasion trace size of the lower specimen. In order to reduce the experimental error, a new YG6 upper specimen was used for each friction test.

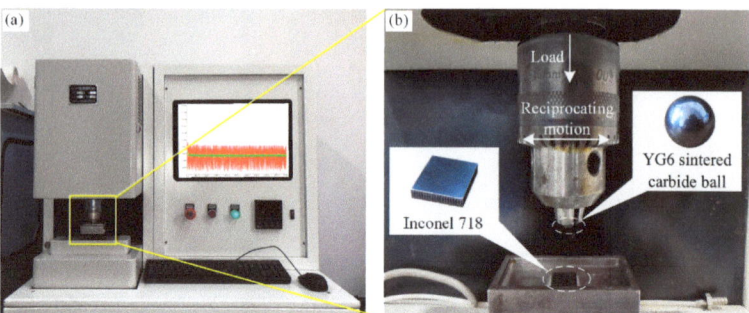

Figure 4. Friction tester and the specimens: (**a**) MWF-500 friction tester; (**b**) lower and upper specimens.

Table 3. Chemical composition of Inconel 718 superalloy.

Chemical Element	Ni	Cr	Nb	Mo	Ti	Al	Co	C	Mn	Si	P	S	B	Cu	Fe
Minimum value	50.0	17.0	4.75	2.8	0.65	0.2	—	—	—	—	—	—	—	—	Margin
Maximum value	55.0	21.0	5.50	3.3	1.15	0.8	1.0	0.08	0.35	0.35	0.015	0.015	0.006	0.3	

Table 4. Mechanical properties of Inconel 718 superalloy.

Yield Strength MPa	Tensile Strength MPa	Elongation %	Poisson Ratio GPa	Thermal Conductivity W/m·K	Density g/cm³	Hardness HV
1110	1310	23.3	206	11.2	8.47	281

Table 5. Chemical composition and mechanical properties of YG6 cemented carbide.

Co Content %	WC Content %	Flexure Strength MPa	Impact Toughness J/cm	Poisson Ratio GPa	Hardness HRA	Density g/cm³
6	94	145	2.6	530	89.5	14.6–15.0

To improve the accuracy of friction test, the friction surface of the Inconel 718 specimen was pre-polished to a roughness of approximately Ra 150 nm. The surface topography of the specimen was shown in Figure 5, where different colors indicate surface fluctuation. During the test, the Inconel 718 specimen was bonded on the bottom of the cargo tank using paraffin wax, the upper specimen was fixed by an elastic chuck, and the friction pairs were lubricated by the lubricants in the cargo tank. Other test conditions include the following: the friction time was 60 min, the distance and frequency of reciprocating motion was 4 mm and 2.5 Hz, the normal load was 80 N, and the testing temperature was room temperature (approximately 20 °C).

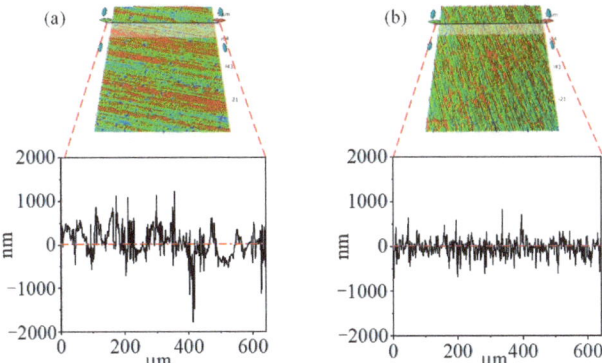

Figure 5. Friction surface of Inconel 718 specimen before and after polishing: (**a**) before polishing; (**b**) after polishing.

3. Results and Discussion

3.1. Viscosity and Wettability of Nanolubricants

3.1.1. Viscosity

The room temperature dynamic viscosities of the nanolubricants are shown in Figure 6. According to Figure 6, for both MoS_2 and SiO_2 nanolubricants, the viscosities increased significantly with the additive percentage. However, due to the prominent thickening effect of the SiO_2 nano-additive, the viscosities of the SiO_2 nanolubricants were much higher than those of the MoS_2 nanolubricants. Although 20 wt% SiO_2 nanolubricants still have certain flowability and can be used for lubrication in low-speed machining processes, their viscosity can no longer be measured using a NDJ-5s rotary viscometer.

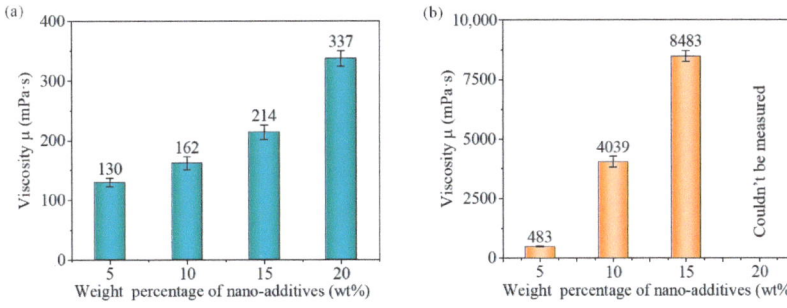

Figure 6. Viscosities of nanolubricants. (**a**) MoS_2; (**b**) SiO_2.

3.1.2. Wettability

Figure 7 shows the contact angles of the base fluid and the nanolubricants, and a small contact angle means that the lubricant has good wettability. The following results can be obtained from Figure 7.

(1) The contact angles of the MoS_2 nanolubricants were smaller than that of the base fluid and decreased gradually with the increase in the additive percentage. This phenomenon indicates that the MoS_2 nanolubricants can easily form an adsorption film on the surface of the Inconel 718 superalloy. Moreover, the higher the additive percentage within the test range is, the better the wettability is.

(2) With the increase in the additive percentage, the contact angles of the SiO_2 nanolubricants presented a completely different variation rule. The contact angle of the 5

wt% SiO$_2$ nanolubricant is smaller than that of the base fluid, while that of the 10 wt% SiO$_2$ nanolubricant was 67.81°, which is much larger than that of the base fluid. With the further increase in the additive percentage, the droplets displayed an irregular shape on the sample surface due to poor fluidity, making it impossible to measure the contact angle. This phenomenon shows that the SiO$_2$ nanolubricants cannot easily form an adsorption film on the surface of the lower specimen, and their wettability becomes bad and can even be lost as the additive percentage increases.

The above results indicate a certain correlation between the viscosity and wettability of nanolubricants. However, due to the differences in physical properties of nano-additives, the wettability of MoS$_2$ and SiO$_2$ nanolubricants exhibited different variation features with their viscosity. With the increase in the additive percentage, the wettability of the MoS$_2$ nanolubricants increased with the viscosity, while the wettability of the SiO$_2$ nanolubricants decreased.

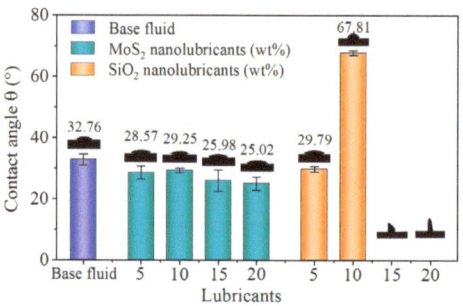

Figure 7. Contact angles of base fluid and nanolubricants.

3.2. Anti-Friction Properties of Different Lubricants

3.2.1. Friction Coefficient

Base Fluid

Figure 8 shows the friction coefficient curve under the lubrication of the base fluid and the abrasion surface morphologies of the Inconel 718 specimen. The repeated tests indicated that the friction coefficient presented a three-stage change trend with the friction time, and the characteristics of the three stages are "low amplitude stable fluctuation", "high amplitude slow decrease", and "low amplitude stable fluctuation", respectively. In order to clarify the reason for the change in friction coefficient, the surface morphologies of the lower specimen were acquired in the middle of each stage. According to Figure 8, there is a close relationship between friction coefficient and abrasion surface morphologies.

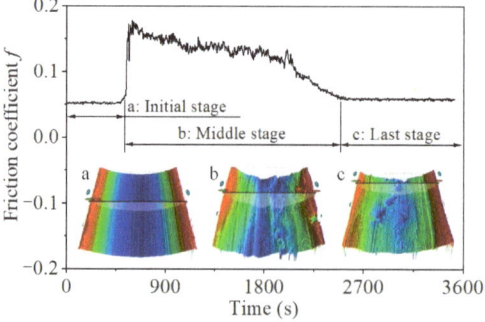

Figure 8. Friction coefficient curve under the lubrication of the base fluid.

(1) In the initial stage, the mean friction coefficient was lowest, approximately 0.052, and the abrasion surface was relatively smooth. In this case, the friction type should be boundary friction, and a stable lubricating film should exist at the friction interface.

(2) In the middle stage, the instantaneous friction coefficient increased sharply and reached 0.178, then showed a slow downward trend. At the same time, some furrows appeared on the abrasion surface. It can be inferred that serious adhesive and abrasive wear happened on the surface of the Inconel 718 specimen, and the friction type should be mixed friction where dry friction and boundary friction coexist. However, the appearance of furrows reduced the contact area of two specimens and increased the unit contact pressure. As a result, the furrows were gradually smoothed out with the rise of friction time, resulting in a gradual decrease in the instantaneous friction coefficient.

(3) In the last stage, the abrasion surface quality was significantly improved, and the frictional process once again exhibited the characteristics of boundary friction; thus, the instantaneous friction coefficient showed a stable fluctuation trend. However, due to the damaged frictional surface, the mean friction coefficient in this stage was 0.0587, which is bigger than that in the initial stage.

Nanolubricants

Figure 9 shows the friction coefficient curves under different nanolubricants.

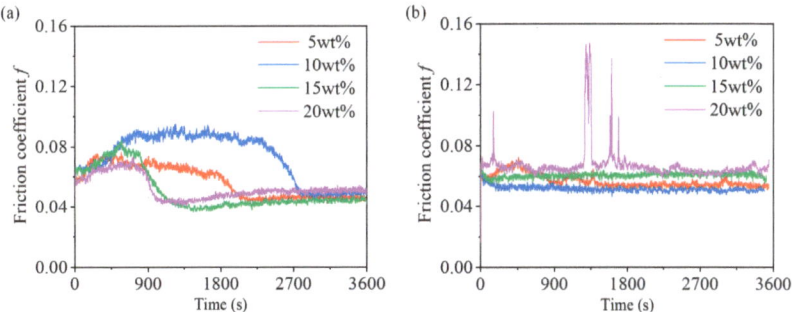

Figure 9. Friction coefficient curves under different nanolubricants: (**a**) MoS$_2$; (**b**) SiO$_2$.

(1) Figure 9a indicates that, with the increase in the test time, the friction coefficient under the MoS$_2$ nanolubricants exhibited a two-stage change trend that is different from that of the base fluid. In the first stage, the friction coefficient increased first and then decreased. In the second stage, a "low amplitude stable fluctuation" feature was observed. In the first stage, the maximum friction coefficients corresponding to the MoS$_2$ nanolubricants of 5 wt%, 10 wt%, 15 wt%, and 20 wt% were 0.077, 0.095, 0.085, and 0.073, respectively, which are all less than that of the base fluid (f = 0.178). According to the duration and maximum friction coefficient of the first stage, the 20 wt% MoS$_2$ nanolubricants have the best anti-friction effect, which should be related to their excellent wettability.

(2) Figure 9b shows that the friction coefficient under the SiO$_2$ nanolubricants was stable as the test time increased and did not show obvious stage characteristics. The maximum friction coefficients were also smaller than that of the base fluid (f = 0.178), and these of the 10 wt% and 15 wt% SiO$_2$ nanolubricants were all less than 0.06. According to the maximum and fluctuation of the instantaneous friction coefficient, when the additive percentage was 5 wt%, 10 wt%, and 15 wt%, the SiO$_2$ nanolubricants could achieve a good anti-friction effect, and the effect of the 10 wt% nanolubricant was

the best. However, when the SiO$_2$ percentage was 20 wt%, the instantaneous friction coefficient increased sharply within a short time. This phenomenon was caused by the increase in the shear resistance due to the excessive additive percentage [41].

In summary, both MoS$_2$ and SiO$_2$ nanolubricants exhibit an excellent anti-friction performance. However, the friction coefficients of the two nanolubricants present different variation characteristics with the test time. This result shows that the soft MoS$_2$ nano-additive and the hard SiO$_2$ nano-additive have different anti-friction mechanisms at the friction interface.

3.2.2. Abrasion Surface Topography of the Lower Specimen

Figure 10 shows the abrasion surface of the Inconel 718 specimen under the lubrication of the base fluid; severe defects appeared on the abrasion surface, including plastic deformation, microcracks, surface exfoliation, and chipping pits. As shown in Figure 10a, plastic deformation appeared at the edge of the abrasion trace and was caused by the friction and extrusion of the upper specimen. As shown in Figure 10c, microcracks, surface exfoliation, and chipping pits mainly occurred in the middle of the abrasion trace, and the closer they were to the center of the abrasion trace, the more severe the damage was. Microcracks appeared in the middle stage of the friction test because of the collapse of the lubricating film and the fatigue failure of the abrasion surface. In contrast, surface exfoliation and chipping pits were formed by the propagation of microcracks and adhesive wear.

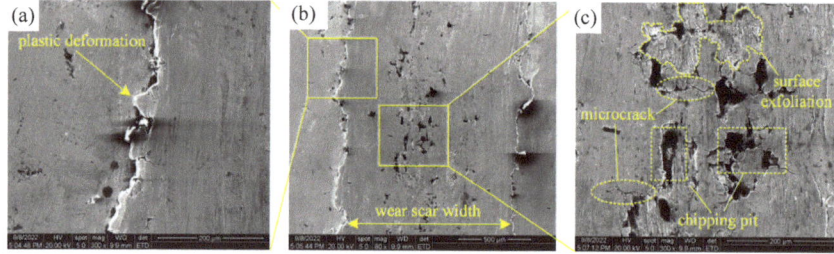

Figure 10. Abrasion surface of the lower specimen under the lubrication of the base fluid: (**a**) edge of the abrasion trace; (**b**) overall view of the abrasion trace; (**c**) middle of the abrasion trace.

Figures 11 and 12 show the abrasion surfaces of the lower specimen under the MoS$_2$ and SiO$_2$ nanolubricants, respectively. Compared with Figure 10, the defect form of the abrasion surfaces changed. Although plastic deformation and chipping pits remained, the topographies of the abrasion surfaces, in most cases, significantly improved.

With MoS$_2$ nanolubricants, furrow and material smearing were the two main topographic characteristics of the abrasion surfaces. Furrow was caused by abrasive wear, which can deteriorate the surface quality. Meanwhile, material smearing was caused by the reciprocating friction and the extrusion of the upper specimen, which can reduce the size of the furrows and improve the surface quality. According to Figure 11a, when the MoS$_2$ percentage was 5 wt%, chipping pits, furrow, and material smearing simultaneously appeared on the abrasion surface, so the abrasion topography was relatively poor. In this situation, the friction coefficient is large. According to Figure 11b, when the MoS$_2$ percentage was 10 wt%, the effect of material smearing was weak, but deep furrows appeared on the abrasion surface, so the corresponding friction coefficient fluctuated with a high amplitude for a long time. When the MoS$_2$ percentage increased from 10 wt% to 20 wt%, the size of the furrows gradually decreased due to the enhancement of the material smearing effect, thereby improving the topography of the abrasion surface and decreasing the duration of the high-amplitude friction coefficient.

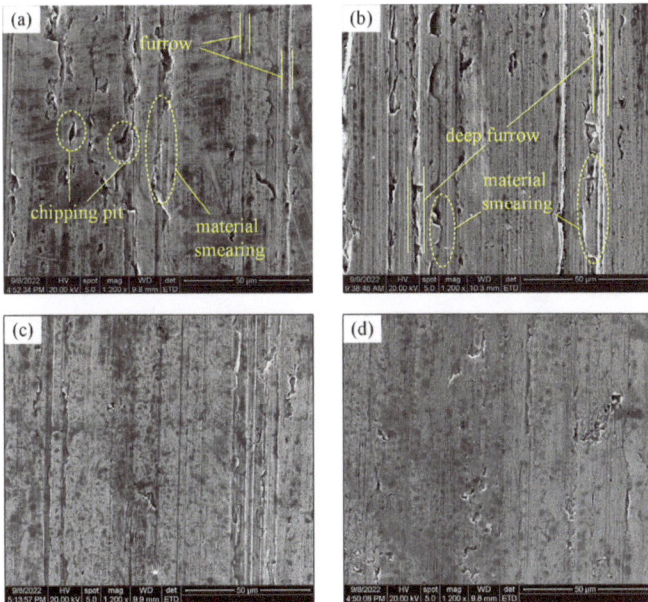

Figure 11. Abrasion surfaces of the Inconel 718 specimen under MoS_2 nanolubricants: (**a**) 5 wt%; (**b**) 10 wt%; (**c**) 15 wt%; (**d**) 20 wt%.

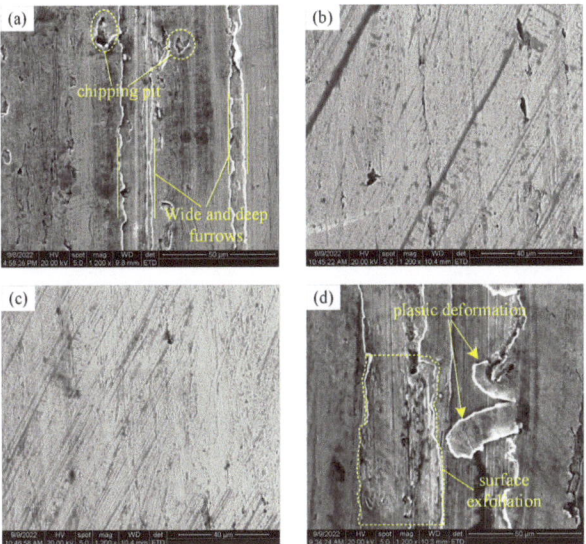

Figure 12. Abrasion surfaces of the lower specimen under SiO_2 nanolubricants: (**a**) 5 wt%; (**b**) 10 wt%; (**c**) 15 wt%; (**d**) 20 wt%.

Compared with the MoS_2 nanolubricants, the SiO_2 nanolubricants have different topographic characteristics of abrasion surfaces. Figure 12a shows that when the SiO_2 percentage was 5 wt%, the abrasion surface had deep furrows and large chipping pits. Accordingly, the friction coefficient was relatively large at the beginning of the test. According to Figure 12b,c, when the SiO_2 percentage was 10 wt% and 15 wt%, the abrasion surfaces were generally smooth in addition to the inclined scratches formed by the rolling of SiO_2 particles. In both cases, the friction coefficient was very stable during the entire test. Figure 12d shows that when the SiO_2 percentage was 20 wt%, the topography of the abrasion surface was very poor, with serious plastic deformation and surface exfoliation, and the corresponding friction coefficient also fluctuated sharply within a short time, which might be caused by the accumulation and agglomeration of excessive nanoparticles.

The above analysis indicates that the abrasion surfaces of the MoS_2 and SiO_2 nanolubricants have different topography characteristics. Moreover, a close internal relationship exists between the abrasion surface and the friction coefficient, and a smooth and steady friction coefficient represents a good abrasion surface quality.

3.2.3. Abrasion Trace Size of the Lower Specimen

In addition to the topography characteristics of the abrasion surface, the abrasion trace size is also an important index for evaluating the anti-friction performance of lubricants. Figure 13 shows the morphologies of abrasion traces under different lubricants, which were measured using a Contour GT-K 3D surface profiler. As shown in Figure 14, the width and depth of the abrasion traces were obtained from the morphology. In order to eliminate the measurement error caused by uneven plastic deformation at the edge of the abrasion trace, the influence of plastic deformation on the abrasion trace size has been eliminated. According to Figures 13 and 14, the following results can be found.

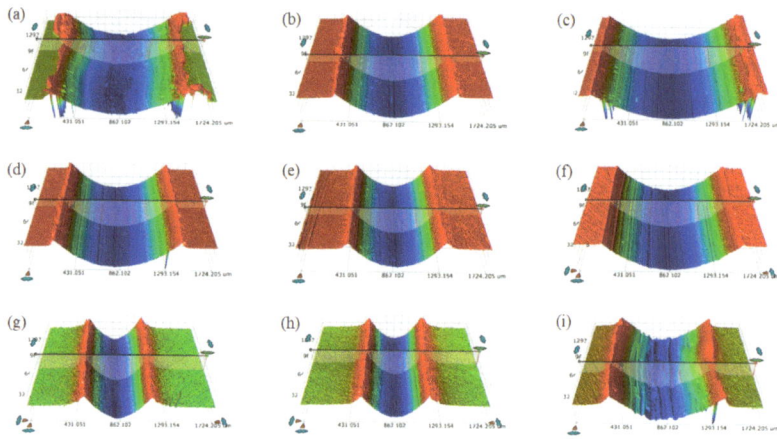

Figure 13. Morphologies of abrasion traces under different lubricants: (**a**) base fluid; (**b**) 5 wt% MoS_2; (**c**) 10 wt% MoS_2; (**d**) 15 wt% MoS_2; (**e**) 20 wt% MoS_2; (**f**) 5 wt% SiO_2; (**g**) 10 wt% SiO_2; (**h**) 15 wt% SiO_2; (**i**) 20 wt% SiO_2.

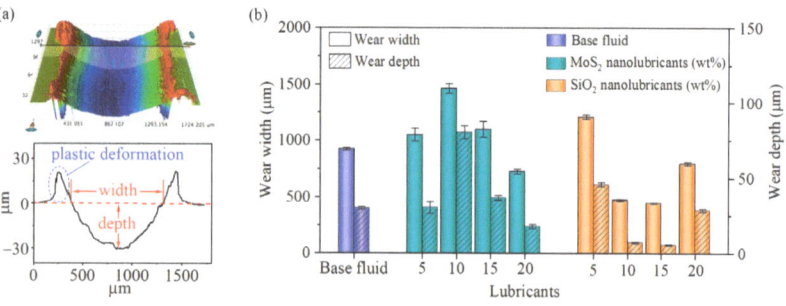

Figure 14. Measurement method and result of abrasion trace size: (**a**) measurement method; (**b**) abrasion trace size.

(1) For MoS$_2$ nanolubricants, the abrasion trace size was smaller than that of the base fluid only when the additive percentage was 20 wt%. For the SiO$_2$ nanolubricants, except for the abrasion trace at 5 wt% additive percentage, the size of the other abrasion traces was smaller than that of the base fluid. The abrasion trace size under the 15 wt% SiO$_2$ nanolubricant was the lowest, with a width and a depth of 51.16% and 80.79% of the base fluid, respectively.

(2) In general, a close relationship exists between the size and surface quality of the abrasion trace. A good surface quality means a small abrasion trace size. For experimental friction pairs, the SiO$_2$ nanolubricant displayed a better anti-wear effect than the MoS$_2$ nanolubricants. Whether the tribological performance is evaluated in terms of friction coefficient, abrasion surface quality, or abrasion trace size, the best results were obtained under the 10 wt% and 15 wt% of SiO$_2$ nanolubricants.

4. Analysis of Action Mechanism for Different Nano-Additives

As shown in Figure 15, during nanofluid lubrication, the action mechanism of nano-additives at the friction interface can be described by four types of effects, namely, the adsorption film, mending, polishing, and rolling effects [42,43].

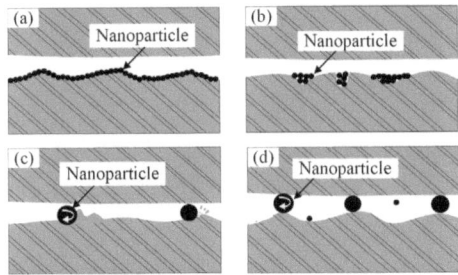

Figure 15. Action mechanism of nano-additives: (**a**) adsorption film effect; (**b**) mending effect; (**c**) polishing effect; (**d**) rolling effect.

Figure 16 shows the element content on the original surface and abrasion surfaces of the Inconel 718 specimen, which is measured by a XFlash Detector 630M energy dispersive spectrometer (Bruker, Billerica, MA, USA). During the measurement, the accelerating voltage was 20 kV, and the magnification was 300. On the basis of the Mo and Si contents, the action mechanism of the MoS$_2$ and SiO$_2$ nano-additives can be further analyzed.

(1) According to Figure 16a,b, the Mo and Si contents on the original surface were 3.70% and 0.31%, while the contents on the abrasion surface under the base fluid were 3.53% and 0.52%, respectively. The difference between the two is very small.

(2) Figure 16c,d show that, under the 5 wt% MoS$_2$ nanolubricant, the Mo contents in the smooth area and severe wear area of the abrasion trace were 8.84% and 12.85%, respectively, which are much higher than the 3.53% under the base fluid, indicating increases of 5.31% and 9.32%, respectively. The significant increase in the Mo content in different areas means that a layer of MoS$_2$ adsorption film formed on the abrasion surface. Meanwhile, the Mo content in the severe wear area is higher than that in the smooth area, indicating that a great amount of MoS$_2$ additives was squeezed into the furrows or chipping pits on the abrasion surface, resulting in the mending effect.

(3) According to Figure 16e,f, under the 5 wt% SiO$_2$ nanolubricants, the Si contents in the smooth area and severe wear area of the abrasion trace were 0.90% and 3.03%, respectively, which are larger than 0.52% under the base fluid, indicating increases of only 0.38% and 2.51%, respectively. This result indicates that the adsorption film and

mending effects of SiO$_2$ additives are not obvious. Therefore, the improvement of the anti-friction performance of SiO$_2$ nanolubricants should be mainly attributed to the rolling and polishing effects of hard SiO$_2$ additives, which transform sliding friction into rolling friction between the friction pairs.

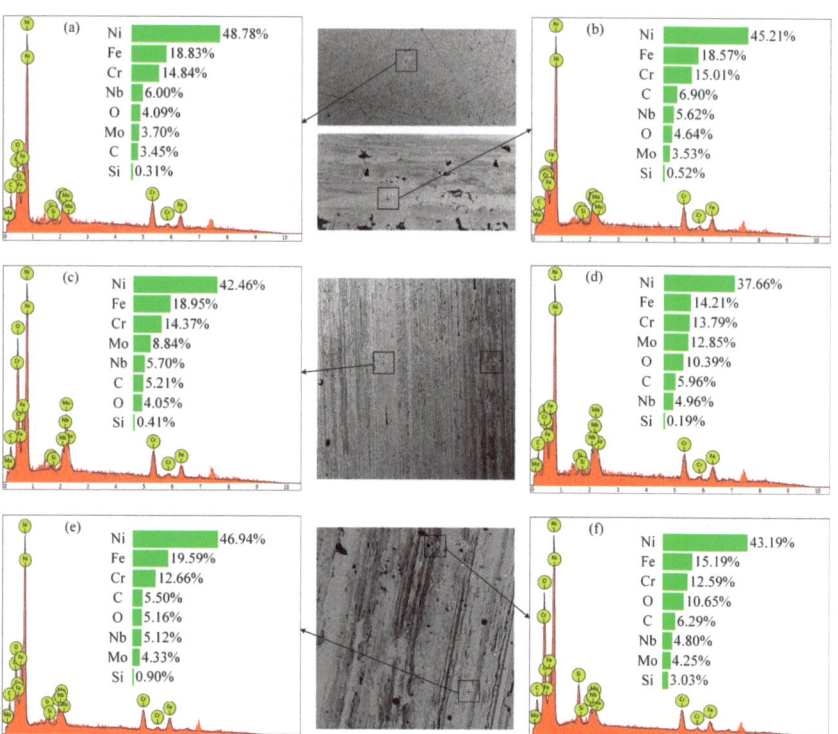

Figure 16. Element content on the original surface and abrasion surfaces of the Inconel 718 specimen: (**a**) original surface; (**b**) abrasion surface under the base fluid; (**c**) smooth area of abrasion surface under 5 wt% MoS$_2$ nanolubricant; (**d**) severe wear area of abrasion surface under 5 wt% MoS$_2$ nanolubricant; (**e**) smooth area of abrasion surface under 5 wt% SiO$_2$ nanolubricant; (**f**) severe wear area of abrasion surface under 5 wt% SiO$_2$ nanolubricant.

5. Conclusions

(1) MoS$_2$ and SiO$_2$ nano-additives could improve the viscosity of mineral oil and change its wettability. And there was a certain correlation between the viscosity and wettability of nanolubricants. Due to the differences in physical properties of the two additives, the wettability of MoS$_2$ and SiO$_2$ nanolubricants exhibited different variation trends with their viscosity. The wettability of the MoS$_2$ nanolubricants increased with the viscosity, while the wettability of the SiO$_2$ nanolubricants decreased.

(2) Although MoS$_2$ and SiO$_2$ nanolubricants had different change characteristics of friction coefficient with the frictional time, their maximum and fluctuation values were all smaller than those of base liquid. For MoS$_2$ nanolubricant, when the additive percentage was 20 wt%, the friction coefficient curve fluctuated the least, and the maximum value was 0.073, which was much smaller than 0.178 of the base fluid. For the SiO$_2$ nanolubricant, the best anti-friction effect was achieved when the additive percentage was 10 wt%, and the maximum friction coefficient was less than 0.06.

(3) Compared with the base fluid, the nanolubricants can improve the surface quality of abrasion traces. However, the abrasion surfaces of the two nanolubricants had

different topography characteristics. Furrow and material smearing were the two main topography features under the MoS_2 nanolubricant. And, with the rise in MoS_2 percentage, the effect of material smearing was enhanced, and the surface quality of the abrasion trace was improved. With the change in SiO_2 percentage, the abrasion surface quality showed a trend of first improvement and then deterioration, and the surface quality with 10 wt% and 15 wt% SiO_2 nanolubricant was better.

(4) A close internal relationship exists between friction coefficient, abrasion surface quality, and abrasion trace size. A steady friction coefficient represents a good abrasion surface quality and a small abrasion trace size. For the experimental friction pairs, the SiO_2 nanolubricant has a better anti-friction effect than MoS_2 nanolubricants. With 10 wt% and 15 wt% SiO_2 nanolubricants, the width and depth of abrasion traces were about half those of the base fluid.

(5) The soft MoS_2 nano-additive and the hard SiO_2 nano-additive have different anti-friction mechanisms. The action mechanism of the MoS_2 nano-additive is the adsorption film effect and mending effect of nanoparticles, while the main action mechanism of the SiO_2 nano-additive should be the polishing effect and rolling effect of nanoparticles.

Author Contributions: Methodology, L.M. and Q.F.; experiments and data analysis, F.L. and Z.M.; software, S.B. and X.M.; writing—original draft preparation, Z.M. and F.L.; writing—review and editing, L.M. and K.Y. All authors have read and agreed to the published version of the manuscript.

Funding: This research was financially supported by the National Natural Science Foundation of China, Grant Number 52175397.

Data Availability Statement: All data generated or analyzed during this study are included in this article.

Conflicts of Interest: The authors declare no conflicts of interest.

References

1. Holmberg, K.; Erdemir, A. Influence of tribology on global energy consumption, costs and emissions. *Friction* **2017**, *5*, 263–284. [CrossRef]
2. Brinksmeier, E.; Meyer, D.; Huesmann-Cordes, A.G.; Herrmann, C. Metalworking fluids—Mechanisms and performance. *CIRP Annals.* **2015**, *64*, 605–628. [CrossRef]
3. Öndin, O.; Kıvak, T.; Sarıkaya, M.; Yıldırım, Ç.V. Investigation of the influence of MWCNTs mixed nanofluid on the machinability characteristics of PH 13-8 Mo stainless steel. *Tribol. Int.* **2020**, *148*, 106323. [CrossRef]
4. Shariatmadar, F.S.; Pakdehi, S.G. Synthesis and characterization of aviation turbine kerosene nanofluid fuel containing boron nanoparticles. *Energy Fuels.* **2016**, *30*, 7755–7762. [CrossRef]
5. Wang, X.; Li, C.; Zhang, Y.; Ali, H.M.; Sharma, S.; Li, R.; Yang, M.; Said, Z.; Liu, X. Tribology of enhanced turning using biolubricants: A comparative assessment. *Tribol. Int.* **2022**, *174*, 107766. [CrossRef]
6. Saidur, R.; Kazi, S.N.; Hossain, M.S.; Rahman, M.M.; Mohammed, H.A. A review on the performance of nanoparticles suspended with refrigerants and lubricating oils in refrigeration systems. *Renew. Sustain. Energy Rev.* **2011**, *15*, 310–323. [CrossRef]
7. Kotia, A.; Rajkhowa, P.; Rao, G.S.; Ghosh, S.K. Thermophysical and tribological properties of nanolubricants: A review. *Heat Mass Transf.* **2018**, *54*, 3493–3508. [CrossRef]
8. Dai, W.; Kheireddin, B.; Gao, H.; Liang, H. Roles of nanoparticles in oil lubrication. *Tribol. Int.* **2016**, *102*, 88–98. [CrossRef]
9. Wang, B.; Qiu, F.; Barber, G.C.; Zou, Q.; Wang, J.; Guo, S.; Yuan, Y.; Jiang, Q. Role of nano-sized materials as lubricant additives in friction and wear reduction: A review. *Wear* **2022**, *490*, 204206. [CrossRef]
10. Jatti, V.S.; Singh, T.P. Copper oxide nano-particles as friction-reduction and anti-wear additives in lubricating oil. *J. Mech. Sci. Technol.* **2015**, *29*, 793–798. [CrossRef]
11. Sia, S.Y.; Sarhan, A.A.D. Morphology investigation of worn bearing surfaces using SiO_2 nanolubrication system. *Int. J. Adv. Manuf. Technol.* **2014**, *70*, 1063–1071. [CrossRef]
12. Sujan, K.; Wang, B.; Hu, M.; Meng, S.; Zhang, L.; Hu, R.; Barber, G.C. Tribological properties of Al_2O_3/WS_2 oil-based composite lubricant utilized on steel-brass frictional couples. *Surf. Topogr. Metrol. Prop.* **2021**, *9*, 015018. [CrossRef]

13. Su, Y.; Gong, L.; Chen, D. An investigation on tribological properties and lubrication mechanism of graphite nanoparticles as vegetable based oil additive. *J. Nanomater.* **2015**, *2015*, 276753. [CrossRef]
14. Yu, R.; Liu, J.; Zhou, Y. Experimental study on tribological property of MoS_2 nanoparticle in castor oil. *J. Tribol.* **2019**, *141*, 102001. [CrossRef]
15. Mo, Y.H.; Tao, D.H.; Wei, X.C. Activation and self-repairing effectiveness of lubrication with nano-tin as additives. *J. Shanghai Univ. Engl. Ed.* **2009**, *13*, 45–50. [CrossRef]
16. Singh, Y.; Singh, N.K.; Sharma, A. Effect of SiO_2 nanoparticles on the tribological behavior of Balanites Aegytiaca (Desert date) oil-based biolubricant. *J. Bio Tribo-corros.* **2021**, *7*, 1–6. [CrossRef]
17. Chu, H.Y.; Hsu, W.C.; Lin, J.F. The anti-scuffing performance of diamond nano-particles as an oil additive. *Wear* **2010**, *268*, 960–967. [CrossRef]
18. Gupta, M.K.; Bijwe, J. A complex interdependence of dispersant in nano-suspensions with varying amount of graphite particles on its stability and tribological performance. *Tribol. Int.* **2020**, *142*, 105968. [CrossRef]
19. Luo, T.; Wei, X.; Huang, X.; Huang, L.; Yang, F. Tribological properties of Al_2O_3 nanoparticles as lubricating oil additives. *Ceram. Int.* **2014**, *40*, 7143–7149. [CrossRef]
20. Peyghambarzadeh, S.M.; Hashemabadi, S.H.; Jamnani, M.S.; Hoseini, S.M. Improving the cooling performance of automobile radiator with Al_2O_3/water nanofluid. *App. Therm. Eng.* **2011**, *31*, 1833–1838. [CrossRef]
21. Mohamed, K.A.A.; Hou, X.J.; Abdelkareem, M.A.A. Anti-wear properties evaluation of frictional sliding interfaces in automobile engines lubricated by copper/graphene nanolubricants. *Friction* **2020**, *8*, 905–916.
22. Kulkarni, D.P.; Vajjha, R.S.; Das, D.K.; Oliva, D. Application of aluminum oxide nanofluids in diesel electric generator as jacket water coolant. *App. Therm. Eng.* **2008**, *28*, 1774–1781. [CrossRef]
23. Buongiorno, J.; Hu, L.W.; Kim, S.J.; Hannink, R.; Truong, B.A.O.; Forrest, E. Nanofluids for enhanced economics and safety of nuclear reactors: An evaluation of the potential features, issues, and research gaps. *Nucl. Technol.* **2008**, *162*, 80–91. [CrossRef]
24. Syafiq, A.M.; Redhwan, A.A.M.; Hazim, A.A.; Aminullah, A.R.M.; Ariffin, S.Z.; Nughoro, W.; Arifuddin, A.; Hawa, A.B.S. An Experimental Evaluation of SiO_2 nano cutting fluids in CNC Turning of Aluminium Alloy AL319 via MQL Technique. *IOP Conf. Ser. Mater. Sci. Eng.* **2021**, *1068*, 012009. [CrossRef]
25. Ni, J.; Cui, Z.; He, L.; Yang, Y.; Sang, Z.; Rahman, M.M. Reinforced lubrication of vegetable oils with nano-particle additives in broaching. *J. Manuf. Process.* **2021**, *70*, 518–528. [CrossRef]
26. Pal, A.; Chatha, S.S.; Sidhu, H.S. Performance evaluation of the minimum quantity lubrication with Al_2O_3-mixed vegetable-oil-based cutting fluid in drilling of AISI 321 stainless steel. *J. Manuf. Process.* **2021**, *66*, 238–249. [CrossRef]
27. Gajrani, K.K.; Ram, D.; Sankar, M.R. Biodegradation and hard machining performance comparison of eco-friendly cutting fluid and mineral oil using flood cooling and minimum quantity cutting fluid techniques. *J. Clean. Prod.* **2017**, *165*, 1420–1435. [CrossRef]
28. Okokpujie, I.P.; Tartibu, L.K. Experimental analysis of cutting force during machining difficult to cut materials under dry, mineral oil, and TiO_2 nano-lubricant. *J. Meas. Eng.* **2021**, *9*, 218–230. [CrossRef]
29. Harsha, A.P.; Khatri, O.P. The Effect of Spherical Hybrid Silica-Molybdenum Disulfide on the Lubricating Characteristics of Castor Oil. *J. Tribol.* **2023**, *145*, 121701-1.
30. Prasad, B.K.; Rathod, S.; Yadav, M.S.; Modi, O.P. Effects of some solid lubricants suspended in oil toward controlling the wear performance of a cast iron. *J. Tribol.* **2010**, *132*, 041602. [CrossRef]
31. Yegin, C.; Lu, W.; Kheireddin, B.; Zhang, M.; Li, P.; Min, Y.; Sue, H.J.; Sari, M.M.; Akbulut, M. The effect of nanoparticle functionalization on lubrication performance of nanofluids dispersing silica nanoparticles in an ionic liquid. *J. Tribol.* **2017**, *139*, 041802. [CrossRef]
32. Basso, C.R.; Crulhas, B.P.; Castro, G.R.; Pedrosa, V.A. A study of the effects of pH and surfactant addition on gold nanoparticle aggregation. *J. Nanosci. Nanotechnol.* **2020**, *20*, 5458–5468. [CrossRef] [PubMed]
33. Bao, L.; Zhong, C.; Jie, P.; Hou, Y. The effect of nanoparticle size and nanoparticle aggregation on the flow characteristics of nanofluids by molecular dynamics simulation. *Adv. Mech. Eng.* **2019**, *11*, 1–17. [CrossRef]
34. Xu, M.H.; Cao, Y.Y.; Gao, S.G. Surface modification of nano-silica with silane coupling agent. *Key Eng. Mater.* **2015**, *636*, 23–27. [CrossRef]
35. Hou, X.; Jiang, H.; Ali, M.K.A.; Liu, H.; Su, D.; Tian, Z. Dispersion behavior assessment of the molybdenum disulfide nanomaterials dispersed into poly alphaolefin. *J. Mol. Liq.* **2020**, *311*, 113303. [CrossRef]
36. Harigaya, Y.; Suzuki, M.; Toda, F.; Takiguchi, M. Analysis of oil film thickness and heat transfer on a piston ring of a diesel engine: Effect of lubricant viscosity. *J. Eng. Gas Turbines Power* **2006**, *128*, 685–693. [CrossRef]
37. Singh, J.; Kumar, D.; Tandon, N. Development of nanocomposite grease: Microstructure, flow, and tribological studies. *J. Tribol.* **2017**, *139*, 052001.
38. Parizad, A.; Shahbazi, K.; Tanha, A.A. SiO_2 nanoparticle and KCl salt effects on filtration and thixotropical behavior of polymeric waterbased drilling fluid: With zeta potential and size analysis. *Results Phys.* **2018**, *9*, 1656–1665. [CrossRef]

39. Sanukrishna, S.S.; Vishnu, S.; Prakash, M.J. Experimental investigation on thermal and rheological behaviour of PAG lubricant modified with SiO_2 nanoparticles. *J. Mol. Liq.* **2018**, *261*, 411–422. [CrossRef]
40. Basha, M.M.; Sankar, M.R. Experimental tribological study on additive manufactured Inconel 718 features against the hard carbide counter bodies. *J. Tribol.* **2023**, *145*, 121702.
41. Sajeeb, A.; Krishnan Rajendrakumar, P. Experimental studies on viscosity and tribological characteristics of blends of vegetable oils with CuO nanoparticles as additive. *Micro Nano Lett.* **2019**, *14*, 1121–1125. [CrossRef]
42. Jason, Y.J.J.; How, H.G.; Teoh, Y.H.; Chuah, H.G. A study on the tribological performance of nanolubricants. *Processes* **2020**, *8*, 1372. [CrossRef]
43. Cui, X.; Li, C.; Ding, W.F.; Chen, Y.; Mao, C.; Xu, X.F.; Liu, B.; Wang, D.Z.; Li, H.N.; Zhang, Y.B.; et al. Minimum quantity lubrication machining of aeronautical materials using carbon group nanolubricant: From mechanisms to application. *Chin. J. Aeronaut.* **2022**, *35*, 85–112. [CrossRef]

Disclaimer/Publisher's Note: The statements, opinions and data contained in all publications are solely those of the individual author(s) and contributor(s) and not of MDPI and/or the editor(s). MDPI and/or the editor(s) disclaim responsibility for any injury to people or property resulting from any ideas, methods, instructions or products referred to in the content.

 *lubricants*

Article

Investigation of Effect of Surface Modification by Electropolishing on Tribological Behaviour of Worm Gear Pairs

Robert Mašović, Suzana Jakovljević *, Ivan Čular, Daniel Miler and Dragan Žeželj

Faculty of Mechanical Engineering and Naval Architecture, University of Zagreb, Ivana Lučića 5, 10000 Zagreb, Croatia; robert.masovic@fsb.unizg.hr (R.M.); ivan.cular@fsb.unizg.hr (I.Č.); daniel.miler@fsb.unizg.hr (D.M.)
* Correspondence: suzana.jakovljevic@fsb.unizg.hr

Abstract: Electropolishing using a high-current density results in a pitting phenomenon, producing a surface texture distinguished by many pits. Apart from the change in surface topography, electropolishing forms an oxide surface layer characterized by beneficial tribological properties. This paper introduces surface texturing in worm gear pairs by electropolishing a 16MnCr5 steel worm surface. Electropolishing produces surface pits 1 μm to 5 μm deep and 20 to 100 μm in diameter. The material characterization of 16MnCr5 steel is compared against the electropolished 16MnCr5 steel based on microstructure, hardness, surface topography and chemical composition. Experimental tests with worm pairs employing electropolished worms are conducted, and the results are compared to conventional worm pairs with ground steel worms. Electropolished worms show up to 5.2% higher efficiency ratings than ground ones and contribute to better running-in of worm gear pairs. Moreover, electropolished worms can reliably support full contact patterns and prevent scuffing due to improved lubrication conditions resulting from the produced surface texture and oxide surface layer. Based on the obtained results, electropolishing presents a promising method for surface texturing and modification in machine elements characterized by highly loaded non-conformal contacts and complex geometry.

Keywords: electropolishing; worm gear pair; surface texturing; steel; lubrication; efficiency

Citation: Mašović, R.; Jakovljević, S.; Čular, I.; Miler, D.; Žeželj, D. Investigation of Effect of Surface Modification by Electropolishing on Tribological Behaviour of Worm Gear Pairs. *Lubricants* **2024**, *12*, 408. https://doi.org/10.3390/lubricants12120408

Received: 15 October 2024
Revised: 3 November 2024
Accepted: 21 November 2024
Published: 24 November 2024

Copyright: © 2024 by the authors. Licensee MDPI, Basel, Switzerland. This article is an open access article distributed under the terms and conditions of the Creative Commons Attribution (CC BY) license (https://creativecommons.org/licenses/by/4.0/).

1. Introduction

The worm gear pair typically comprises a steel worm as the driving member and a bronze worm wheel as the driven member. Worm gear pairs are often used in power transmission applications due to their high transmission ratios in a compact design, heavy shock loading capability, and ability to self-lock and reduce noise and vibration due to the dominant sliding motion. On the other hand, some disadvantages derive from the worm gear pair's geometry and working principle: lower efficiency compared to other gear types, high sensitivity to assembly and manufacturing errors, and frictional heat generation paired with unfavourable lubrication conditions [1–4]. The latter is a frequent research topic since lubrication conditions are directly related to worm gear pair efficiency and overall life span. Relatively unfavourable oil entraining geometry of most common worm gear designs limits film-forming ability, load capacity, and efficiency compared to other gear types. These limitations can lead to a contact separated into two parts with contact temperatures in the middle of the contact zone being substantially higher than temperatures at the outer regions of the worm wheel tooth flank. Consequently, these temperature differences cause oil film thinning or even lubrication breakdown [5–7].

Researchers investigated new material pairs to improve lubrication conditions in worm gear pairs. Benedetti et al. [8] investigated steel–steel tribo-pairs with different coatings. Coated steel surfaces showed higher wear resistance compared to bronze, which is conventionally used as a worm wheel material. Fontanari et al. [9,10] studied wear damage mechanisms in steel–bronze and alloy steel–cast iron pairs under a mixed lubrication regime. Cast iron showed a lower wear rate, indicating that alternative worm wheel materials may perform better than bronze. Mašović et al. [11] investigated electropolished steel–bronze pairs. Electropolishing created surface texture with many pits/dimples, improving the lubrication and reducing friction by up to 30%. Furthermore, several studies comparing lubricating oils were conducted. The general conclusion was that mineral oils result in approximately 3–5% lower efficiency than synthetic oils [12,13]. Lastly, some researchers have investigated new geometry types on a theoretical level. Such geometries are usually characterized by improved load-carrying capabilities that should improve lubricating conditions in worm gear pairs [14–16].

Surface or material modification is commonly employed to improve the lubrication conditions and tribological performance of contacting surfaces of machine elements [17,18]. Typically, surfaces are modified by altering surface topography through surface texturing or employing surface coatings. Surface texturing produces surface micro-cavities with several beneficial functions: entrapment of wear debris, secondary lubrication, friction and wear reduction, and increased oil film thickness [17]. Applying surface texturing in gears is challenging due to their complex geometry, narrow tolerances, high loads accompanied by non-conformal contacts, and strict surface roughness requirements. Therefore, a conventional approach for improving lubrication conditions in gears is primarily surface grinding or superfinishing [19]. Nonetheless, some studies focused on the surface texturing of gears or gear steel. Gupta et al. used chemical etching and laser texturing on spur gear tooth flanks to produce a dimpled surface [20,21]. Such gears presented reduced wear and a decrease in vibrations. Petare et al. [22] laser-textured helical and straight bevel gears. The findings indicated higher wear resistance and increased microhardness compared to untextured gears. Nakatsuji and Mori [23,24] applied electropolishing to medium carbon steel gears. The produced surfaces were characterized by pits and micropores with oxidized and phosphoric surface compounds. The benefits of such surfaces were evident in the better oil film creation which improved pitting and scuffing durability. Li et al. [25] shot peened specimens made of gear steel, thus producing a dimpled surface with friction-reduction properties.

Based on the literature overview, various studies were conducted to improve lubrication conditions in worm gear pairs, as well as to apply surface texturing in other types of gears. However, surface modification in the form of surface texturing in worm gear pairs has not yet been attempted. This paper investigates the tribological behaviour of steel–bronze worm gear pairs employing surface-modified steel worms by electropolishing. The work conducted in this paper is based on the results of electropolished steel–bronze sliding tests presented in earlier research conducted by the authors [11]. The effects of electropolishing on surface topography, material hardness, microstructure, and chemical composition are presented and discussed. Worm gear pairs employing electropolished steel worms are compared to a conventional worm gear pairs with ground steel worms. Obtained benefits regarding better worm gear pair efficiency and operational characteristics suggest that electropolishing can be a promising surface modification method for non-conformal worm pair contact characterized by unfavourable lubrication conditions.

2. Materials and Methods

2.1. Materials

Investigated worm gear pairs comprise case-hardened 16MnCr5 steel worms and CuSn12 worm wheels, which are also considered reference materials for worm pairs by the ISO/TS 14521 standard [26]. The 16MnCr5 steel was supplied in hot-rolled round bars while the CuSn12 bronze was centrifugally cast. The chemical composition (wt.%) of the materials is presented in Table 1. Since this research focuses primarily on surface modification of 16MnCr5 steel, the properties of 16MnCr5 are covered in more detail. The microstructure of the case-hardened 16MnCr5 steel worm is presented in Figure 1. The specimens were ground (1200 grit sandpaper), polished, and etched with 3% nital for

20 seconds before images were taken using an optical microscope. The 16MnCr5 steel specimen surface shows fine martensite, while the core shows low-carbon martensite at 200× magnification. Additional analyses of 16MnCr5 steel were conducted using a scanning electron microscope (SEM-Tescan Vega 3 Easyprobe, from Tescan, Brno, Czech Republic) and by employing energy dispersive spectroscopy (EDS-Bruker B-Quantax EDS Detector, from Bruker, Karlsruhe, Germany). The details of these analyses are presented in the "Results" section.

Table 1. Chemical composition of the materials, wt.%.

	Fe	C	Si	Mn	P	S	Cr	Ni	Mo	As	Al	Cu
16MnCr5	97.05	0.19	0.31	1.11	0.018	0.01	1.01	0.08	0.01	0.034	0.033	0.15
	Cu	Zn	S	Co	Cr	Pb	Sn	Ni	Si	-	-	-
CuSn12	86.15	0.36	0.14	0.08	0.08	0.67	12.10	0.4	0.2	-	-	-

(a) (b)

Figure 1. Microstructure and hardness of a 16MnCr5 worm: (**a**) core (200×); (**b**) case-hardened surface layer (200×).

2.2. Electropolishing Setup

The surface of 16MnCr5 steel worms was modified via electropolishing. The same electropolishing solution and similar setup were already successfully employed in our previous work [11]. Before electropolishing, the steel worms were cleaned in an ultrasonic bath with ethanol (96%) for 10 min. The electropolishing solution was a mixture of 34% sulphuric acid, 42% phosphoric acid, and 24% water [27]. The solution was heated at 50 ± 2 °C without agitation using a Končar flask heater (Končar, Zagreb, Croatia). After electropolishing, the steel worms were rinsed and dried. The cathode, in the form of a cylindrical shell, was made of AISI 304 stainless steel. The distance between the worm and the cathode was approximately 3 cm. The bearing surfaces of the worm shaft should not be electropolished. Therefore, they were insulated using polytetrafluorethylene (PTFE) tape (Figure 2). Electropolishing parameters are provided in Table 2. It should be noted that the abbreviation "EP" will be used for the remainder of the paper to distinguish electropolished from conventional ground-only worms.

Table 2. Electropolishing parameters for 16MnCr5 steel worms.

	Potential (V)	Current Density (A/dm^2)	Time (min)
Worm EP-1	6.1	20	5
Worm EP-2	4.7	15	5

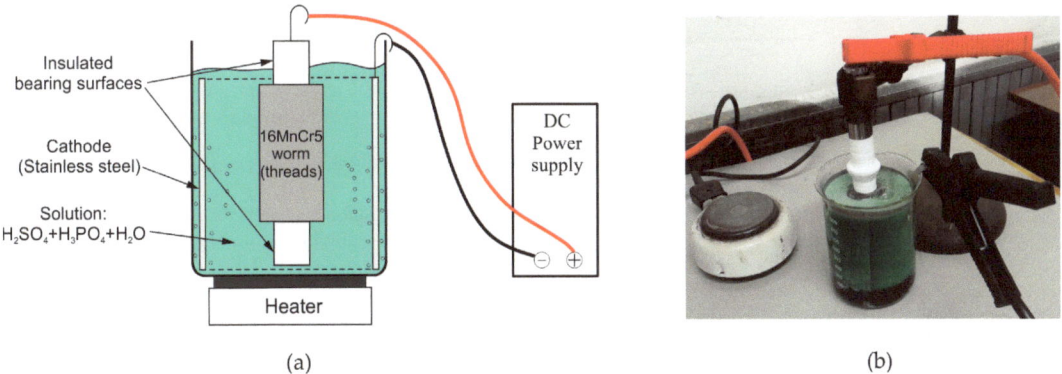

Figure 2. Electropolishing setup of a steel worm: (**a**) schematic; (**b**) actual image.

2.3. Worm Pair Experimental Testing

Tested worm pairs were a part of a commercially available gearbox (Figure 3). Rolling bearings were used for worm and worm wheel shafts. A new set of bearings was used for each test run. Continuous oil lubrication was supplied through the oil inlet on the top of the gearbox directly onto the worm wheel. The worm was positioned beneath the worm wheel.

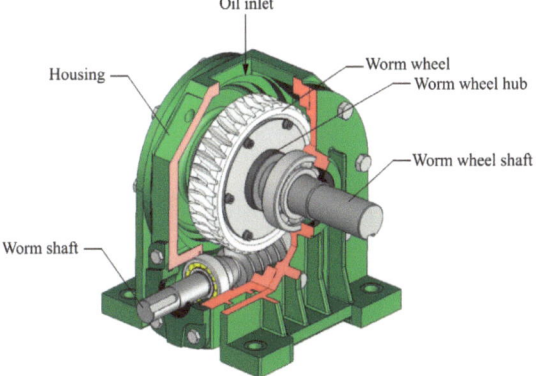

Figure 3. CAD model of the gearbox.

The worm surface was finely ground to the average surface roughness of R_a = 0.25 μm and then electropolished to modify the surface. Bronze worm wheels were produced by the gear hobbing process without any additional surface finishing (R_a = 1.0 μm). The average surface hardness of CuSn12 bronze wheels was 110 HV. The detailed worm gear pair geometry is provided in Table 3.

The employed lubricating oil was Castrol Alpha SP 150 mineral oil. The same lubricating oil was used in the previous research conducted by the authors [11]. This oil is recommended for industrial gearboxes with forced circulation, splash, or bath lubrication. The oil contains extreme-pressure (EP) additives for good thermal and load-carrying stability. The EP additives are compatible with both ferrous and non-ferrous materials. The selected oil conforms with AGMA 9005-E02 [28] and DIN 51517-3 [29] standards. The properties of the oil are provided in Table 4.

Table 3. Worm pair geometry.

Worm		Worm Wheel	
Number of threads, z_1 (-)	2	Number of teeth, z_2 (-)	36
Thread direction	right	Thread direction	right
Module, m (mm)	4	Module, m (mm)	4
Pitch diameter, d_{w1} (mm)	36	Pitch diameter, d_{w2} (mm)	144
Pressure angle, α (°)	20°	Pitch, p (mm)	12.566
Axial pitch, p_x (mm)	12.566	Reference diameter, d_{w2} (mm)	144
Lead, P (mm)	25.132	Profile shift, $x_2 \cdot m$ (mm)	0
Lead angle, γ_{m1}, (°)	12.529°		
Tooth thickness, s_{m1} (mm)	6.134		
Center distance, a (mm)	90		
Profile type	ZN		

Table 4. Main properties of the Alpha SP150 lubricating oil.

Density at 20 °C (kg/m³)	Kinematic Viscosity (mm²/s)		Viscosity Index (-)	Open Flash Point (°C)	Pour Point (°C)
	40 °C	100 °C			
890	150	14.5	>95	223	−21

All worm gear pairs were dimensionally inspected according to the DIN 3974 standard [30] using an ATOS optical 3D scanner. In addition, the tooth thickness of worm wheels was also measured. The accuracy of the scanner was verified by the acceptance test values based on VDI/VDE 2634 [31] provided in Table 5. Since all worms and worm wheels were produced in the same batch, the quality grades were relatively similar and are reported as average values in Table 6. Grade 1 indicates the highest accuracy and tight tolerances, whereas grade 12 indicates the lowest accuracy.

Table 5. Acceptance test values for ATOS 5 400 MV 320.

Parameter	Maximum Deviation (mm)	Allowable Limit (mm)
Probing error form (sigma)	0.001	0.004
Probing error (size)	0.004	0.015
Sphere spacing error	−0.008	0.012
Length measurement error	−0.006	0.027

Table 6. Quality grades.

Deviation	Worm		Wheel	
Single pitch deviation (axial)	f_{px}	9	-	-
Single pitch deviation	-	-	f_{p2}	11
Adjacent pitch difference	f_{ux}	3	f_{u2}	11
Total pitch deviation	F_{pz}	9	-	-
Total cumulative pitch deviation	-	-	F_{p2}	9
Total profile deviation	$F_{\alpha 1}$	11	$F_{\alpha 2}$	12
Runout	F_{r1}	2	F_{r2}	12

The experimental stand and a corresponding schematic representation are shown in Figure 4. A DC motor/generator (GEN) working in generator mode was used to provide the load that could be varied by the excitation current (ECC) supplied to the generator windings. The rotational speed of the driving electric motor (EM) was regulated by the frequency inverter (FI). Two shaft torque transducers (TT1 and TT2) were used to measure input and output torque from the worm pair gearbox (WP-GB). An additional gear multiplier (MP) was installed to increase the generator input shaft rotational speed. Electrical energy from the generator was supplied to electrical appliances in the test room, e.g., heaters. Oil circulation was carried out using two oil pumps (OP). The oil temperature of 60 °C was controlled by heating inside the oil tank (OT + OH) and by cooling through the oil chiller (OC). The oil was constantly filtered (OF) and fully replaced after each test run. A running-in of each worm wheel was conducted before each test run under gradually increasing load until an acceptable contact pattern was achieved.

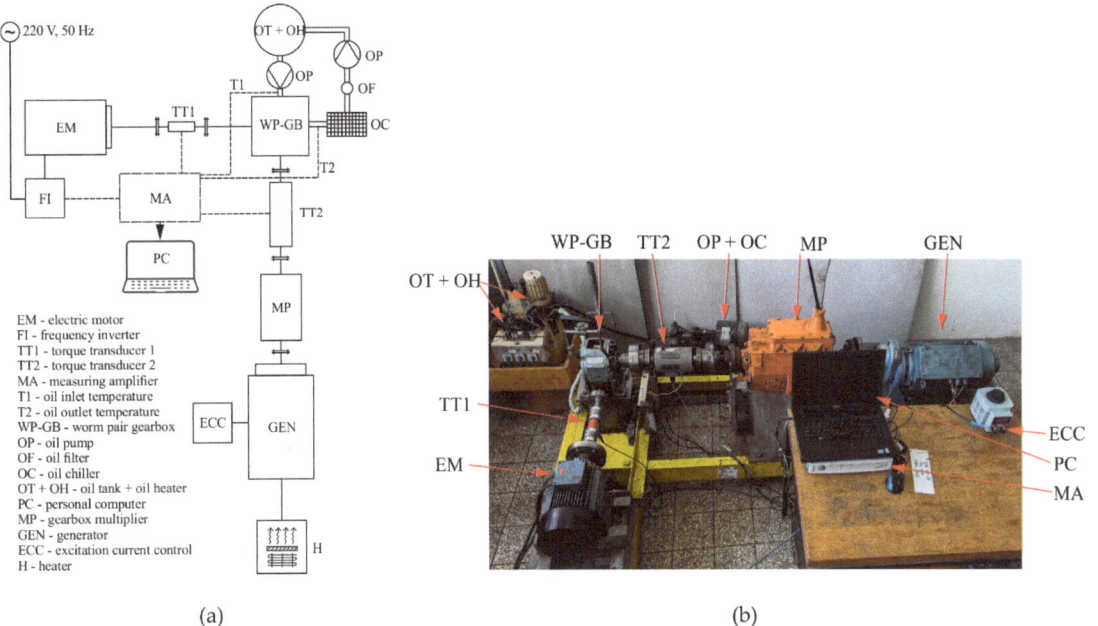

Figure 4. Worm pair experimental setup: (**a**) schematic representation; (**b**) experimental stand.

3. Results

In this section, the effects of electropolishing on material microstructure, hardness profile, surface topography, and chemical composition of 16MnCr5 steel worms are presented and discussed, followed up with the results of worm pair experimental testing and comparison with conventional worm pairs.

3.1. Microstructure and Chemical Composition

A comparison of ground and electropolished worm microstructure and chemical composition is presented in Figure 5. According to SEM images, the material martensite microstructure remained unchanged after electropolishing, as seen in Figure 5a,b. The difference in chemical composition at the surface level was found based on EDS line analysis. In Figure 5c, the position where iron significantly reduces represents the surface of the ground worm. Other elements remained at relatively low levels, regardless of the position.

On the other hand, on the surface of the electropolished worm, the reduction of iron coincides with the increase in oxygen and silicon. The average chemical composition is provided in Table 7. The presence of oxygen can be attributed to the formation of an oxide surface layer, common in mild steel electropolishing in H_2SO_4 solutions [11,32,33]. An increase in silicon can be explained by the formation of a viscous salt film on the surface of the electropolished workpiece by diffusion of acceptor anions from the electrolyte or incomplete dissolution of the films produced or precipitated on the metal surface [34]. An additional reason may be found in the uneven development of crystallographic etching.

Etching may be present when electropolishing conditions are employed which do not provide perfect polishing conditions.

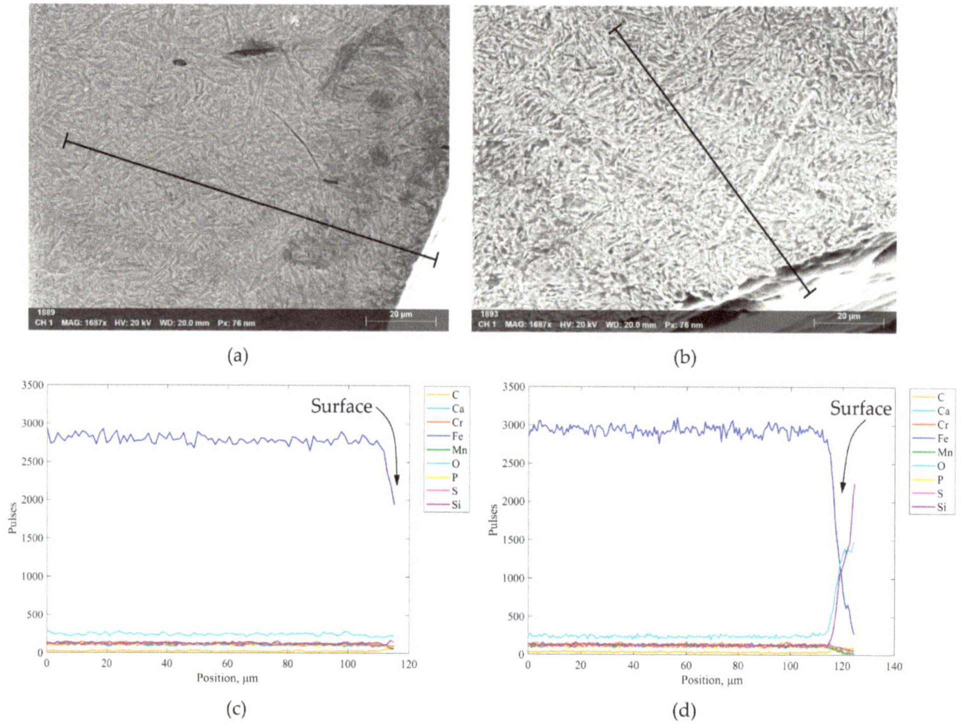

Figure 5. Microstructure and chemical composition of a steel worm: (**a**) SEM of a ground worm cross-section; (**b**) SEM of an electropolished worm cross-section; (**c**) EDS line analysis of a ground worm; (**d**) EDS line analysis of an electropolished worm.

Table 7. EDS line analysis of ground and electropolished worm, wt.%.

Element	Fe	O	P	S	Si	Ca	Mn	Cr	C
Ground worm	97.24	-	0.01	0.01	0.23	0.02	1.38	1.11	-
Electropolished worm	92.69	3.52	0.02	0.03	1.25	0.07	1.27	1.15	-

In this research, the goal was to avoid producing a smoothened and bright surface but rather a surface texture through the pitting phenomenon. Pitting occurs when electropolishing potentials are higher than optimal for conventional electropolishing. Gas evolution occurs when oxygen bubbles are trapped on the workpiece surface. Bubbles represent a place with lower resistance where current density sharply rises, increasing the dissolution rate and consequently generating a pitting hole [35]. Another explanation for pitting can be found in the broken bubble effect. When bubbles in the viscous layer diffuse to the solution, the bubbles tend to break due to the change in surface tension between the viscous layer and the solution. Broken bubbles create a shorter path for electrical current in the viscous layer, and this effect can increase the current density by as much as two times. Consequently, higher current density within the broken bubble increases the local dissolution rate and produces the pitting hole [36].

3.2. Hardness

Subsurface hardness profiles of investigated ground and electropolished worms are shown in Figure 6. A worm case-hardening depth of approximately 0.7 mm (±0.03 mm) was determined according to ISO 18203 standard using a limit of 550 HV [37]. Moreover, the case-hardening depth was in accordance with the recommended values for worm module $m = 4$ mm [38]. Presented profiles correspond to subsurface hardness with the closest measuring point to the surface at 20 µm. A minor decline in hardness can be observed up to the depth of 150 µm. It is important to note that this measured decline in subsurface hardness is inside the measurement uncertainty of the surface hardness measuring method and therefore should not be considered evidence of reduced subsurface hardness in electropolished steel. Also, it is possible that minor microstructural differences among measured samples influenced the presented subsurface hardness values. At larger depths from the surface, the difference in hardness becomes negligible.

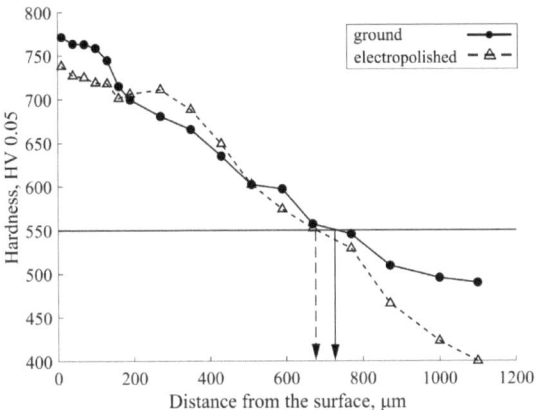

Figure 6. Hardness profiles of investigated ground and electropolished worms.

Additionally, surface hardness measurement of ground and electropolished worms was conducted. The results showed 870 HV for the ground and 560 HV for the electropolished surface. Additional measurements of electropolished steel worms were conducted after careful removal of the oxide surface layer using 1200-grit sanding paper to investigate the effect of the oxide surface layer on the surface hardness. The results showed a surface hardness of 848 HV. The literature also provides evidence of reduced surface hardness explained by two electropolishing effects. Since electropolishing is a material removal mechanism, a thin case-hardened surface layer of material might be removed from the surface which can affect the measured surface hardness [24]. Secondly, when electropolishing mild steel, the workpiece is passivated by the formation of an oxide and/or phosphate surface layer which is relatively softer than the base material [11,32,33]. Based on the earlier EDS analysis, the thickness of the oxide surface layer in 16MnCr5 steel worms was up to a couple of microns. The results showed that by removing the mentioned oxide surface layer, the surface hardness value of electropolished steel (848 HV) was similar to that of the ground steel (870 HV). According to the presented results, it can therefore be concluded that the formation of a soft oxide surface layer during electropolishing primarily contributes to lower values of surface hardness in electropolished 16MnCr5 steel worms.

3.3. Surface Topography

Surface profile measurements were performed using a Mitutoyo SJ-500 measuring instrument according to the ISO 4287 standard [39]. Examples of surface profiles with corresponding surface parameters of worm wheel and ground and electropolished worms are presented in Figure 7. Electropolished worms were characterized by many pits. Worm

EP-1 had deeper pits (≈2–5 μm) compared to shallower pits on worm EP-2 (≈1–2 μm). This observation can be attributed to different current densities as worm EP-1 was electropolished with a current density of 20 A/dm^2 compared to 15 A/dm^2 employed on worm EP-2. An additional noticeable difference was found in pit diameters. Lower current density employed on worm EP-2 enlarged existing pits/valleys of ground surface that were approximately 1 μm deep, resulting in 50–100 μm pit diameters. On the other hand, pit diameters on worm EP-1 surface were in the range of 20–50 μm.

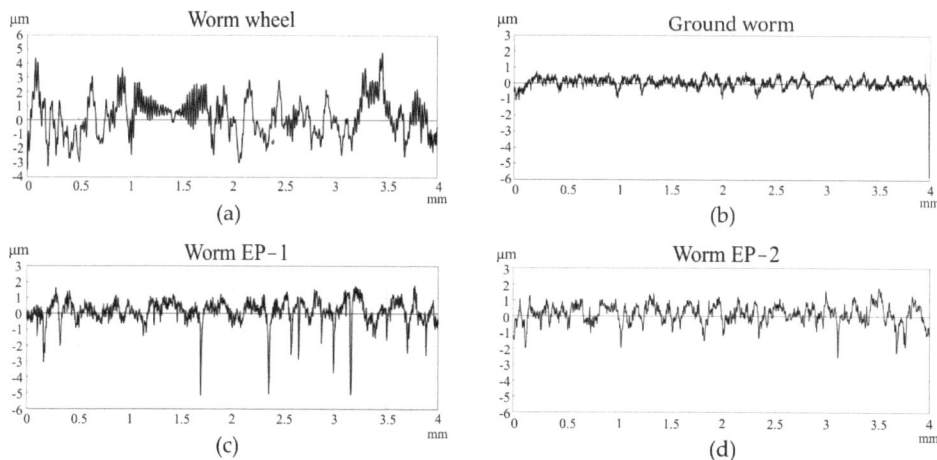

Figure 7. Surface profiles: (**a**) worm wheel; (**b**) ground worm; (**c**) worm EP-1; (**d**) worm EP-2.

Regarding surface texture properties, the goal was to produce a surface characterized by pits and negative R_{sk} and higher R_{ku} surface parameters. Pitted or dimpled surfaces are often characterized by negative R_{sk} and higher R_{ku} values; such surface textures have already shown friction-reduction characteristics [11,40–42]. Pits (or dimples) act as oil reservoirs, providing secondary lubrication and reducing friction. Furthermore, the significant increase in R_a was aimed to be avoided because it usually results in higher friction and wear. In the presented case, the average surface roughness increased from R_a = 0.25 for ground worms to $R_a \approx$ 0.50 for electropolished worms. This observation results in two types of surface textures that were applied and tested on worm pairs. The first one is the electropolished worm surface (worm EP-1), which has characteristics similar to surfaces mentioned in the literature, with surface parameters R_{sk} = −0.996 and R_{ku} = 7.228 and deeper but smaller pits. The second one is the electropolished worm surface (worm EP-2) with slightly different characteristics, namely R_{sk} = −0.008 and R_{ku} = 3.554, but shallower and larger pits. Surface parameter values are presented in detail in Table 8.

Table 8. Average values of surface parameters for worm wheels, ground, and electropolished worms.

	R_a (μm)	R_q (μm)	R_{sk}	R_{ku}	R_k (μm)	R_{pk} (μm)	R_{vk} (μm)
Worm wheel	1.09	1.35	0.63	2.82	3.37	1.43	1.14
Ground worm	0.25	0.33	−1.05	17.68	0.74	0.17	0.45
Worm EP-1	0.55	0.77	−0.99	7.23	1.43	0.48	1.54
Worm EP-2	0.48	0.6	0.01	3.55	1.53	0.40	0.90

Ground and electropolished worm surfaces are shown in Figure 8. Ground worm's surface is characterized by grinding marks, while the electropolished surface has many irregular pits/dimples distributed stochastically. The pit area density was 11% in the regions with more pits. In contrast, the pit area density in sparser regions reduced to 9% of the total surface area. In previous research [11], electropolishing 16MnCr5 steel discs with a current density of 30 A/dm^2 yielded a similar albeit slightly higher pit area density of 12%. Generally, 5% to 20% pit area density was reported as beneficial regarding friction reduction [43–47].

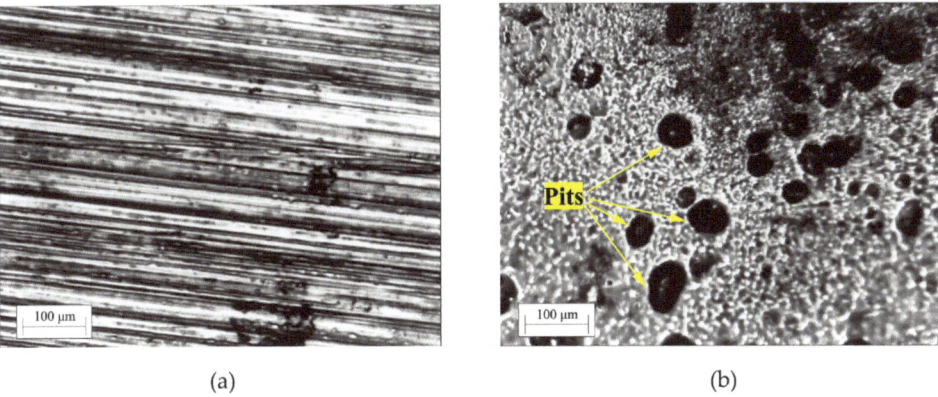

Figure 8. Worm surface: (**a**) ground worm; (**b**) worm EP-2.

3.4. Worm Gear Pair Experimental Tests

Experimental testing conditions, such as nominal rotational worm speed, nominal working load, and oil temperature, were selected to replicate realistic working conditions for employed worm gear pair geometry and materials. Also, working conditions were selected to satisfy surface durability (i.e., pitting resistance) and wear load capacity calculation procedures according to ISO/TS 14521 [26]. Tests were carried out for $N_L = 2 \cdot \times 10^6$ worm wheel load cycles. Tested worm pair combinations, testing conditions, and values calculated according to ISO/TS 14521 are presented in Table 9. Worm pair designations that will be used for the remainder of this paper can be found in the rightmost column of Table 9.

Table 9. Tested worm pair combinations and testing conditions.

Worm Material	Worm Wheel Material	Designation
16MnCr5, ground	CuSn12	WP-1
16MnCr5, ground	CuSn12	WP-2 *
16MnCr5, ground and electropolished	CuSn12	WP-EP-1
16MnCr5, ground and electropolished	CuSn12	WP-EP-2
Nominal worm shaft speed	1480 rpm	
Worm wheel nominal load, T_2	300 Nm	
Oil inlet temperature	60 °C	
Worm wheel load cycles, N_L	2×10^6	
Mean contact stress, σ_{Hm}	323 N/mm^2	
Limiting contact stress, σ_{HG}	415 N/mm^2	
Pitting resistance-safety factor, S_H	1.285	
Worm wheel tooth flank loss, δ_{Wn}	0.517 mm	
Limiting value of tooth flank loss, $\delta_{Wlim,n}$	1.171 mm	
Wear load capacity-safety factor, S_W	2.265	

* Occurrence of scuffing.

The comparison of the overall gearbox efficiency is presented in Figure 9. The average efficiencies of WP-1, WP-EP-1, and WP-EP-2 were 84.9%, 85.8%, and 90.1%, respectively. The worm gear pair WP-2 will be discussed separately due to the occurrence of scuffing.

Both worm gear pairs with electropolished worms showed improved efficiencies compared to WP-1. As mentioned, the electropolished surface had many pits that had multiple functions. Pits act as oil reservoirs, providing secondary lubrication, increasing oil film thickness, and enhancing heat dissipation which is attributed to an increase in overall efficiency. Also, pits can improve the tribological behaviour of contacting surfaces by entrapping abrasive particles [48]. In the case of worm gear pairs, the latter function can be particularly useful as worm wheel wear is usually one or two orders of magnitude larger than wear in steel gears in general. This phenomenon is especially pronounced during the running-in phase and early stages of the operation.

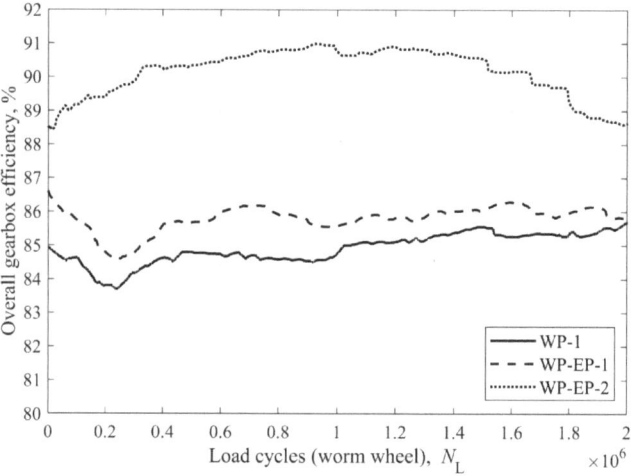

Figure 9. Comparison of worm gearbox efficiency.

A more noticeable increase in efficiency rating was found for WP-EP-2, where the worm surface was characterized by shallower but larger pits. The explanation for such a phenomenon lies in the geometry of the pits. In the locations of deeper pits in the contact zone, the lubricant experiences a severe decrease in pressure, leading to a lower viscosity that prevents the lubricant from completely separating the contacting surfaces [49]. Also, a countereffect of textured surfaces in non-conformal contacts (as in gears and worm pairs) can occur. An occurrence of cavitation inside deeper cavities in highly loaded non-conformal contacts can result in an increase in the coefficient of friction compared to an untextured surface [50]. On the other hand, the benefit of shallower pits can be explained by their smaller volume, which is more easily filled with oil to build additional hydrodynamic pressure and thus provide the separation of contacting surfaces [40].

An example of surface profiles after the test is shown in Figure 10. By comparing the surface profiles of the electropolished worm before and after the test, it can be concluded that minimal to no wear occurred on the electropolished surface of the worm as its surface topography remained unchanged. The values of R_a, R_{sk}, and R_{ku} were similar to those before the test (Figure 7). In addition, the surface pits produced by electropolishing did not deteriorate. The softer oxide surface layer was damaged and removed during worm gear pair operation, presumably already in the running-in phase [11,51]. The appearance of worm threads after electropolishing and after the test can be seen in Figure 11.

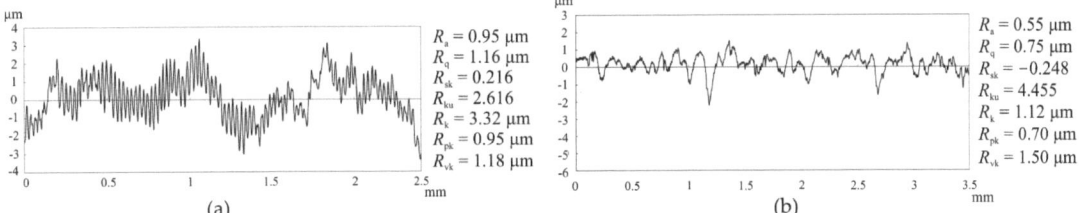

Figure 10. Surface profiles of worm pair after the test: (**a**) worm wheel; (**b**) electropolished worm.

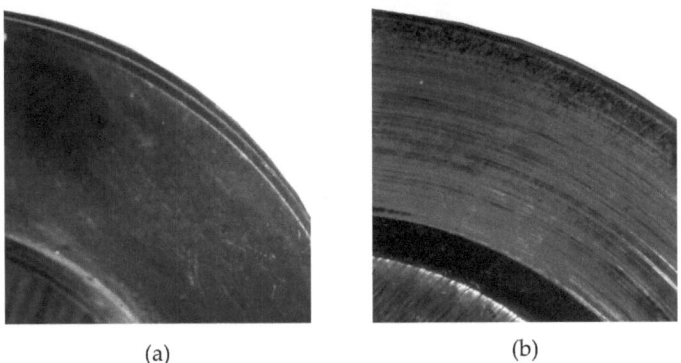

Figure 11. Worm surface: (**a**) after electropolishing; (**b**) after the test.

In contrast to conventional steel–steel gear material pairs, the steel–bronze pair favours the surface texture on a harder steel worm as most of the wear takes place on a softer bronze worm wheel. A similar finding was reported by Kasem et al. [48]. The authors suggested that surfaces textured with small and shallow dimples could reduce friction in lubricated contact with one contacting body having high wear resistance. Regardless of surface parameters used in the design of surface textures, the presented results imply that in the case of worm gear pairs, the focus should be placed primarily on the pit/dimple geometry and then complemented by surface parameters as a promising guide towards successful surface texturing application.

The additional phenomenon of electropolished steel surfaces in contact with bronze is higher bronze wear and therefore faster running-in [11]. By observing efficiency plots in Figure 9, faster running-in can be observed for WP-EP-2. At the same time, up to $N_L = 2.5 \times 10^5$, both WP-1 and WP-EP-1 had a period of decreasing efficiency, indicating a prolonged running-in process until the steady-state running was reached at approximately $N_L = 4 \times 10^5$. Higher wear was evident in both WP-EP-1 and WP-EP-2 compared to WP-1.

A visual representation of worm wheel wear after running-in and at the end of the test is given in Figure 12. By inspecting the lower right section of the worm wheel tooth flank, the ridge on the WP-EP-2 worm wheel is considerably larger compared to the other two worm wheels. Generally, the difference in wear between "after running-in" and "end-of-test" figures is negligible. This observation suggests that, as anticipated, most wear occurred during the running-in phase due to the nature of wear in sliding contacts [52]. The average tooth thicknesses of a worm wheel based on 36 measurements per worm wheel ($z_2 = 36$) are shown in Table 10. Tooth thickness was measured at the middle of the flank on the worm wheel pitch diameter.

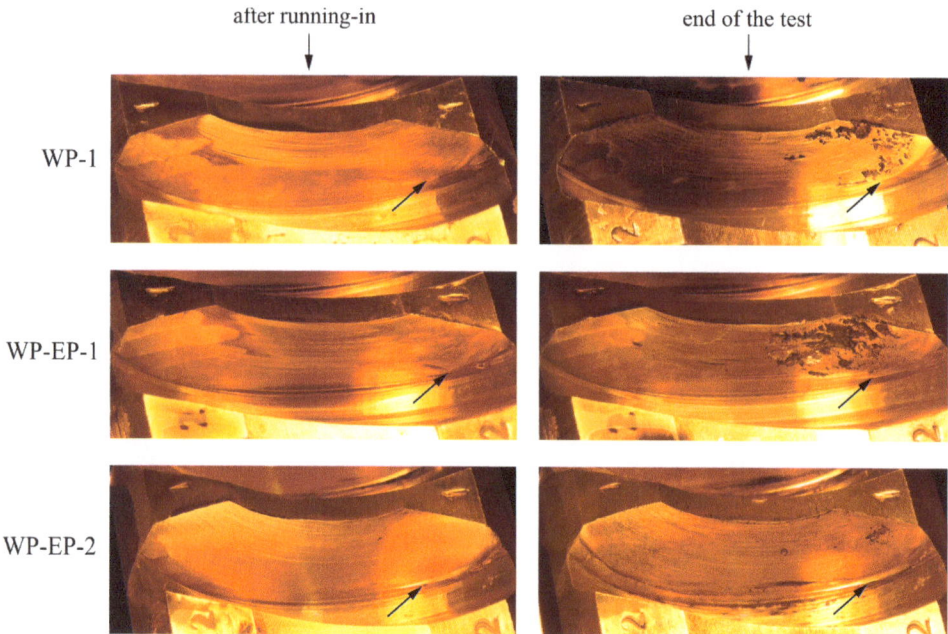

Figure 12. Worm wheel wear after the running-in phase and at the end of the test.

Table 10. Average worm wheel tooth thickness before and after the test.

Tooth Thickness, s_{m2}	Before the Test, mm	End of the Test, mm	Difference, mm
WP-1	6.150	6.099	0.051
WP-EP-1	6.174	6.107	0.067
WP-EP-2	6.086	5.832	0.254

Based on the tooth thickness measurement, the worm wheel in WP-EP-2 experienced the most wear, which can be observed in Figure 12. Larger wear during the running-in phase resulted in a larger initial contact pattern than worm wheels in WP-1 and WP-EP-1. Contact patterns can be recognized by abrasive grooves in the direction of sliding and matted surface appearance.

More clear representation of contact patterns can be found in Figure 13. Contact pattern adjustment in worm gear pairs is usually carried out by axial alignment of the worm wheel shaft. Larger contact patterns contribute to more uniform contact stress distribution, which can affect the uniformity and thickness of oil film. This observation also explains the difference in pitting, as the worm wheel in WP-EP-1 experienced the most pitting while having the smallest contact pattern among tested worm pairs. The relationship between contact patterns, pitting, and worm pair efficiency is relatively complex. In the case of 16MnCr5-CuSn12 worm pairs, Huber [53] concluded that neither larger contact patterns nor variation in load or pitting have an explicit or clear effect on the overall efficiency of worm gear pairs. In addition, pitting can provide additional lubrication in the contact zone as pits can act as oil reservoirs, resulting in pitted worm wheels having a similar efficiency rating to non-pitted ones [13,54].

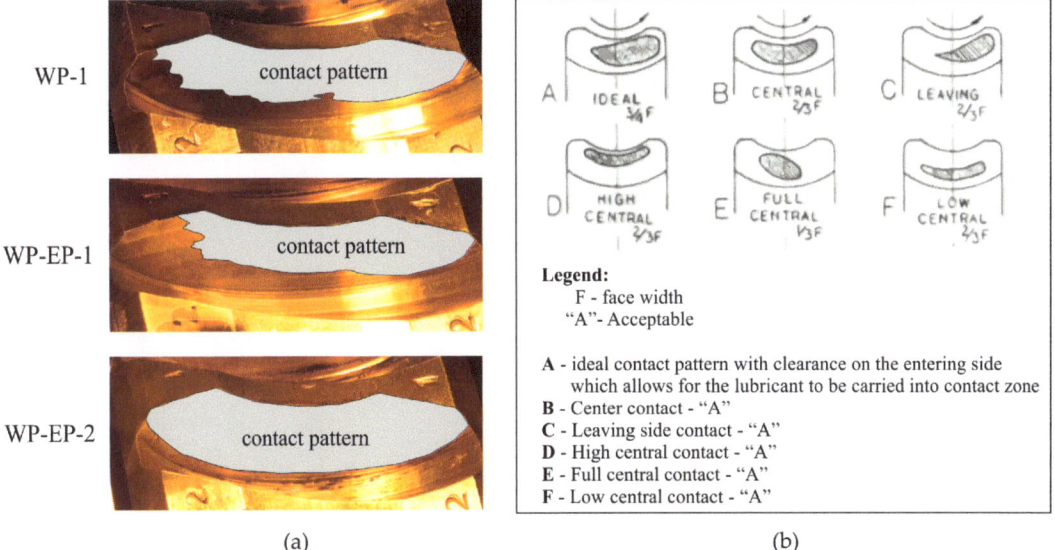

Figure 13. Worm wheel contact patterns: (**a**) tested worm wheels; (**b**) acceptable contact patterns according to [55].

As per Figure 13, the contact pattern established in WP-EP-2 can be considered a full contact pattern. In general, a full contact pattern in worm pairs is avoided as it limits the oil from entering the contact zone, promoting the onset of scuffing and, consequently, worm gear pair failure [55]. Instead, incomplete but acceptable contact patterns primarily positioned on the leaving side of the worm wheel tooth flank are frequently present (Figure 13). Although the full contact pattern was established in WP-EP-2, no problems with scuffing occurred. Moreover, the highest overall efficiency and lowest pitting were recorded for WP-EP-2. This finding suggests that electropolishing modifies the worm surface to provide improved or additional lubrication that can consistently and reliably support full contact patterns in worm gear pairs.

An additional test run was carried out to study the behaviour of conventional worm gear pair WP-2 under a full contact pattern. The results are presented in Figure 14. The test run ended prematurely due to two periods of intense scuffing. In worm gear pairs, scuffing occurs primarily as a consequence of improper lubricating conditions. Due to the dominant sliding motion in the mesh, the oil is constantly scrapped off from contacting surfaces. Therefore, an adequate amount of constant oil supply is necessary for worm gear pairs to function properly. When a full contact pattern is established, the contact spans across the whole worm wheel tooth flank surface, starting from entering and spreading towards the leaving side of the flank. If a contact is established on the entering side of the worm wheel tooth flank, the oil cannot properly supply the contact zone and the temperature in the contact zone will rise significantly. The rise in contact temperatures in turn reduces the oil viscosity, oil film thickness, and desorption of the lubricant from the surface. The sudden increase in heat in the contact zone leads to the micro-welding of asperities in contact. Finally, as surfaces leave the contact, the welded material is torn apart, promoting the adhesive wear of contacting surfaces.

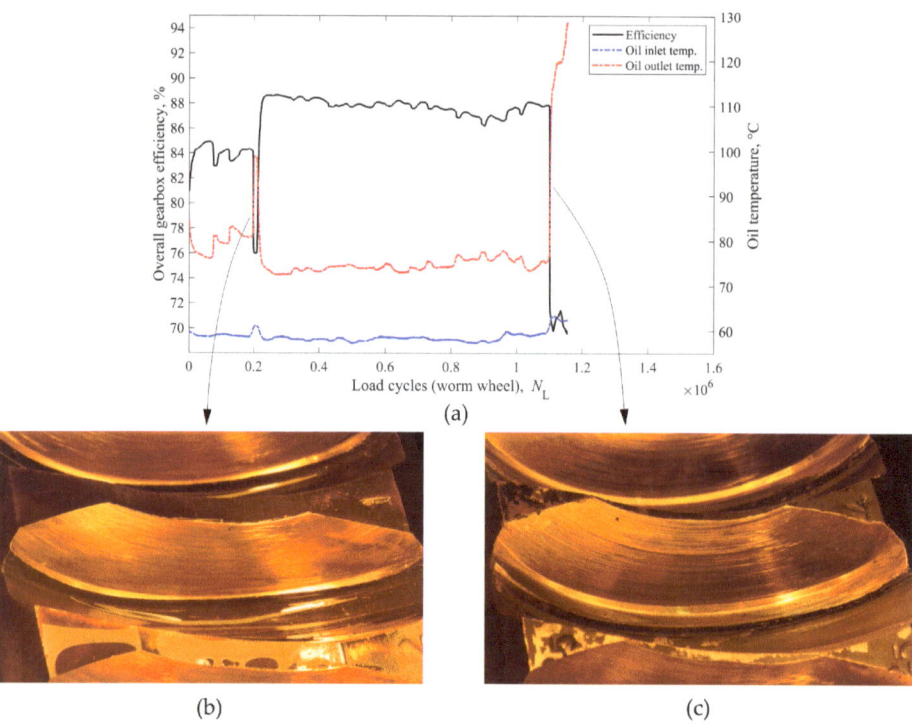

Figure 14. Scuffing of worm pair WP-2: (**a**) efficiency and oil temperatures; (**b**) worm wheel tooth flank wear at $N_L = 2 \times 10^5$; (**c**) worm wheel tooth flank wear at $N_L = 1.1 \times 10^6$.

In the example of WP-2, the first scuffing period was noticed at $N_L = 2 \times 10^5$, while the second was observed at $N_L = 1.1 \times 10^6$, causing worm pair failure. The overall efficiency was relatively high between the two scuffing periods, accompanied by high sliding wear. The high wear rate prevented pitting formation as the material was constantly removed from the worm wheel tooth flank. The high wear rate also sharply decreased worm wheel tooth thickness (Figure 14b). By the end of the test, worm wheel teeth were completely worn out and pointed (Figure 14c). Scuffing periods were characterized by low efficiency and a high rise in oil outlet temperature. The oil outlet temperature was approximately 74 °C in other tested worm pairs. In contrast, during two periods of severe scuffing observed in WP-2, the oil temperature was 81 °C during the first period and almost 130 °C before the test was aborted. These findings indicate that the operation of conventional worm gear pairs with an established full contact pattern is not sustainable as improper lubrication conditions can lead to the breakdown of lubrication and scuffing.

While this research focused primarily on the feasibility of surface texturing in worm gear pairs, the question of how different surface texture parameters influence the tribological behaviour of worm gear pairs remains open. Various surface textures can be produced and investigated by varying electropolishing conditions regarding overall efficiency rating and wear behaviour. The main difference and disadvantage of surface texture produced by electropolishing is its stochastic nature and inability to control the pit/dimple geometry and structure. On the other hand, electropolishing is a relatively simple, fast, and affordable process compared to other common surface texturing methods (e.g., laser surface texturing, chemical etching), especially when application on complex geometry is considered. Therefore, comparing different textures produced by electropolishing to optimize surface texture parameters for highly loaded non-conformal contacts as found in gears will be of future interest.

Furthermore, as steel worms experience little wear, there is a possibility to apply two-step electropolishing to "recycle" steel worms. Firstly, conventional electropolishing can be used to remove wear marks and then, in the second step, electropolishing with altered parameters can be employed to produce surface texture. In this way, the need for producing new worms is reduced. A potential disadvantage of this approach is the degraded dimensional accuracy of such worms as electropolishing removes surface material, leading to increased clearance and backlash in worm gear pairs. Nonetheless, this proposed approach remains open for future investigations.

4. Conclusions

In this paper, material characterization of electropolished 16MnCr5 steel was conducted, followed by an investigation of the tribological behaviour of electropolished 16MnCr5 steel worms paired with bronze CuSn12 worm wheels. The focus was on assessing efficiency rating, worm wheel wear, and surface topography. The results were compared to a conventional worm gear pair with a ground 16MnCr5 steel worm. The main results can be summarized as follows:

- Electropolishing modifies the surface by producing surface texture and an additional oxide surface layer. Surface texture is characterized by pits 1 to 5 μm deep and 20 to 100 μm in diameter with area densities from 9% to 11%.
- The formation of a softer oxide surface layer during electropolishing primarily contributes to lower surface hardness values in electropolished 16MnCr5 steel worms. However, no deterioration of the produced surface texture was noticed after the test runs.
- Worm pairs employing electropolished worms showed a higher efficiency rating, up to 5.2%, compared to conventional worm gear pairs with ground worms. This phenomenon can be attributed to improved lubrication conditions due to surface pits providing additional lubrication and promoting hydrodynamic pressure build-up.
- Improved lubrication conditions can reliably support full contact patterns and prevent scuffing in worm pairs with electropolished worms.
- Electropolished worms result in larger worm wheel wear, which can assist in faster running-in and better contact patterns.

This research presents the initial attempt at applying and assessing the effects of surface texturing in worm pairs. As such, some challenges and questions regarding the presented work remain open. As the density and geometry of the produced pits cannot be precisely controlled, different electropolishing parameters should be investigated to achieve different pit geometry, followed by an investigation of their tribological performance. Although the literature explains several of the electropolishing effects investigated in this research, the influence of electropolishing solution on surface topography and chemical composition is not yet fully known, as mild steel is not commonly electropolished. Lastly, besides gears and highly loaded non-conformal contacts, the application of electropolishing in conformal contacts is not yet investigated and will be the aim of future work.

Author Contributions: Conceptualization, R.M., S.J. and D.Ž.; methodology, R.M., S.J. and D.Ž.; validation, R.M. and D.Ž.; formal analysis, R.M., I.Č. and D.M.; investigation, R.M., S.J. and D.Ž.; resources, S.J. and D.Ž.; data curation, R.M., I.Č. and D.M.; writing—original draft preparation, R.M.; writing—review and editing, R.M., S.J., I.Č., D.M. and D.Ž.; visualization, R.M. and S.J.; supervision, S.J. and D.Ž. All authors have read and agreed to the published version of the manuscript.

Funding: This research received no external funding.

Data Availability Statement: The data presented in this study are available on request from the corresponding author.

Conflicts of Interest: The authors declare no conflicts of interest.

References

1. Opalić, M.; Žeželj, D.; Vučković, K. A New Method for Description of the Pitting Process on Worm Wheels Propagation. *Wear* **2015**, *332–333*, 1145–1150. [CrossRef]
2. Crosher, W.P. *Design and Application of the Worm Gear*; ASME Press: New York, NY, USA, 2002; ISBN 0-7918-0178-0.
3. Litvin, F.L.; Fuentes, A. *Gear Geometry and Applied Theory*, 2nd ed.; Cambridge University Press: New York, NY, USA, 2004; ISBN 978-0-511-23000-4.
4. Razdevich, S.P. *Dudley's Handbook of Practical Gear Design and Manufacture*, 3rd ed.; CRC Press: Boca Raton, FL, USA, 2016; ISBN 9781498753104.
5. Sharif, K.J.; Kong, S.; Evans, H.P.; Snidle, R.W. Contact and Elastohydrodynamic Analysis of Worm Gears Part 1: Theoretical Formulation. *Proc. Inst. Mech. Eng. Part C J. Mech. Eng. Sci.* **2001**, *215*, 817–830. [CrossRef]
6. Sharif, K.J.; Kong, S.; Evans, H.P.; Snidle, R.W. Contact and Elastohydrodynamic Analysis of Worm Gears Part 2: Results. *Proc. Inst. Mech. Eng. Part C J. Mech. Eng. Sci.* **2001**, *215*, 831–846. [CrossRef]
7. Kong, S.; Sharif, K.; Evans, H.P.; Snidle, R.W. Elastohydrodynamics of a Worm Gear Contact. *J. Tribol.* **2001**, *123*, 268–275. [CrossRef]
8. Benedetti, M.; Fontanari, V.; Torresani, E.; Girardi, C.; Giordanino, L. Investigation of Lubricated Rolling Sliding Behaviour of WC/C, WC/C-CrN, DLC Based Coatings and Plasma Nitriding of Steel for Possible Use in Worm Gearing. *Wear* **2017**, *378–379*, 106–113. [CrossRef]
9. Fontanari, V.; Benedetti, M.; Straffelini, G.; Girardi, C.; Giordanino, L. Tribological Behavior of the Bronze-Steel Pair for Worm Gearing. *Wear* **2013**, *302*, 1520–1527. [CrossRef]
10. Fontanari, V.; Benedetti, M.; Girardi, C.; Giordanino, L. Investigation of the Lubricated Wear Behavior of Ductile Cast Iron and Quenched and Tempered Alloy Steel for Possible Use in Worm Gearing. *Wear* **2016**, *350–351*, 68–73. [CrossRef]
11. Mašović, R.; Miler, D.; Čular, I.; Jakovljević, S.; Šercer, M.; Žeželj, D. The Effect of Steel Electropolishing on the Tribological Behavior of a Steel–Bronze Pair in the Mixed and Boundary Lubrication Regimes. *Lubricants* **2023**, *11*, 325. [CrossRef]
12. Muminović, A.; Repčić, N.; Žeželj, D. The Efficiency of Worm Gears Lubricated with Oils of Mineral and Synthetic Bases. *Trans. Famena* **2013**, *37*, 65–72.
13. Mautner, E.M.; Sigmund, W.; Stemplinger, J.P.; Stahl, K. Efficiency of Worm Gearboxes. *Proc. Inst. Mech. Eng. Part C J. Mech. Eng. Sci.* **2016**, *230*, 2952–2956. [CrossRef]
14. Litvin, F.L.; Gonzalez-Perez, I.; Yukishima, K.; Fuentes, A.; Hayasaka, K. Design, Simulation of Meshing, and Contact Stresses for an Improved Worm Gear Drive. *Mech. Mach. Theory* **2007**, *42*, 940–959. [CrossRef]
15. Sohn, J.; Park, N. Modified Worm Gear Hobbing for Symmetric Longitudinal Crowning in High Lead Cylindrical Worm Gear Drives. *Mech. Mach. Theory* **2017**, *117*, 133–147. [CrossRef]
16. Simon, V.V. Characteristics of a New Type of Cylindrical Worm-Gear Drive. *J. Mech. Des.* **1998**, *120*, 139–146. [CrossRef]
17. Rosenkranz, A.; Grützmacher, P.G.; Gachot, C.; Costa, H.L. Surface Texturing in Machine Elements—A Critical Discussion for Rolling and Sliding Contacts. *Adv. Eng. Mater.* **2019**, *21*, 1900194. [CrossRef]
18. Todić, A.; Djordjević, M.T.; Arsić, D.; Džunić, D.; Lazić, V.; Aleksandrović, S.; Krstić, B. Influence of Vanadium Content on the Tribological Behaviour of X140CrMo12-1 Air-Hardening Steel. *Trans. FAMENA* **2022**, *46*, 15–22. [CrossRef]
19. Britton, R.D.; Elcoate, C.D.; Alanou, M.P.; Evans, H.P.; Snidle, R.W. Effect of Surface Finish on Gear Tooth Friction. *J. Tribol.* **2000**, *122*, 354–360. [CrossRef]
20. Gupta, N.; Tandon, N.; Pandey, R.K. An Exploration of the Performance Behaviors of Lubricated Textured and Conventional Spur Gearsets. *Tribol. Int.* **2018**, *128*, 376–385. [CrossRef]
21. Gupta, N.; Tandon, N.; Pandey, R.K.; Vidyasagar, K.E.C.; Kalyanasundaram, D. Tribological and Vibration Studies of Textured Spur Gear Pairs under Fully Flooded and Starved Lubrication Conditions. *Tribol. Trans.* **2020**, *63*, 1103–1120. [CrossRef]
22. Petare, A.; Deshwal, G.; Palani, I.A.; Jain, N.K. Laser Texturing of Helical and Straight Bevel Gears to Enhance Finishing Performance of AFF Process. *Int. J. Adv. Manuf. Technol.* **2020**, *110*, 2221–2238. [CrossRef]
23. Nakatsuji, T.; Mori, A. Tribological Properties of Eiectrolytically Polished Surfaces of Carbon Steel. *Tribol. Trans.* **1998**, *41*, 179–188. [CrossRef]
24. Nakatsuji, T.; Mori, A. Pitting Durability of Electrolytically Polished Medium Carbon Steel Gears—Succeeding Report. *Tribol. Trans.* **1999**, *42*, 393–400. [CrossRef]
25. Li, W.; Lu, L.; Zeng, D. The Contribution of Topography Formed by Fine Particle Peening Process in Reducing Friction Coefficient of Gear Steel. *Tribol. Trans.* **2020**, *63*, 9–19. [CrossRef]
26. *ISO/TS 14521*; Gear—Calculation of Load Capacity of Worm Gears. International Organization for Standardization: Geneva, Switzerland, 2020.
27. *ASTM-E1558*; Standard Guide for Electrolytic Polishing of Metallographic Specimens. ASTM International: West Conshohocken, PA, USA, 1999.
28. *ANSI/AGMA 9005-E02*; Industrial Gear Lubrication. American Gear Manufacturers Associations: Washington, DC, USA, 2014.
29. *DIN 5157-3*; Lubricants—Lubricating oils—Part 3: Lubricating oils CLP. Minimum Requirements: Berlin, Germany, 2018.
30. *DIN 3974-1*; Toleranzen Für Schneckengetriebe-Verzahnungen—Teil 1: Grundlagen. Deutsches Institut für Normung: Berlin, Germany, 1995.

31. *VDI/VDE 2634*; Optical 3D-Measuring Systems—Multiple View Systems Based on Area Scanning. VDI/VDE-Gesellschaft Mess- und Automatisierungstechnik: Berlin, Germany, 2008.
32. Gabe, D.R. Electropolishing of Mild Steel in Phosphoric and Perchloric Acid Containing Electrolytes. *Corros. Sci.* **1973**, *13*, 175–185. [CrossRef]
33. Nakatsuji, T.; Mori, A. The Tribological Effect of Electrolytically Produced Micro-Pools and Phosphoric Compounds on Medium Carbon Steel Surfaces in Rolling—Sliding Contact. *Tribol. Trans.* **2001**, *44*, 173–178. [CrossRef]
34. Landolt, D. Fundamental Aspects of Electropolishing. *Electrochim. Acta* **1987**, *32*, 1–11. [CrossRef]
35. Han, W.; Fang, F. Fundamental Aspects and Recent Developments in Electropolishing. *Int. J. Mach. Tools Manuf.* **2019**, *139*, 1–23. [CrossRef]
36. Lee, S.J.; Chen, Y.H.; Hung, J.C. The Investigation of Surface Morphology Forming Mechanisms in Electropolishing Process. *Int. J. Electrochem. Sci.* **2012**, *7*, 12495–12506. [CrossRef]
37. *ISO 18203*; Steel—Determination of the Thickness of Surface-Hardened Layers. International Organization for Standardization: Geneve, Switzerland, 2016.
38. Errichello, R.; Milburn, A. Optimum Carburized and Hardened Case Depth. *Gear Technol. January/Febr.* **2020**, 58–65.
39. *ISO 4287:1997*; Geometrical Product Specifications (GPS)—Surface Texture: Profile Method—Terms, Definitions and Surface Texture Parameters. International Organization for Standardization: Geneva, Switzerland, 1997.
40. Sedlaček, M.; Gregorčič, P.; Podgornik, B. Use of the Roughness Parameters S Sk and S Ku to Control Friction—A Method for Designing Surface Texturing. *Tribol. Trans.* **2017**, *60*, 260–266. [CrossRef]
41. Sedlaček, M.; Podgornik, B.; Vižintin, J. Planning Surface Texturing for Reduced Friction in Lubricated Sliding Using Surface Roughness Parameters Skewness and Kurtosis. *Proc. Inst. Mech. Eng. Part J J. Eng. Tribol.* **2012**, *226*, 661–667. [CrossRef]
42. Akamatsu, Y.; Tsushima, N.; Goto, T.; Hibi, K. Influence of Surface Roughness Skewness on Rolling Contact Fatigue Life. *Tribol. Trans.* **1992**, *35*, 745–750. [CrossRef]
43. Yue, H.; Schneider, J.; Deng, J. Laser Surface Texturing for Ground Surface: Frictional Effect of Plateau Roughness and Surface Textures under Oil Lubrication. *Lubricants* **2024**, *12*, 22. [CrossRef]
44. Schneider, J.; Braun, D.; Greiner, C. Laser Textured Surfaces for Mixed Lubrication: Influence of Aspect Ratio, Textured Area and Dimple Arrangement. *Lubricants* **2017**, *5*, 32. [CrossRef]
45. Ali, S.; Kurniawan, R.; Moran, X.; Ahmed, F.; Danish, M.; Aslantas, K. Effect of Micro-Dimple Geometry on the Tribological Characteristics of Textured Surfaces. *Lubricants* **2022**, *10*, 328. [CrossRef]
46. Ghaei, A.; Khosravi, M.; Badrossamay, M.; Ghadbeigi, H. Micro-Dimple Rolling Operation of Metallic Surfaces. *Int. J. Adv. Manuf. Technol.* **2017**, *93*, 3749–3758. [CrossRef]
47. Ronen, A.; Etsion, I.; Kligerman, Y. Friction-Reducing Surface-Texturing in Reciprocating Automotive Components. *Tribol. Trans.* **2001**, *44*, 359–366. [CrossRef]
48. Zhang, H.; Pei, X.; Jiang, X. Anti-Wear Property of Laser Textured 42CrMo Steel Surface. *Lubricants* **2023**, *11*, 353. [CrossRef]
49. Mourier, L.; Mazuyer, D.; Ninove, F.-P.; Lubrecht, A.A. Lubrication Mechanisms with Laser-Surface-Textured Surfaces in Elastohydrodynamic Regime. *Proc. Inst. Mech. Eng. Part J J. Eng. Tribol.* **2010**, *224*, 697–711. [CrossRef]
50. Rosenkranz, A.; Szurdak, A.; Gachot, C.; Hirt, G.; Mücklich, F. Friction Reduction under Mixed and Full Film EHL Induced by Hot Micro-Coined Surface Patterns. *Tribol. Int.* **2016**, *95*, 290–297. [CrossRef]
51. Ueda, M.; Spikes, H.; Kadiric, A. Influence of Black Oxide Coating on Micropitting and ZDDP Tribofilm Formation. *Tribol. Trans.* **2022**, *65*, 242–259. [CrossRef]
52. Khonsari, M.M.; Ghatrehsamani, S.; Akbarzadeh, S. On the Running-in Nature of Metallic Tribo-Components: A Review. *Wear* **2021**, *474–475*, 203871. [CrossRef]
53. Huber, G. *Untersuchungen Über Flankentragfähigkeit Und Wirkungsgrad von Zylinderschneckengetrieben (Evolventenschnecken)*; TU München: Munich, Germany, 1978.
54. Sievers, B.; Gerke, L.; Predki, W.; Pohl, M. Verschleiß- Und Grübchen Tragfähigkeit von Bronze- Schneckenrädern. *Giesserei* **2011**, *98*, 36–45.
55. Janninck, W. Contact Surface Topology of Worm Gear Teeth. *Gear Technol. March/April* **1988**, 31–47.

Disclaimer/Publisher's Note: The statements, opinions and data contained in all publications are solely those of the individual author(s) and contributor(s) and not of MDPI and/or the editor(s). MDPI and/or the editor(s) disclaim responsibility for any injury to people or property resulting from any ideas, methods, instructions or products referred to in the content.

Article

Research on the Sealing Performance of Segmented Annular Seals Based on Fluid–Solid–Thermal Coupling Model

Zhenpeng He [1,2], Lanhao Jia [3], Jiaxin Si [4,*], Ning Li [4,5], Hongyu Wang [3], Baichun Li [1], Yuhang Guo [3], Shijun Zhao [3] and Wendong Luo [2]

1 Aeronautical Engineering Institute, Civil Aviation University of China, Tianjin 300300, China; hezhenpeng@tju.edu.cn (Z.H.); baichun_li@hotmail.com (B.L.)
2 Aeronautical Mechanical & Electrical Engineering Institute, Chongqing Aerospace Polytechnic, Chongqing 400021, China; luowendong@cqht.edu.cn
3 Sino-European Institute of Aviation Engineering, Civil Aviation University of China, Tianjin 300300, China; 2022122013@cauc.edu.cn (L.J.); 2023122021@cauc.edu.cn (H.W.); 2022122012@cauc.edu.cn (Y.G.); 200540135@cauc.edu.cn (S.Z.)
4 AECC Hunan Aviation Powerplant Research Institute, Zhuzhou 412002, China; lining-608@163.com
5 Science and Technology on Helicopter Tranmission Laboratory, Zhuzhou 412002, China
* Correspondence: 19158706887@163.com

Abstract: High-speed segmented annular seals are often subjected to friction and wear, and the groove design on the sealing surface can effectively suppress this loss. For the purpose of improving the sealing performance, the segmented annular seal models of three structures are established, and the accuracy of the calculation model is verified by comparing with the previous results. Through fluid–solid–thermal coupled analysis, the flow field characteristics, opening characteristics, and leakage characteristics of the segmented annular seal under high working condition parameters were studied. The results show that the setting of the shallow groove forms the hydrodynamic effect by squeezing and hindering the flow of fluid in the clearance. The increase in rotational speed and pressure difference can promote the increase in the opening force, while the temperature has no significant effect on the opening of the seal. Seals with ladder-like grooves have the best opening performance, and seals without shallow grooves are already difficult to open under conditions of high pressure difference and large spring forces. Temperature and pressure difference are the main factors affecting the leakage of the seal, while the influence of the rotation speed is small. When the sealed pressure increases from 0.15 MPa to 0.4 MPa, the maximum increase in the leakage of the seal with specific groove design is 4.657 times the original. As the temperature rises from 420 K to 620 K, the maximum decrease in the three structures is up to 22.9%. Among the seals of the three structures, seals with ladder-like grooves have medium leakage. This research will contribute to the improvement of research methods for the sealing performance of segmented annular rings, especially for the evaluation of groove design and opening characteristics.

Keywords: segmented annular seal; groove design; fluid-solid-thermal coupling; opening characteristics; leakage characteristics

Citation: He, Z.; Jia, L.; Si, J.; Li, N.; Wang, H.; Li, B.; Guo, Y.; Zhao, S.; Luo, W. Research on the Sealing Performance of Segmented Annular Seals Based on Fluid–Solid–Thermal Coupling Model. *Lubricants* 2024, 12, 407. https://doi.org/10.3390/lubricants12120407

Received: 5 November 2024
Revised: 20 November 2024
Accepted: 20 November 2024
Published: 22 November 2024

Copyright: © 2024 by the authors. Licensee MDPI, Basel, Switzerland. This article is an open access article distributed under the terms and conditions of the Creative Commons Attribution (CC BY) license (https://creativecommons.org/licenses/by/4.0/).

1. Introduction

An aeroengine contains various rotating machine systems that have a large number of dynamic–static contact surfaces, leading to significant leakage issues. Therefore, efficient sealing devices are required to reduce the high-pressure fluid leakage of the engine on the mechanical contact surfaces. It has been shown that the performance of the sealing devices in the aeroengine has a significant impact on the efficiency of engine components and the overall operation of the engine [1]. Meanwhile, with the rapid advancement of the aviation industry, aircraft engines are evolving toward higher speeds, higher temperatures, and greater pressure differentials. Consequently, the operating conditions for dynamic seals in high-parameter aircraft engines are becoming increasingly stringent, and traditional sealing methods can no longer meet the requirements. As an advanced seal with excellent performance, segmented annular seals have garnered widespread attention.

The history of segmented annular seals dates back to the 1970s when NASA [2] invented the segmented annular seal based on the face seal. It is a type of radial seal that has self-compensation capabilities under high working condition parameters. Through the design of a split ring, the torque of the segments during wear is reduced, and the friction of the segments is uniform. Compared to traditional floating ring seals, segmented annular seals have lower leakage and higher stability [3–5]. Due to the excellent performance of the segmented annular seals, the seals have been widely used in the main bearing chamber seal of aeroengines. However, the friction and wear between the inner surface of the segments and the runway have increasingly constrained the seal's service life. To improve the sealing performance, a segmented annular seal with shallow grooves on the inner surface was developed [6]. This sealing device makes full use of the hydrodynamic effect, enabling high-speed and frictionless sealing, significantly enhancing the reliability and service life of the seals. The balance between achieving optimal sealing clearance and minimizing the risk of rotor rubbing has become a challenge that scholars need to address.

The research methods for segmented annular seals initially focused on independent fields and experiments. Oike et al. [7] solved the opening force and leakage rate of the graphite seal with Rayleigh ladder-like grooves based on the incompressible fluid Reynolds equation, and the results were verified through experiments. Arghir et al. [8] conducted a numerical analysis to study the effects of operating parameters and geometric parameters on seal leakage and frictional power consumption. The results indicated that the depth of the shallow grooves is the primary factor affecting the seal's leakage rate, while the influence of groove length and groove width are inconspicuous. Bai et al. [9,10] studied the hydrodynamic characteristics of shallow grooves of the segmented annular seals based on gas lubrication theory. The finite difference method was used to analyze the influence of operating parameters and groove geometric parameters on the gas film pressure and leakage rate. The results showed that increasing the groove width, groove number, and groove depth enhanced the hydrodynamic effect of the seal. Additionally, as the ratio of groove width to groove radius increased, the dimensionless average pressure initially increased and then decreased. Li et al. [11] studied the variation in the leakage of seals by combining numerical calculations with experimental methods, focusing on key parameters such as pressure, rotational speed, and the circumferential spring preload of the seal. With the advancement of technology, the limitations of the independent field model are becoming more and more significant, and the research based on it cannot meet the needs of scholars. Wang et al. [12] established a model for the static and rotordynamic characteristics based on the local differential quadrature (LDQ) method. The results show that structural parameters of the shallow grooves are the key parameters that influence the lift force, stiffness coefficient, and damping coefficient of segmented annular seals, while have little influence on the leakage.

Under the action of high temperature and high pressure airflow, the segments will undergo obvious deformation [13]. To further investigate the performance of the seal under high parameter conditions, the multi-physics coupling model has become an effective method. Yun et al. [14] considered the uneven distribution of spring force and performed structure–heat coupling simulation calculations on the main components of the circumferential sealing device. The results showed that the uneven distribution of spring force leads to an increase in clearance and leakage. Chen et al. [15] constructed a fluid–solid coupling model whose fluid domain included the main seal between the segments and the runway. The main mechanical parameters affecting the segments' deformation were analyzed by using orthogonal experimental design. Furthermore, the influence of joint shapes, joint clearances, and the groove profile of the auxiliary seal interface on the deformation of segments were discussed. Ma et al. [16] considered the influence of the temperature field and established numerical calculation models for the solid domain and flow field of segmented annular seals. By optimizing the parameters of the shallow grooves based on numerical analysis, the seals achieve stable frictionless operation at high speed and maintain a low leakage rate, resulting in the reduction of the temperature rise and the frictional wear of the seal. Yan et al. [17] proposed a fluid–solid–thermal coupling model that included the auxiliary seal between the segments and the case. The characteristics of the flow field, temperature field, and structural field of the segment annular seals were analyzed.

In recent years, for the purpose of improving the performance of the segmented annular seal, several studies on structural optimization design have been made. Arghir et al. [18–20] are devoted to the development of a kind of rotor micro-textured segmented annular seal and they have done various kinds of research. It was found that the textured rotor can increase the leakage, while decreasing the torque and temperature. Fan et al. [21] established a numerical model based on two-phase flow theory to analyze the flow field characteristics and leakage performance of three types of rotor micro-textured segmented annular seals. The results showed that the helical groove structure exhibited the best sealing performance and hydrodynamic effect, effectively reducing the friction and wear of the graphite segments. Ren et al. [22] designed a kind of segmented annular seal with triangular grooves and established the numerical solution model of fluid–solid–thermal coupling. Through the calculation under the condition of high working condition parameters, it is found that the seal with triangular grooves has a large opening force and a small leakage, and the sealing performance is good.

According to the summary of the literature, the current simulation study of segmented annular seals mainly focuses on the independent field model of flow or structure. Some progress has been made by the utilization of multi-physics coupling models, while the established models are relatively simple. The influence of segment deformation on the opening performance and leakage characteristics under fluid–solid–thermal coupling action is not considered comprehensively. Meanwhile there is still room for further optimization of the sealing structure. In this paper, various forms of fluid–solid–thermal coupling models of the segmented annular seal are constructed, and the flow field characteristics, opening characteristics, and leakage characteristics without grooves, with rectangular grooves, and with ladder-like grooves under different working conditions are analyzed. Finally, the structure with the best performance is selected. The purpose of the study is to provide a basis for the design and optimization of segmented annular seals with a high linear speed and large pressure difference, and lay a foundation for the improvement of the sealing performance of aeroengines.

2. Numerical Model

2.1. Working Principle

2.1.1. Theoretical Model

A typical segmented annular seal with shallow grooves is shown in Figure 1; it is composed of several graphite segments, a case, a circumferential spring, and axial springs. In the radial direction, the segments are pressed by the circumferential spring against the rotor. To prevent the radial displacement of the segments, the compressed springs press the segments to the case. The segments are connected in turn through lap joints. There is a certain circumferential assembly clearance between the joints to compensate for the wear and manufacturing errors of the segments during the working process. The inner surface of segments and the outer surface of the runway form the main seal, which is the main leakage path of the fluid.

2.1.2. Force Analysis

The force schematic of the segments is shown in Figure 2. In the axial direction, F_{as} is the axial spring force, F_{m2} is the force of the casing on a segment. In the radial direction, F_{m1} is the runway support, F_p is the dynamic pressure buoyancy, F_{cs} is the circumferential spring force, F_a is the circumferential closing force generated by the medium, and F_f is the friction of the casing on a segment, where $F_a + F_{cs} + F_f$ is known as the opening resistance, and F_p is known as the opening force.

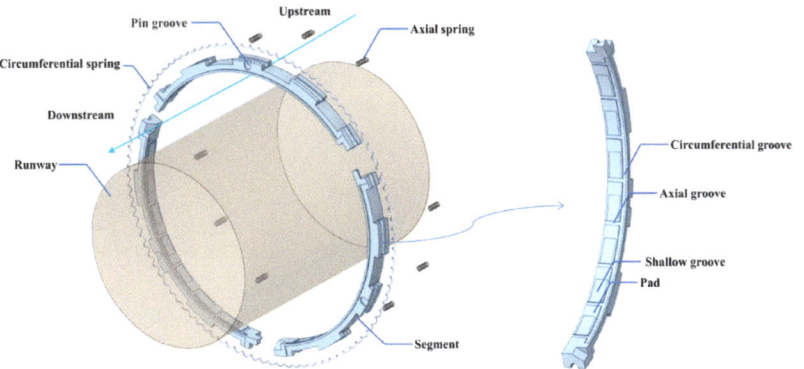

Figure 1. Segmented annular seal with enhanced lift effects.

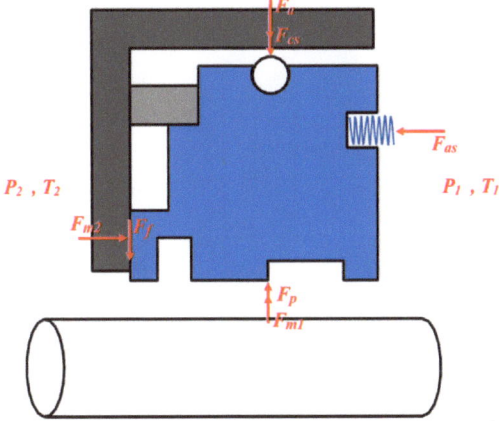

Figure 2. Force schematic of a segmented annular seal.

Segments in a steady state satisfy the following relations:

$$F_a + F_{cs} + F_f = F_p + F_{m1} \qquad (1)$$

$$F_f = \mu F_{m2} \qquad (2)$$

where μ is the friction coefficient and the value is taken as 0.1 in this study.

The opening process of a segmented annular seal is shown in Figure 3. Initially, when the rotor is at rest, the segments are pressed against the runway by the opening resistance. The opening force increases and the runway support decreases as the rotation speed increases. When the runway support force drops to zero, the seal opens. When the seal clearance increases, the dynamic pressure effect is weakened, the opening force decreases, and the seal tends to close; when the clearance decreases, the dynamic pressure effect increases, the opening force increases, and the seal tends to open. Ultimately, a dynamic balance of opening force and opening resistance is achieved with a steady clearance.

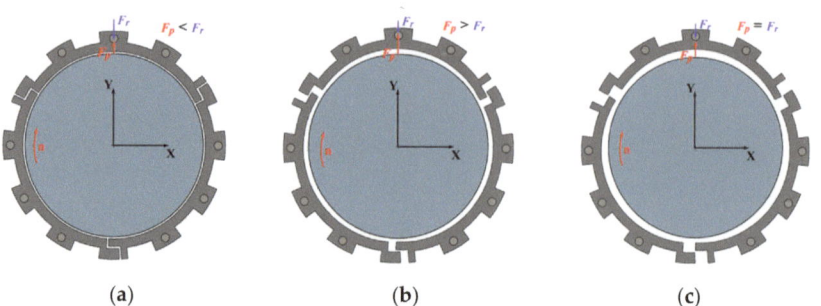

Figure 3. Opening process of a segmented annular seal: (**a**) initial state of the seal; (**b**) rising state of the seal; and (**c**) balance state of the seal.

2.2. Method of Analysis

2.2.1. Method of Flow Field Characteristics Analysis

In this study, the fluid in the clearance is considered as an ideal compressible gas, and the governing equations are shown as the following equation [23]:

$$\begin{cases} \frac{\partial \rho}{\partial t} + U \cdot \nabla \rho + \rho \nabla \cdot U = 0 \\ \rho \frac{DU}{Dt} = \rho f + \nabla \cdot (F_{ij} e_i e_j) \\ \frac{\partial (E + \frac{U^2}{2})}{\partial t} + U \cdot \nabla (E + \frac{U^2}{2}) = f \cdot U + \dot{q} + \frac{\lambda}{\rho} \cdot \Delta t_{temp} + \frac{1}{\rho} \nabla \cdot (U \cdot F_{ij} e_i e_j) \end{cases} \quad (3)$$

where p is the pressure on the fluid microelement, t is the time, U is the velocity vector, ρ is the density, F_{ij} is the component of stress tensor, e_i and e_j are the components unit stress, E is the internal energy, $\dot{q}$ is the generation in unit mass of fluid micro-mass, λ is the Fourier thermal conductivity, and Δt_{temp} is the temperature gradient.

The fluid in the clearance is set as Newtonian fluid, which maintains a constant viscosity regardless of the applied shear rate, and its constitutive equation is as follows [24]:

$$\begin{cases} P_{ij} = -p\delta_{ij} + 2\mu(S_{ij} - \frac{1}{3}S_{kk}\delta_{ij}) \\ S_{kk} = S_{11} + S_{22} + S_{33} \end{cases} \quad (4)$$

where P_{ij} is the component of viscous stress tensor, δ_{ij} is the isotropic part of stress, U is the velocity vector, μ is the viscosity, and S_{ij} is the component of deformation rate tensor.

The state of fluid flow in the sealing clearance can be judged by the Reynolds number obtained from the following relation [23]:

$$\text{Re} = \frac{2\rho v h}{\mu} \quad (5)$$

where h is the thickness of seal clearance.

The Reynolds numbers of the fluids involved in this study were calculated to be small and in accordance with the specifications for the use of laminar flow models. Due to the high operating speed of the runway, the groove structure of the main sealing surface is more complex, and the fluid is very prone to turbulence and vortex. Therefore, the $k - \varepsilon$ model is used in this study. The $k - \varepsilon$ model takes into account the effects of a low Reynolds number and shear flow corrections, which can better handle near-wall motions [25].

2.2.2. Method of Opening Characteristics Analysis

In order to clearly evaluate the opening characteristics of a segmented annular seal, the study of the opening characteristics of a segmented annular seal is divided into two parts: opening force calculation and opening speed calculation. Opening force is the numerical expression of the dynamic pressure effect of clearance, and the opening rotation speed is the runway speed that makes the segments float. A segmented annular seal with excellent performance requires a small opening resistance and a low opening rotation speed, so as to ensure that the seal can open even at a low working rotation speed and avoid the friction and wear of the segments.

The opening force can be calculated by the average pressure; the method is shown as follows in Equation (6) [26]:

$$\begin{cases} p_{av} = \dfrac{1}{A} \int p dA = \dfrac{1}{A} \sum_{i=1}^{n} p_i |A_i| \\ F_P = p_{av} \cdot S \end{cases} \qquad (6)$$

where p_{av} is the average pressure acting on the inner surface of the segments and S is the area of the inner surface.

The calculation of the opening rotation speed must first determine the clearance when the segments are in contact with the runway. The concept of contact gas film thickness is introduced in this paper [27,28]. For whether the two ends are in contact, the criteria are as follows: when the minimum gas film between the two sides is greater than or equal to the thickness of the contact gas film, it is considered that the two sides are not in contact; otherwise, the two sides are in contact. For a contact surface as shown in Figure 4, the definition of the contact air film thickness is related to the roughness of the sealing face, which can be expressed as follows:

$$h_c = 3.75 \sqrt{R_1^2 + R_2^2} \qquad (7)$$

where R_1 and R_2 represent the roughness of the inner surface of the seal ring and the outer surface of the runway, with both R_1 and R_2 taken as 0.2 μm in the study.

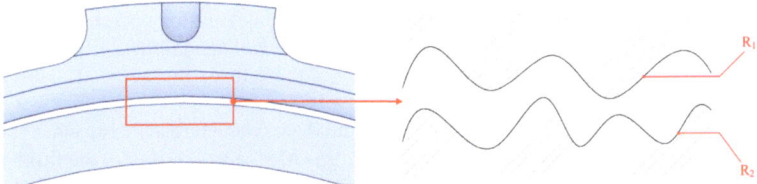

Figure 4. Contact surface of segments and runway.

The calculation method of the opening rotation speed is shown below.

1. Establish the fluid domain model according to contact gas film thickness;
2. Calculate the opening resistance according to structural parameters, Equations (1) and (2);
3. Set the opening rotation speed until the opening force is slightly larger than the opening resistance.

2.2.3. Method of Leakage Characteristics Analysis

The influence of segment deformation should be considered in the analysis of leakage characteristics. The solid domain focuses on the segments and runway affected by external forces, temperature, and rotation speed. The equilibrium state of the solid domain under fluid pressure can be solved by the following equation [29]:

$$M\ddot{u} + C\dot{u} + Ku = F \qquad (8)$$

where M is the mass matrix, C is the damping matrix, K is the stiffness matrix, $\ddot{u}$, $\dot{u}$, and u are the acceleration vector, velocity vector, and displacement vector, respectively, and F is the load vector generated by the fluid in the clearance.

Thermal deformation is a key factor affecting the movement of the segments, which can be solved by Equation (9) as follows:

$$f = \alpha_T \nabla T \qquad (9)$$

where α_T is the coefficient of thermal expansion.

For the fluid–solid–thermal coupling interface, the stress, deformation, heat flow density, and temperature of the solid and fluid should be satisfied when equal, and the fluid–solid–thermal coupling control equations are as follows [17]:

$$\begin{cases} \tau_f \cdot n = \tau_s \cdot n \\ d_f = d_s \\ q_f = q_s \\ T_f = T_s \end{cases} \qquad (10)$$

where n is the normal direction of the coupling surface, τ_f and τ_s are the stresses on the respective coupling surfaces of the fluid and solid domains, and d_f and d_s are the deformations on the respective coupling surfaces of the fluid and solid domains.

Through the fluid–solid–thermal coupling calculation, the axial deformation of the solid domain can be obtained. Then, the new fluid domain is constructed. Since the hydrodynamic pressure exerted by the fluid domain on the solid domain is a variable varying with the clearance thickness, multiple iterations are required until the deformation of the segments converges. Finally, the leakage after convergence is extracted.

2.2.4. Fluid–Solid–Thermal Coupling Calculation Flowchart for a Segmented Annular Seal

This study starts from the fluid domain and carries out the flow field characteristics analysis of the influence of structural optimization on the clearance. Then, aiming at the opening force and opening rotation speed, the opening ability of the seal is quantitatively studied. At last, the leakage is calculated to represent the leakage characteristics. Through the study of these three characteristics, the performance of the seal can be analyzed comprehensively.

The flowchart of the three characteristics analysis is shown in Figure 5. By combining Fluent software and Ansys Mechanical module, the fluid–solid–thermal coupling model of the segmented annular seal is constructed in Ansys Workbench platform 2022R1. First, the fluid domain of the segmented annular seal is calculated to obtain the pressure field and temperature field; then, the flow field characteristics and opening characteristics are analyzed. The solid domain model is established and the opening resistance of the segments is calculated. Then, the pressure field and temperature field are used as boundary conditions to couple with the structural field by grid interpolation. Through iterative solutions until convergence, the calculation results of fluid–solid–thermal coupling can be obtained.

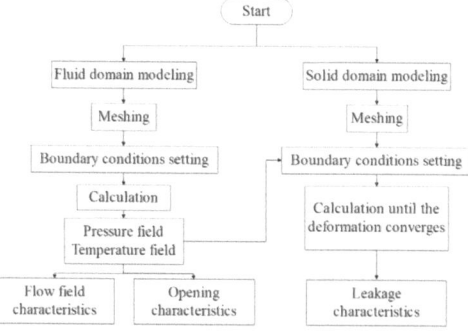

Figure 5. Fluid–solid–thermal coupling calculation flowchart for a segmented annular seal.

2.3. Calculation Model

The principal structure of the computational domain is depicted in Figure 6, encompassing a solid domain and a fluid domain. The solid domain contains segments and a runway and the fluid domain is the clearance between the segments and the runway. This paper focuses on the flow field characteristics and leakage characteristics of the seal, so the clearance at the lap joint is ignored.

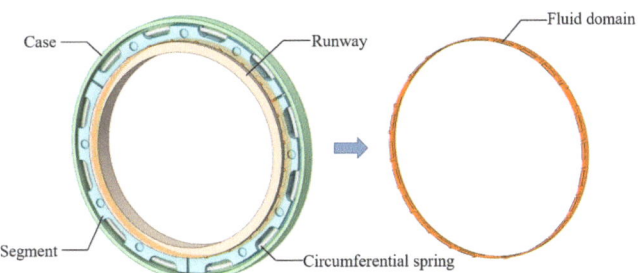

Figure 6. Construction of segmented annular seals.

Considering the operability of practical processing, the following shapes of grooves are introduced in this study: segments without shallow grooves, segments with rectangular grooves, and segments with ladder-like grooves. The three structures of the segments are shown in Figure 7. For the opening rotation speed calculation, the initial seal clearance is calculated by Equation (7), which is 1.06 μm. For the other research, the initial seal clearance is set as 3 μm to monitor a seal opening. The parameters of the shallow grooves are obtained from the literature [16] to ensure a good opening performance and to avoid a large effect of frictional wear on the grooves. The structural parameters are listed in Table 1.

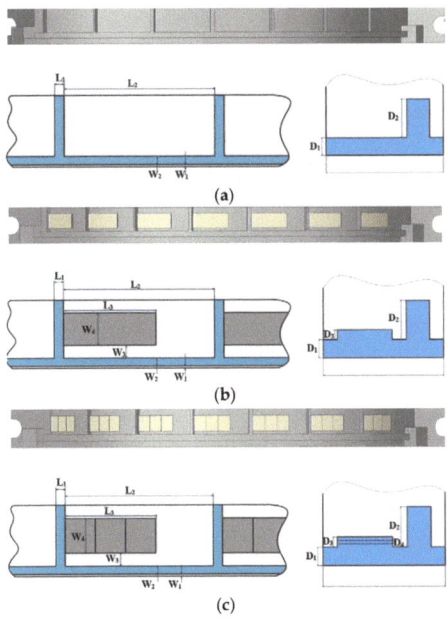

Figure 7. Structures of different segmented annular seals: (**a**) structure of a segmented annular seal without grooves; (**b**) structure of a segmented annular seal with rectangular grooves; and (**c**) structure of a segmented annular seal with ladder-like grooves.

Table 1. Set parameter value.

Parameter	Value
Rotor diameter D/(mm)	100
Segmented number	3
Step number	3
Circumferential width of axial groove L_1/(mm)	1.265
Circumferential width of pad L_2/(mm)	11.5
Circumferential width of shallow groove L_3/(mm)	7.8
Width of shallow groove W_1/(mm)	1.5
Width of circumferential groove W_2/(mm)	1.5
Width between circumferential groove and shallow groove, W_3/(mm)	0.55
Width of shallow groove, W_4/(mm)	3
Initial seal clearance D_1/(μm)	1.06, 3
Depth of axial and circumferential groove, D_2/(mm)	0.6
Depth of shallow groove, D_3/(mm)	0.03

Segments, as the main sealing part studied in this paper, are made of carbon graphite. The materials of the runway and case are structural steel. The material properties are shown in Table 2.

Table 2. Material properties [30].

Material Properties	Carbon Graphite	Structural Steel
Density ρ/(kg/m^3)	2100	7800
Modulus of elasticity E/(GPa)	14	210
Poisson's ratio	0.25	0.3
Thermal conductivity k/(W·m^{-1}·K)	434	900
Coefficient of thermal expansion β/(10^{-6} °C)	6	11

3. Grid Independence Verification and Model Verification

Taking the segmented annular seal without shallow grooves as an example, the specific mesh division is shown in Figure 8. The fluid domain grid is divided by sweep method and the obtained meshes are all hexahedral meshes. The inlet and outlet of the fluid film are set as the pressure inlet and pressure outlet, and the inner surface is set as the rotating wall. The solid domain is divided into tetrahedral meshes, whose boundary conditions are set as follows: the rotation speed is applied to the runway and the fixed support constraint is set on its inner surface. The spring force of the circumferential spring and the compressed spring are equivalent to the radial force and the axial force applied to the segments. The circumferential displacement constraint is applied equivalently at the anti-rotation pin.

Aiming at the leakage and the deformation of the segments, the grid independence verification analysis of the fluid domain and the solid domain is carried out. As can be seen in Figure 9a,b, the leakage begins to converge when the grid quantity of fluid domain reaches 1.87 million. The leakage begins to converge when the grid quantity of the solid domain reaches 1.8 million. In order to ensure the accuracy and efficiency of the calculation, the finalized grid quantity of the fluid domain and solid domain are 2.26 million and 2.52 million, respectively, where the error of leakage and maximum deformation is less than 1%.

In order to verify the accuracy of the calculation method, the segmented annular seal model and boundary conditions in reference [31] are used. According to the grid division method and boundary condition setting method above, the solution of the sealing flow field is carried out. As can be seen in Figure 9c, the calculated data are in good agreement with the literature data. The leakage deviation increases with the increase in pressure difference and the maximum deviation of leakage is 5.54%. The results indicate that the calculation method is feasible.

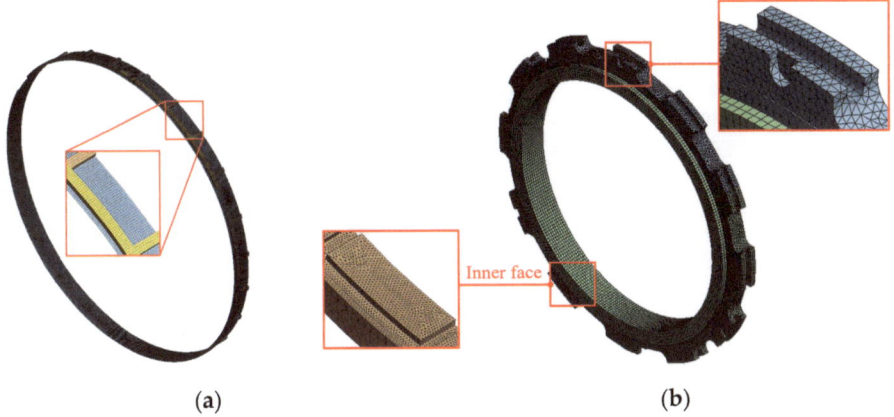

Figure 8. Meshing of a seal without a shallow groove: (**a**) meshing of fluid domain of a segmented annular seal and (**b**) meshing of solid domain of a segmented annular seal.

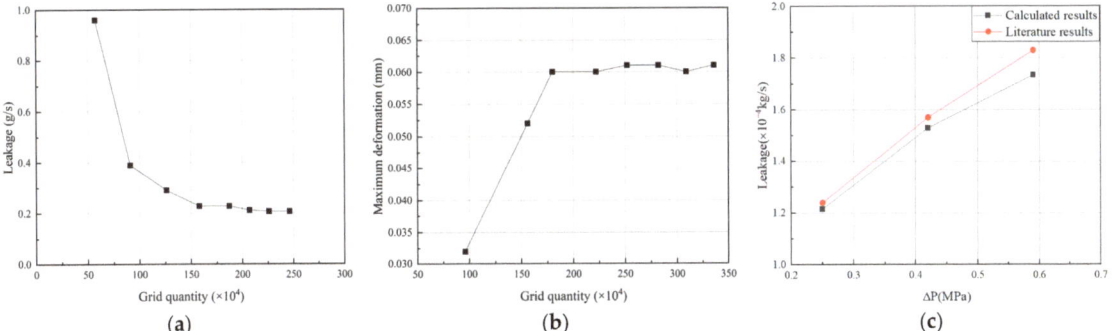

Figure 9. Grid independence verification and accuracy verification: (**a**) fluid domain; (**b**) solid domain; and (**c**) accuracy verification.

4. Results

4.1. Fluid Field Characteristics Analysis

The flow field pressure distribution of the fluid clearance directly determines the opening force of the seal, which in turn affects the sealing performance. Therefore, this study first investigates the effect of grooves on the fluid field characteristics of the segmented annular seal. The corresponding parameter values are shown in Table 3.

Table 3. Condition parameters.

Parameter	Value
Rotating speed $n/(\text{r/min})$	$0\text{--}2.4 \times 10^5$
Inlet temperature $T/(\text{K})$	420–620
Sealed pressure $P_0/(\text{MPa})$	0.1–0.5
Outlet pressure $P_1/(\text{MPa})$	0.1

4.1.1. Pressure Distribution Cloud Analysis

Figure 10 shows the pressure distribution of the fluid domain of three segmented annular seals with an inlet pressure of 3×10^5 Pa, and outlet pressure of 1×10^5 Pa. From the figure, it can be seen that the pressure in most areas of the flow field is consistent with the inlet pressure setting. However, the pressure drops suddenly at the outlet, forming a low pressure zone. Due to the isolation effect of the lip, the development of a low pressure zone is restrained. As a result, the low pressure zone is concentrated in the clearance at the lap joint and less in other regions. The high pressure zone of the fluid domain is located between two axial grooves. Taking one of the pads and grooves as a unit for analysis, the pressure distribution is shown in Figure 11. From Figure 11a, it can be seen that the high pressure zone A_1 is located at the end of the axial groove. This is because when the fluid is driven by the runway from the axial groove, the fluid is squeezed due to the decrease in the clearance, thus forming a slight high pressure zone. The maximum pressure is 0.322 MPa, which is only 7.33% higher than the inlet pressure. As Figure 11b demonstrated, the high pressure zone A_2 of the flow field is located at the end of the rectangular groove, and the setting of the rectangular groove obviously enhances the hydrodynamic effect of the fluid. The maximum pressure is increased by 15.67% compared with the inlet pressure. It can be seen from Figure 11c that the multiple extrusions of the ladder-like groove can further enhance the hydrodynamic pressure effect; the maximum pressure of the flow field is increased by 22.67% up to the inlet pressure.

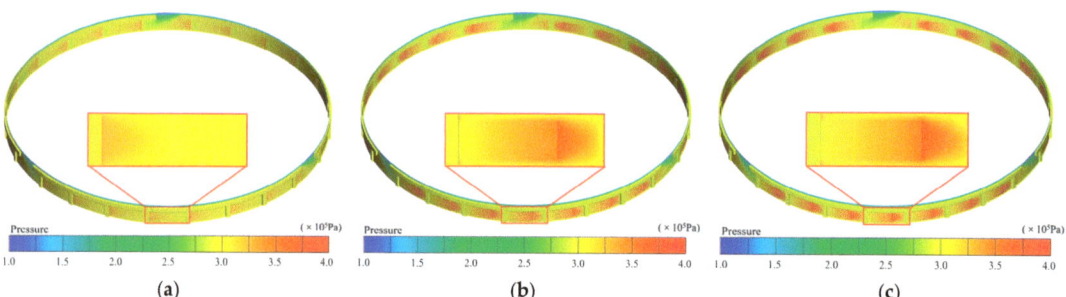

Figure 10. Pressure distribution cloud of the whole fluid domain: (**a**) pressure distribution cloud of a segmented annular seal without grooves; (**b**) pressure distribution cloud of a segmented annular seal with rectangular grooves; and (**c**) pressure distribution cloud of a segmented annular seal with ladder-like grooves.

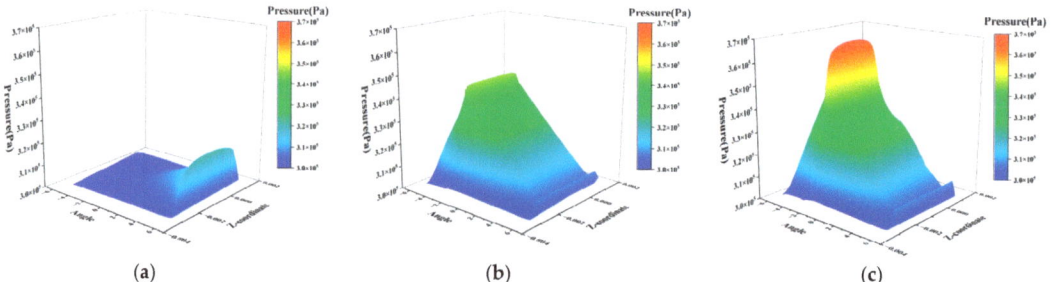

Figure 11. Pressure distribution cloud of a unit of the fluid domain: (**a**) pressure distribution of a segmented annular seal without grooves; (**b**) pressure distribution of a segmented annular seal with rectangular grooves; and (**c**) pressure distribution of a segmented annular seal with ladder-like grooves.

4.1.2. Velocity Distribution Cloud Analysis

The velocity distribution within the clearance with an inlet pressure of 3×10^5 Pa and outlet pressure of 1×10^5 Pa is given in Figure 12. The flow velocity gradually decreases in the radial direction from the inner surface to the outer surface. Compared with Figure 12a,b, it can be seen under the action of the rectangular grooves; the fluid flow is first concentrated in the shallow grooves and then diverges, thereby enhancing the hydrodynamic effect of the fluid by squeezing and hindering the flow. From Figure 12c, the setting of ladder-like grooves makes the fluid motion more complicated, and the squeezing effect between the fluids is also enhanced, thus further enhancing the hydrodynamic effect of the fluid.

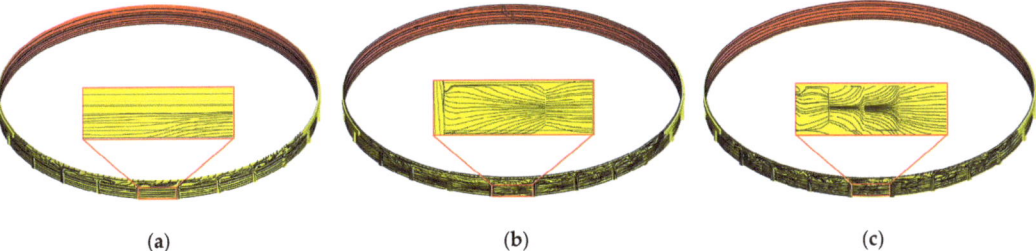

Figure 12. Velocity distribution cloud of a segmented annular seal: (**a**) velocity distribution cloud of a segmented annular seal without grooves; (**b**) velocity distribution cloud of a segmented annular seal with rectangular grooves; and (**c**) velocity distribution cloud of a segmented annular seal with ladder-like grooves.

4.1.3. Temperature Distribution Cloud Analysis

Figure 13 shows the temperature distribution clouds of flow fields of three types of segmented annular seals. It can be seen that the overall distribution of the flow field temperature is relatively uniform. The temperature does not change significantly at the shallow grooves, and the low temperature zone is only generated near the outlet and at the lap joint. Therefore, it can be considered that the design of shallow grooves has no effect on the temperature of the flow field.

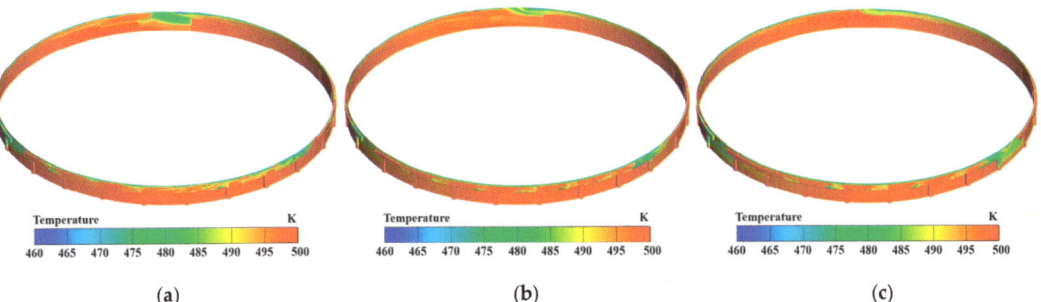

Figure 13. Temperature distribution cloud of a segmented annular seal: (**a**) temperature distribution cloud of a segmented annular seal without grooves; (**b**) temperature distribution cloud of a segmented annular seal with rectangular grooves; and (**c**) temperature distribution cloud of a segmented annular seal with ladder-like grooves.

Overall, the design of the shallow grooves has a great influence on the pressure distribution in the fluid domain of a segmented annular seal. Among the three structures, the ladder-like grooves can effectively enhance the dynamic pressure effect and increase the maximum pressure. In order to quantitatively analyze the effect of the shallow grooves, the opening characteristics of the seal will be carried out.

4.2. Opening Characteristics Analysis

The opening characteristic analysis includes two parts: opening force calculation and opening speed calculation. The opening force is the key parameter of seal opening, which is affected by many operating parameters, and the opening rotation speed can be used to measure the overall opening characteristics of a segmented annular seal.

4.2.1. Effect of Operating Parameters on Opening Force

Figure 14a shows the effect of rotation speed on the opening force with a temperature of 500 K, inlet pressure of 3×10^5 Pa, and outlet pressure of 1×10^5 Pa. From the figure, it can be seen that the increase in rotational speed enhances the hydrodynamic effect, resulting in the increase in the opening force. When the rotation speed is zero, the opening force of the three types of seals is approximately equal, and as the rotation speed increases, the difference between the opening forces of the different groove types also increases gradually. At a speed of 18,000 r/min, the opening force of seals with rectangular grooves is increased by 1.32% compared to seals without grooves, and the opening force of seals with ladder-like grooves is increased by 4.6% compared to seals with rectangular grooves.

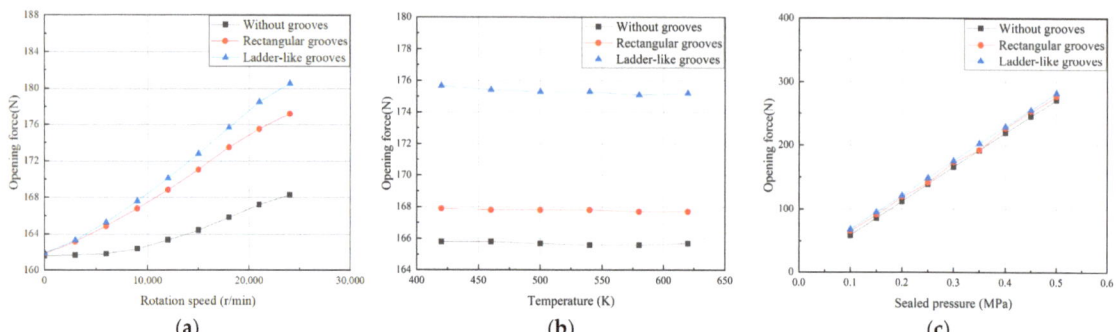

Figure 14. Effect of different condition parameters on opening force: (**a**) effect of rotation speed on opening force; (**b**) effect of temperature on opening force; and (**c**) effect of sealed pressure on opening force.

Figure 14b illustrates the effect of sealed pressure on the opening force with a rotation speed of 18,000 r/min, outlet pressure of 1×10^5 Pa, and temperature of 300 K. It can be seen that the opening force increases linearly when the inlet pressure increases. The increase in pressure difference accelerates the fluid, which in turn allows more fluid to enter the shallow grooves and enhances the hydrodynamic effect. At a sealed pressure of 5×10^5 Pa, the opening force of the seal with rectangular grooves increased by 2% compared with that of the segmented annular seal without grooves, and the opening force of seals with ladder-like grooves increased by 2.1% compared to seals with rectangular grooves.

Figure 14c demonstrates the effect of temperature on the opening force. It can be seen that the change in temperature had no effect on the seal opening force. When the temperature reached 620 K, the opening force of seals with rectangular grooves increased by 1.98% compared with that of seals without grooves, and the opening force of seals with ladder-like grooves increased by 2% compared with that of seals with rectangular grooves. Combined with the research in 3.1.3, it is considered that the temperature does not affect the opening characteristics of the seal.

4.2.2. Effect of Sealed Pressure and Spring Force on Opening Rotation Speed

In this section, the opening rotation speed is calculated to evaluate the opening characteristics of a segmented annular seal. According to the previous study, the opening force varies considerably with sealed pressure, and the spring force is directly related to the opening resistance of the seal. Therefore, the effect of these two parameters will be studied.

Figure 15a shows the variation in opening rotation speed with the sealed pressure. It can be seen that the opening rotation speed increases as the sealed pressure increases. This is because both the opening force and opening resistance of the seal are affected by the pressure of the medium. With the increase in the pressure difference, the opening force and opening resistance of the seal increase accordingly. Due to the strong dynamic pressure effect of the segmented annular seal with ladder-like grooves, when the pressure difference rises from 0 to 3×10^5 Pa, the opening speed of the seal only increases by 8600 r/min. However, the opening rotation speed of segmented annular seal without grooves reaches 20,560 r/min, which makes it difficult to open.

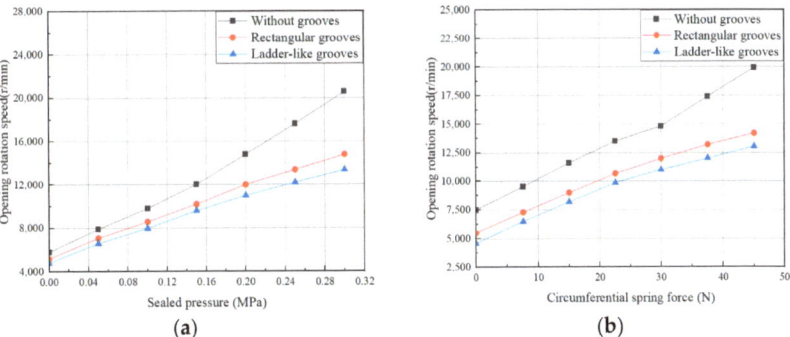

Figure 15. Effect of different parameters on opening rotation speed: (**a**) effect of sealed pressure on opening rotation speed and (**b**) effect of circumferential spring force on opening rotation speed.

Figure 15b illustrates the variation in opening rotation speed with the circumferential spring force. Due to the spring force directly affecting the opening resistance of the seal, the opening rotation speed increases as the spring force increases. The opening rotation speed of the seal with rectangular grooves and with ladder-like grooves stays under 15,000 r/min; when the circumferential spring force reaches t45 N, the opening rotation speed of the seal with rectangular grooves increases by 8.8% compared to the seal with ladder-like grooves. Meanwhile the opening rotation speed of the seal without shallow grooves reaches 19,900 r/min, which makes it difficult to open.

4.3. Leakage Characteristics Analysis

According to the previous study, segmented annular seals with shallow grooves have a larger opening force and increase the clearance, which is an important parameter in leakage characteristics. In this section, the fluid–solid–thermal coupling method is used to solve the sealing clearance and calculate the leakage. In order to make the seal easy to open and the deformation of the segment more obvious, the circumferential spring force is set to 7.5 N.

The deformation of three kinds of segments is shown in Figure 16. It can be seen that the deformation of the lap joint area of the segment is the largest and the deformation of the middle area is the smallest. According to the research in Section 4.1.1, there is a low pressure zone in the fluid domain below the lap joint, and the large pressure difference on both sides of the lap joint promotes its deformation. Due to the discontinuous characteristics of the joint interface, there exists a deformation difference of 2.4% at the lap joints on two sides of the segments. The total deformation of the three kinds of segments is almost the same, and the differences are not more than 1%. Then, the influence of the radial variation in the segments and the runway is considered to solve and analyze the leakage.

Figure 17a shows the effect of rotation speed on the leakage of a segmented annular seal. It can be seen that the leakage decreases slowly with the increase in rotational speed. This is because the sheer force of the runway to the sealed fluid increases when the rotation speed increases, which makes the fluid vortex phenomenon more intense and hinders the leakage of the fluid along the sealing clearance. Among the three structures, the leakage of a segmented annular seal with ladder-like grooves decreases the fastest with the rotation speed. When the speed reaches 18,000 r/min, the leakage of the seal with ladder-like grooves increases by 4.21% relative to the seal without grooves, and the leakage of the seal with rectangular grooves increases by 3.40% relative to seals with ladder-like grooves.

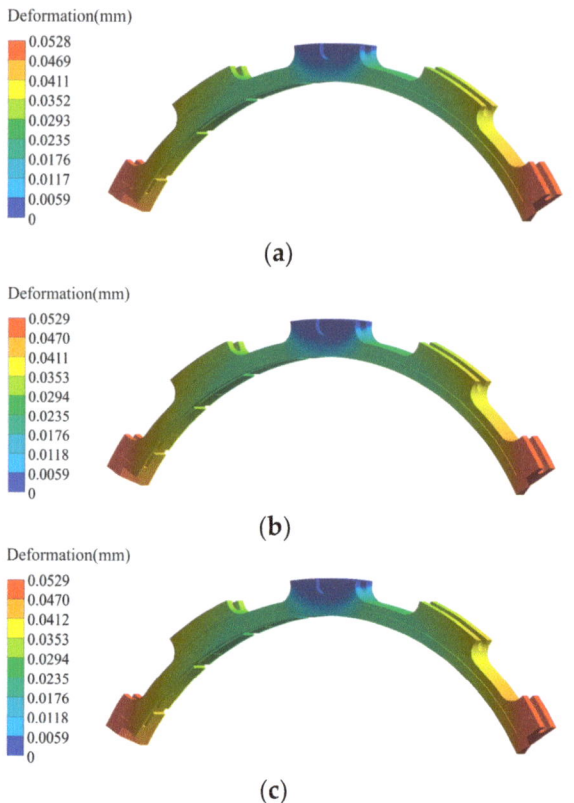

Figure 16. Deformation of three kinds of segments: (**a**) deformation of a segment without grooves; (**b**) deformation of a segment with rectangular grooves; and (**c**) deformation of a segment with ladder-like grooves.

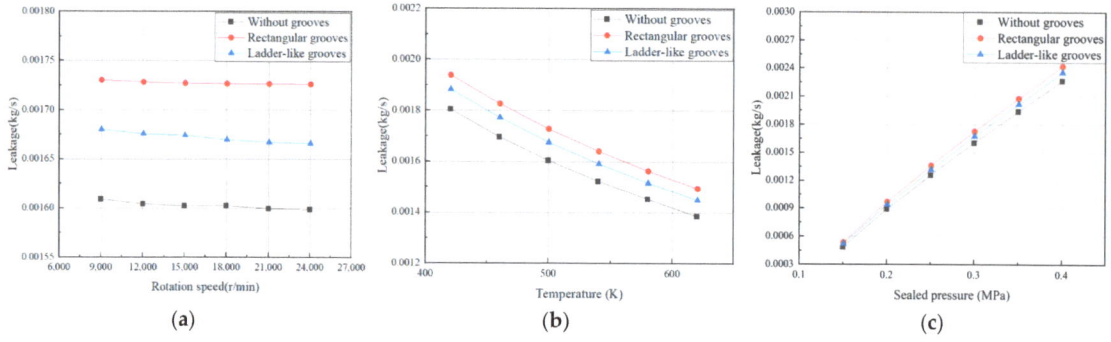

Figure 17. Effect of different condition parameters on leakage: (**a**) effect of rotation speed on leakage; (**b**) effect of temperature on leakage; and (**c**) effect of sealed pressure on leakage.

Figure 17b illustrates the effect of temperature on the leakage of a segmented annular seal. It can be seen that, with the increase in temperature, the leakage of seals with three kinds of shallow groove structure decreases. This is because with the increase in temperature, the viscosity of the fluid increases and the flow rate decreases. As the temperature rises from 420 K to 620 K, the maximum decrease in the three structures is up to 22.9%. At all temperatures, a segmented annular seal with rectangular grooves has the highest leakage. When the temperature reaches 620 K, the leakage of a seal with rectangular grooves increases by 4.62% relative to a seal without grooves, and the leakage of a seal with ladder-like grooves increases by 3.20% relative to seals with rectangular grooves.

Figure 17c demonstrates the effect of sealed pressure on the leakage of a segmented annular seal. Fluid flow is affected by pressure difference and always flows from high-pressure areas to low-pressure areas. As the pressure difference increases, the fluid flow is accelerated and the leakage increases significantly. The maximum increase in the leakage of the seal with a specific groove design is 4.657 times that of the original as the sealed pressure increases from 0.15 MPa to 0.4 MPa. When the pressure reaches 0.4 MPa, seals with rectangular grooves had increased leakage by 3.92% relative to seals without grooves, and seals with ladder-like grooves had increased leakage by 2.81% relative to seals with rectangular grooves.

It can be seen from Figure 17 that sealed pressure is the biggest influencing factor of seal leakage, while the influence of rotation speed is small. The seal without grooves has the lowest leakage and the seal with rectangular grooves has the highest leakage; the seal with ladder-like grooves can maintain medium leakage under the premise of good opening performance, which is proven to have good sealing performance.

5. Conclusions

This paper presented the sealing performance of three types of segmented annular seals, a fluid–solid–thermal coupling model is established to obtain the flow field characteristics, opening characteristics, and leakage characteristics of segmented annular seals. The effects of different shallow grooves and the condition parameters on the performance of segmented annular seals are systematically presented and discussed. Proper groove design can play a role in optimizing the opening performance of segmented annular seals by enhancing the opening force. However, better opening performance will also lead to larger leakage. The selection of the three types of seals can be determined according to the actual use requirements. The key conclusions are summarized as follows:

(1) The setting of shallow grooves can effectively enhance the hydrodynamic effect by squeezing the fluid, and the enhancement effect of the ladder-like grooves is more significant than that of the rectangular grooves. However, the change in the groove type has little effect on the fluid temperature field.

(2) For the seals with arbitrarily shaped shallow grooves in this paper, the sealed pressure has the most significant influence on the opening force of the seal; the greater the sealed pressure, the greater the opening force. The increase in rotational speed will also promote the opening force. However, the temperature has no obvious effect on the opening force.

(3) Both the increase in the leakage pressure and the increase in the circumferential spring force will lead to an increase in the opening speed. Under the conditions of a high speed and a large spring force, the opening rotation speed of the seal without shallow grooves reaches around 20,000 r/min, which makes it difficult to open.

(4) The total deformation of the segment increases gradually from the middle to the lap joint, while the shallow groove design has little effect on the total deformation.

(5) An increase in the sealed pressure will lead to an increase in the leakage. When the sealed pressure increases from 0.15 MPa to 0.4 MPa, the maximum increase in the leakage of the seal with specific groove design is 4.657 times that of the original. Nevertheless, the leakage decreases with the increase in temperature and rotation speed.

(6) Among the seals with three groove structures, the seal with ladder-like grooves has the best opening performance and is easy to open and maintain a frictionless seal under high parameter conditions. Compared to the seal without shallow grooves, the leakage of the seal with ladder-like grooves has only a small increase.

These results highlight the excellent sealing performance of seals with specific shallow grooves, and provide a method to analyze the sealing performance of segmented annular seals and an idea of structural optimization design. In order to further study and apply the segmented annular seals, there are two goals for future work: the establishment of more complex models and structural optimization design. In the establishment of the model, the influence of variable spring force can be considered to study the deformation and displacement of the segments under complex stress conditions. Meanwhile, fluid shear heat and friction heat need to be considered to comprehensively analyze the temperature field of the segments. In terms of structural optimization design, more grooves with excellent performance should be discovered on the basis of this study.

Author Contributions: Conceptualization, Z.H., L.J. and J.S.; Methodology, Z.H., L.J., H.W., Y.G., B.L. and S.Z.; Validation, L.J.; Data curation, L.J.; Writing—original draft, L.J.; Writing—review and editing, Z.H., L.J., J.S., N.L., H.W., Y.G. and W.L.; Visualization, L.J.; Supervision, Z.H., J.S. and N.L.; Funding acquisition, Z.H. All authors have read and agreed to the published version of the manuscript.

Funding: This research was funded by XX national science and technology major project, the Tianjin Municipal Science and Technology Bureau Science and Technology Plan Project grant number 23JCY-BJC00110, the Fundamental Research Funds for the Central Universities grant number 3122023045, and the Chongqing Natural Science Foundation Project grant number CSTB2024NSCQ-MSX0388.

Data Availability Statement: The original contributions presented in the study are included in the article; further inquiries can be directed to the corresponding author.

Acknowledgments: The authors sincerely acknowledge the financial support mentioned above, which made it possible to continue this study.

Conflicts of Interest: Author J.S. and author N.L. are employed by the company AECC Hunan Aviation Powerplant Research Institute, and author W.L. is employed by the company Chongqing Aerospace polytechnic. The remaining authors declare that the research was conducted in the absence of any commercial or financial relationships that could be construed as a potential conflict of interest.

References

1. Mayhew, E.; Bill, R.; Voorhees, W. Military Engine Seal Development-Potential for Dual Use. In Proceedings of the 30th Joint Propulsion Conference and Exhibit, Indianapolis, IN, USA, 27–29 June 1994.
2. Ellen, R.; Mayhew, E. *Air Force Seal Activity USA: Aero Propulsion & Power Directorate Wright Laboratory*; NASA: Washington, DC, USA, 1980.
3. Sorokina, N.E.; Redchitz, A.V.; Ionov, S.G.; Avdeev, V.V. Different exfoliated graphite as a base of sealing materials. *J. Phys. Chem. Solids* **2006**, *67*, 1202–1204. [CrossRef]
4. Song, Y.Z.; Zhai, G.T.; Song, J.R.; Li, G.S.; Shi, J.L. Seal and Wear Properties of Graphite from MCMBs/Pitch-Based Carbon/Phenolic-Based Carbon Composites. *Carbon Int. J. Spons. Am. Carbon Soc.* **2006**, *44*, 2793–2796. [CrossRef]
5. Baklanova, N.I.; Zima, T.M.; Boronin, A.I.; Kosheev, S.V.; Titov, A.T.; Isaeva, N.V.; Graschenkov, D.V.; Solntsev, S.S. Protective ceramic multilayer coatings for carbon fibers. *Surf. Coat. Technol.* **2006**, *201*, 2313–2319. [CrossRef]
6. Burcham, R.E. *High-Speed Cryogenic Self-Acting ShaftSeals for Liquid Rocket Turbopumps*; NASA-CR-168194; NASA: Washington, DC, USA, 1983.
7. Oike, M.; Nagao, R. Characteristics of a Shaft Seal System for the LE-7 Liquid Oxygen Turbopump. In Proceedings of the 31st Joint Propulsion Conference and Exhibit, San Diego, CA, USA, 10–12 July 1995.
8. Arghir, M.; Mariot, A. Theoretical Analysis of the Static Characteristics of the Carbon Segmented Seal. *J. Tribol. Trans. Asme* **2017**, *139*, 062202. [CrossRef]
9. Yang, H.W.; Bai, S.X. Gas Hydrodynamic Lubrication Performance of Split Floating Ring Seals with Rayleigh Step Grooves. *J. Propuls. Technol.* **2022**, *43*, 87–94.
10. Bai, S.X.; Chu, D.D.; Ma, C.H.; Yang, J.; Bao, S.Y. Thermo-Hydrodynamic Effect of Gas Split Floating Ring Seal with Rayleigh step Grooves. *Materials* **2023**, *16*, 2283. [CrossRef]
11. Li, Q.Z.; Li, S.X.; Zheng, Y.; Ma, W.J.; Zhuang, S.G. Sensitive parameters affecting performance of three-petal high-speed floating-ring seal. *J. Beijing Univ. Aeronaut. Astronaut.* **2020**, *46*, 571–578.
12. Wang, S.; Sun, D.; Yang, Z.; Xu, W.F.; Zhao, H.; Wen, S.F. Investigation of static and rotordynamic characteristics of the segmented annular seals based on the differential quadrature method. *Tribol. Int.* **2024**, *196*, 109657. [CrossRef]
13. Dahite, S.; Arghir, M. Thermogasodynamic Analysis of the Segmented Annular Seal. *J. Tribol.* **2021**, *143*, 072301. [CrossRef]
14. Yun, R.D.; Chen, Z.Y.; Liu, Y.; Zhang, J.Y. Effects of Circumferential Spring Force Distribution on Sealing Performance of Circumferential Seal. *J. Propuls. Technol.* **2021**, *42*, 1361–1371.

15. Chen, Z.X.; Peng, X.D.; Zhao, W.J.; Jiang, J.B. Deformation Characteristics and Control Strategies of a Dynamic Segmented Circumferential Seal. *J. Tribol.* **2023**, *43*, 143–156.
16. Ma, R.M.; Zhao, X.; Li, S.X.; Chen, X.Z.; Zhao, H.C. Seal characteristics and friction and wear of dynamic pressure split floating ring seal. *J. Propuls. Technol.* **2022**, *43*, 210099.
17. Yan, Y.T.; Wei, R.; Hu, G.Y.; Wang, F.C.; Zhang, L.J. Circumferential Seal Characteristics with Thermal-Fluid-Structure Multiphysics Field Coupling. *J. Aeroengine Power* **2020**, *35*, 305–317.
18. Arghir, M.; Dahite, S. Numerical Analysis of Lift Generation in a Radial Segmented Gas Seal. In Proceedings of the ASME Turbo Expo 2019: Turbomachinery Technical Conference and Exposition, Phoenix, AZ, USA, 17–21 June 2019.
19. Dahite, S.; Arghir, M. Numerical modelling of a segmented annular seal with enhanced lift effects. *Mech. Syst. Signal Process.* **2021**, *152*, 107455. [CrossRef]
20. Fourt, E.; Arghir, M. Experimental Analysis of the Leakage Characteristics of Three Types of Annular Segmented Seals. *J. Eng. Gas Turbines Power* **2023**, *145*, 91–105. [CrossRef]
21. Fan, R.F.; Zhao, H.; Ren, G.Z.; He, Y.; Sun, D.; Wang, X.Y. Influence of rotor micro-texture on oil leakage flow characteristics of circular graphite seal. *J. Aerosp. Power* **2024**, *40*, 20240102.
22. Ren, G.Z.; Li, Y.P.; Sun, D.; Zhao, H.; Wen, S.F.; Wang, X.Y. Numerical Study on Leakage Characteristics of Dynamic Pressure Split Floating Ring Seal with Fluid-solid-thermal Coupling. *J. Aeroengine Power* **2022**, *37*, 114–123.
23. Zhang, Z.S.; Cui, G.X. *Fluid Mechanics*, 2nd ed.; Tsinghua University Press: Beijing, China, 2015; pp. 108–146.
24. Hu, T.X.; Zhou, K.; Wang, X.Y.; Ning, L.; Zou, H.Y. Numerical calculation and experiment on leakage characteristics of floating ring seal. *J. Aerosp. Power* **2020**, *35*, 888–896.
25. Yakhot, V.; Orszag, S.A. Renormalization group analysis of turbulence: I basic theory. *J. Sci. Comput.* **1986**, *1*, 3–51. [CrossRef]
26. Cai, Y. Research on Multi-Field Coupling and Sealing Performance of a Single Carbon-Graphite Circumferential Seal. Master's Thesis, Sichuan University, Chengdu, China, 2021.
27. Gulich, J.F. *Centrifugal Pumps*; Springer: New York, NY, USA, 2007.
28. Peng, X.D.; Li, J.Y.; Sheng, S.E.; Yin, X.N.; Bai, S.X. Effect of surface roughness on performance prediction and geometric optimization of a spiral-groove face seal. *Tribology* **2007**, *6*, 567–572.
29. Ren, H.; Li, F.C.; Du, L. Numerical calculation for the effect of FSI on marine propeller strength. *J. Wuhan Univ. Technol.* **2015**, *39*, 144–147.
30. Zhou, H.N.; Li, G.Q.; Wang, C.F. Multi physical field coupling characteristics of high speed graphite seals. *Lubr. Eng.* **2024**, *49*, 49–59.
31. Allen, G.P. *Self-Acting Lift-Pad Geometry for Circumferential Seals-a Non-Contacting Concept*; NASA-TP-1583C.1; NASA: Washington, DC, USA, 1980.

Disclaimer/Publisher's Note: The statements, opinions and data contained in all publications are solely those of the individual author(s) and contributor(s) and not of MDPI and/or the editor(s). MDPI and/or the editor(s) disclaim responsibility for any injury to people or property resulting from any ideas, methods, instructions or products referred to in the content.

Article

On Lubrication Regime Changes during Forward Extrusion, Forging, and Drawing

Man-Soo Joun [1,*], Yun Heo [2], Nam-Hyeon Kim [2] and Nam-Yun Kim [2]

[1] Research Center for Advanced Research in Future, School of Mechanical and Aerospace Engineering, Gyeongsang National University, Gyeongnam, Jinju 52828, Republic of Korea
[2] School of Mechanical and Aerospace Engineering, Gyeongsang National University, Gyeongnam, Jinju 52828, Republic of Korea; dbsdl609@gnu.ac.kr (Y.H.); kd3j89@gnu.ac.kr (N.-H.K.); skadbs9592@gnu.ac.kr (N.-Y.K.)
* Correspondence: msjoun@gnu.ac.kr

Abstract: The tribological phenomena concerning the lubrication regime change (LRC) during bulk metal forming are comprehensively studied. A multi-step cold forward extrusion process shows the evolution of LRC and reveals the shortcomings of the traditional Coulomb friction law. The previous works of the specific author's research group on friction are reviewed, focusing on the LRC during bulk metal forming. Various LRC phenomena from various examples are revealed. It has been found that the drawing and forward extrusion processes are vulnerable to LRC because of significant sliding motion at the material–die interface, and that when the strain hardening of the material is slight, the influence of friction increases, and as a result, the influence of LRC increases excessively. The new findings also include the impact of LRC on the macroscopic phenomena of the process and the reason for the sharp increase in friction coefficient via LRC, which is validated by the work of Wilson. This paper aims to make engineers and researchers think much of the tribology with lubricant in bulk metal forming with a focus on the dependence of tribological phenomena on the state of the lubricants and the irrationality of traditional friction law, especially in the forging of materials with a low strain hardening capability.

Keywords: lubrication regime change; variable friction coefficient; forward extrusion; forging; drawing; adhesive wear

Citation: Joun, M.-S.; Heo, Y.; Kim, N.-H.; Kim, N.-Y. On Lubrication Regime Changes during Forward Extrusion, Forging, and Drawing. *Lubricants* **2024**, *12*, 352. https://doi.org/10.3390/lubricants12100352

Received: 13 September 2024
Revised: 27 September 2024
Accepted: 8 October 2024
Published: 14 October 2024

Copyright: © 2024 by the authors. Licensee MDPI, Basel, Switzerland. This article is an open access article distributed under the terms and conditions of the Creative Commons Attribution (CC BY) license (https://creativecommons.org/licenses/by/4.0/).

1. Introduction

Process development is a critical element in the forging industry [1–7]. The success of forging process development depends on price and quality competitiveness. Even three decades ago, many trial and errors were indispensable in developing a forging process that satisfied these comprehensive requirements. Currently, forging simulators are leading the development of the metal forming industry in the direction that eliminates trial production through high accuracy. Even ten years ago, calculation time was the most critical factor in evaluating a forging simulator [8,9].

However, the importance of simulation accuracy is still growing year after year [10]. Factors affecting the analysis results include the material model, numerical model of the process, mesh quality, material properties represented by flow characteristics, tribological characteristics, etc. Among them, flow behaviors [11–17] and tribological characteristics [18–32] are the actual issues from the application researcher's viewpoint since, from a macroscopic perspective, bulk metal forming is a fierce battle between the flow characteristics of the material and friction characteristics at the material–die interface.

During the recent two decades, flow characterization technology has steadily improved with the development of materials testing equipment and the advancements in the combined experiment and finite element (FE) method (FEM) approach to characterizing the flow behavior of metallic materials. On the other hand, research on friction in metal forming is limited compared to its importance. The problem of friction in steel forging has not been greatly highlighted. The forging of forgeable steel is relatively easy compared to the forging of non-ferrous materials. This is due to the excellent strain hardening characteristics of forgeable steels. Forgeable steel has a relatively high elongation, and thus, its strain hardening is high. Materials with high elongation and strain hardening capability tend to spread plastic deformation widely. For this reason, the dominance of flow characteristics in steel forging is clear.

On the contrary, in the case of materials with a low strain hardening capability, like aluminum alloys, the influence of flow behaviors on the macroscopic phenomena of the process is somewhat reduced, but the impact of friction dramatically increases. The low strain hardening of the material induces friction, affecting much of the plastic deformation of the material. This feature also appears in the hot forging of aluminum alloys [33]. As a result, from the perspective of a forging engineer accustomed to steel with an apparent strain hardening capability, the friction during the forging of low strain hardening material like aluminum alloys is a big issue. This also governs the die wear and material's surface roughness [34–39].

Even though four and a half decades ago, Wilson [40] cynically criticized the Coulomb friction law with constant friction coefficient in bulk metal forming, the literature survey shows that many researchers employed the constant shear friction law. Wilson conducted a pioneering study on the LRC during bulk metal forming, starting from a thick film lubrication regime towards a boundary lubrication regime via thin film and mixed lubrication regimes. Wilson emphasized that the constant shear friction can describe only the thick film lubrication regime that may seldom occur during bulk metal forming. He calculated the friction coefficient by linearly interpolating the friction coefficients in the thick film and boundary lubrication regimes.

The need for research on friction during metal forming is increasing due to the ongoing need for forging various metals such as aluminum and magnesium alloys, as well as the need for scientific forging process development in accordance with net-zero requirements. A few studies that meet the current demands have been conducted on the non-constant friction coefficient during sheet metal forming. Westeneng [41] presented a new contact model for dealing with the friction of a boundary layer in the deep drawing processes depending on its chemical structure, thickness, pressure, temperature, and sliding speed. Berthier [42] presented that more than one velocity accommodation mechanism between two rubbing surfaces can exist at once and that mechanisms can vary depending on the contact and time. Hol et al. [43] presented the physically based friction model to account for the change in surface topography and the development of friction in the boundary lubrication regime during sheet metal forming. They revealed that the friction coefficients vary in space and time and depend on local process conditions, such as the material's nominal contact pressure and plastic strain. Shisode et al. [44] combined the flattening and asperity ploughing models to present a new multi-scale boundary friction model for deep drawing processes. They determined the friction coefficient at each node based on the die contact surface, nodal contact pressure, and nodal equivalent strain.

In bulk metal forming, especially forging, the contact surface undergoes a significant LRC. In the cold forging of steel, the surface of most materials is coated with a lubricant. The lubricant coating can also be used in the hot forging of aluminum alloys [33]. The state of lubrication changes rapidly depending on the state of the coated lubricant film, that is, the degree of damage to the coated surface. Suppose the coating remains intact in the thick film lubrication regime. In that case, the frictional stress is bound to be very small (for example, friction coefficient: 0.016 [40]) because it blocks direct contact between the material and the die. Direct contact between the material and die causes micro-bonding, fracture, separation of asperities, and the creation of debris, increasing frictional stress and generating friction heat, ultimately accelerating adhesive wear [45]. This creates a vicious cycle. The friction phenomenon is then greatly affected by the degree of the destruction of the coated lubricant film, especially in forging. Therefore, friction is not simply expressed by pressure and a constant friction coefficient. Recent experimental and analytical research results [33,46,47] have revealed that friction is significantly affected at least by the condition of the friction surface, that is, whether the lubricant film is damaged. It should be noted that, as Wilson [40] presented, the Coulomb friction law with a constant friction coefficient is helpful in predicting metal flow only in the boundary lubrication regime, like in the hot forging of steel.

In this study, by synthesizing the lubrication phenomena during bulk metal forming identified previously through experimental and numerical methods, we examine the dependence of friction on the state of the contact surface in bulk metal forming, including multi-step cold forward extrusion, hot forging of an aluminum piston, automatic multistage cold forging (AMSCF) of an aluminum yoke, and round-to-half circle drawing. The multi-step cold forward extrusion process is newly designed to emphasize and visualize the LRC phenomenon. The other examples are reviewed focusing on the LRC. Based on the various LRC phenomena, the inadequacy of the constant shear friction law and the traditional Coulomb friction law with a constant friction coefficient is emphasized. The forging simulator used in this study is AFDEX, an APA (Altair Partner Alliance) SW [10].

2. Problems and Countermeasures of Traditional Friction Laws

As shown in Figure 1, the frictional stress σ_t defined at the material–die contact surface is in a functional relationship with the normal stress σ_n and acts in the illustrated direction to prevent relative motion between the two objects. For this functional relationship, the Coulomb and constant shear friction laws are widely used in metal forming mechanics [19]. Hybrid friction is a combination of two friction laws, which follow the Coulomb friction law at low friction (with a constant friction coefficient μ) and the constant shear friction law at high friction (with an increased friction factor m' defined in Figure 1).

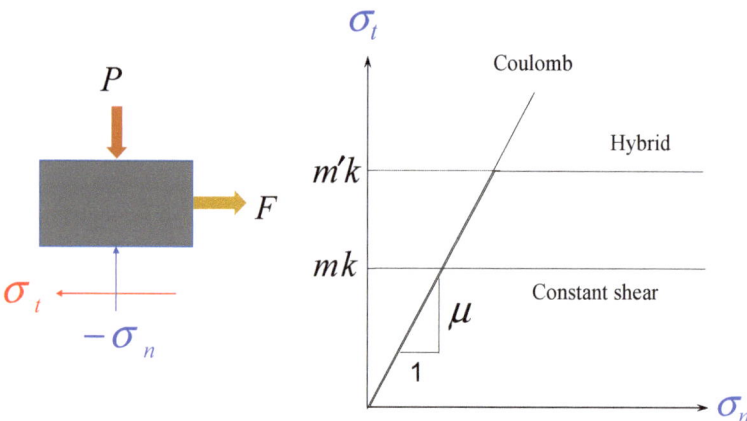

Figure 1. Traditional laws of friction.

Ring compression tests often come up when discussing the Coulomb friction law (with friction coefficient) or constant shear friction (with friction factor). Although ring compression tests help understand friction, they do not always provide a reasonable basis for evaluating it. The results of tracking the height of the ring specimen and the change in its inner diameter, that is, the experimental friction calibration curves, imply the friction characteristics. The friction calibration curves of the Coulomb and constant shear friction laws are similar, and equivalent values can be assumed to exist. Based on this fact, many researchers believe that the two friction laws are equivalent. However, this is generally not true. The change in normal stress at the contact surface in the ring compression test is relatively small compared to that in the forging. Therefore, even if the Coulomb friction law is used, the frictional stress does not significantly differ depending on the location.

Thus, the two friction calibration curves should be similar to each other. Contrarily, the friction state of the contact surface during forging can be significantly different from time to time and from position to position [40]. The Coulomb friction law or its variant is thus better than the constant shear friction law. Notably, the functional relationship between frictional and normal stresses is significant, especially when studying wear.

Assuming that there is a linear relationship between the normal and frictional stresses at the contact surface, the Coulomb friction law is expressed by the following equation:

$$\sigma_t = -\mu \sigma_n g(v_t - \bar{v}_t) \tag{1}$$

where the function $g(v_t - \bar{v}_t)$ reflects the effect of the relative velocity $v_t - \bar{v}_t$ on frictional stress σ_t, and the following function is widely used:

$$g(v_t - \bar{v}_t) = -\frac{2}{\pi}\tan^{-1}\frac{(v_t - \bar{v}_t)}{a} \tag{2}$$

where a is a positive constant that is small compared to $|\bar{v}_t|$.

In the sticking contact surface belonging to the plastic region, the shear stress must satisfy not only the equation of equilibrium, but also the yield criterion. On the other hand, the frictional stress acting on the contact surface where slipping occurs in the plastic region must satisfy not only the equation of equilibrium and yield criterion, but also the friction law. Since the Coulomb friction law in Equation (1) also deals with sticking conditions, there is no mathematical problem in expressing all types of contact stresses, including the frictional stress in the sliding contact surface and the tangential stress in a sticking region. As a result, Equation (2) effectively keeps the frictional stress from being greater than k when the Coulomb friction law is used. If $|\mu\sigma_n|$ exceeds k, then $|v_t - \bar{v}_t|$ should be small enough to meet the yield criterion. Thus, sticking should occur in the extreme case that $|\mu\sigma_n|$ is sufficiently large. What should be noted here is that the tangential stress in a sticking condition generally does not reach the yield stress in pure shear (k) [19].

The traditional Coulomb friction law, which considers the friction coefficient as a constant, also poses some problems in metal forming. First, there is a fundamental problem regarding the simple linearity of normal and frictional stresses. Second, the lubrication regime changes at the contact surface due to the extreme tribological states during metal forming being inevitable. Therefore, the friction coefficient is bound to change during metal forming.

Wilson and Cazeault's work [48] found that the friction coefficient during a wax-lubricated strip drawing was almost zero in the thick film lubrication regime. In contrast, it increased to around 0.16 in the boundary lubrication regime. The friction coefficient varied drastically with the die angle in the strip drawing, especially around the die semi-angle ranging from 15 °C to 20 °C, implying that the lubrication regime change caused a drastic change in the tribological condition of friction and die wear. This fact emphasizes that the friction coefficient is the function of not only a material, die, and lubricant property, but also the process geometry and sliding speed [40]. Notably, the latter governs the mechanical and tribological state of the lubricant, which is concerned with the lubrication regime change.

To solve the problem of the traditional Coulomb friction law described above, Lee et al. [33] formulated the friction coefficient in Equation (1) as a function of state variables such as pressure, surface strain of the material, and temperature. To improve practicality, a separable variable function consisting of the influence of each state variable, that is, the weight, was used, as shown in the following equation:

$$\mu = \mu_0 W_E(\varepsilon) W_T(T) W_P(P) \tag{3}$$

where μ_0 is a friction constant and $W_E(\varepsilon)$, $W_T(T)$, and $W_P(P)$ are weighting functions of surface strain, temperature, and pressure, respectively. Each weighting function was defined as a piecewise linear function. For example, Figure 2 shows a piecewise linear function of $W_E(\varepsilon)$.

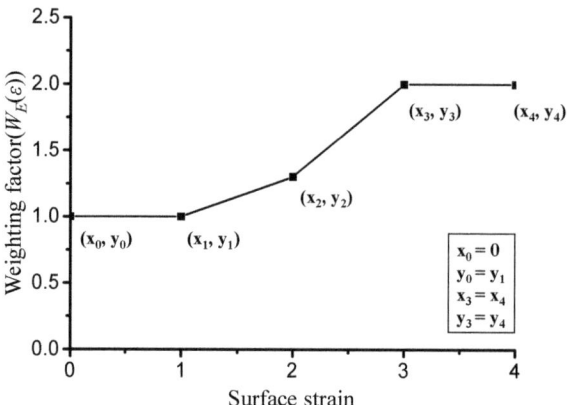

Figure 2. Definition of design variables and initial guess.

3. Evidence of LRC and Its Effect on the Processes and Products

3.1. Case 1: Cold Forward Extrusion of Pure Aluminum Using Step Die

The multi-step forward cold extrusion with poor-lubricated surface, defined in Figure 3, is a typical example where a friction coefficient linked to the state of the contact surface must be used. When a constant friction coefficient over a value is used for the multi-step cold forward extrusion process with a moderate reduction of area, plastic deformation near the skin at the inlet container may occur when the lubricant film on the surface is thin. Joun et al. [19] solved this problem by using a varying friction coefficient. The previous method of imposing the friction coefficient on the specific area or region is impractical.

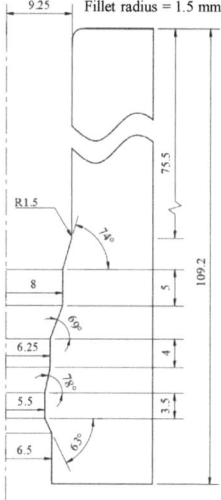

Figure 3. Multi-step forward cold extrusion process.

The friction force inevitably generated from the exit side increases the material's hydrostatic pressure, leading to the increase in contact pressure at the material–die interface inside the container. Notably, it causes a vicious cycle of increased friction and extreme plastic deformation in the case of a significant friction coefficient. This phenomenon differs from plastic deformation in the inlet container, even though the friction at the exit is so substantial that burnt skin can be observed in the actual process. However, this phenomenon has nothing to do with the fundamental problem of the Coulomb friction law.

To visualize this phenomenon, a simulation was conducted for the multi-step cold forward extrusion process using the elastoplastic finite element method. The analysis information is as follows:

- Flow stress of the material: $\sigma = 50.3(1 + 20\varepsilon)^{0.26}$ MPa;
- Young's modulus of the material: 90,000; Poisson's ratio: 0.3;
- Rigid dies;
- Ram's speed: 1 mm/s;
- Frictions:
  Case (1) Constant friction coefficient $\mu = 0.1$;
  Case (2) Surface strain-dependent friction coefficient defined by the piecewise linear function of the following vertices:

$$(\varepsilon, \mu) : (0, 0.01); (0.3, 0.03); (0.5, 0.05); (1.0, 0.1); (2.0, 0.2)$$

The initial finite element mesh system with 5500 uniform quadrilaterals (each quadrilateral size: 0.24 (x-direction) × 0.43 (y-direction) mm) was employed. The remeshing function was turned off to accurately monitor the skin-shearing occurring in Case (1) directly from the FE mesh system. The upper material's effective strain and grid distortion at the final stroke are shown in Figure 4. The maximum strain reached 1.8, implying that the friction coefficient increased to over 0.18 in the third step near the exit. It was found that the upper material in Case (1) started to be plastically deformed when the leading edge of the material touched the third step, causing the skin-shearing of the upper material. On the contrary, the upper material in Case (2) maintained rigid-body translation throughout the process even though the maximum friction coefficient exceeded 0.18 in the third step.

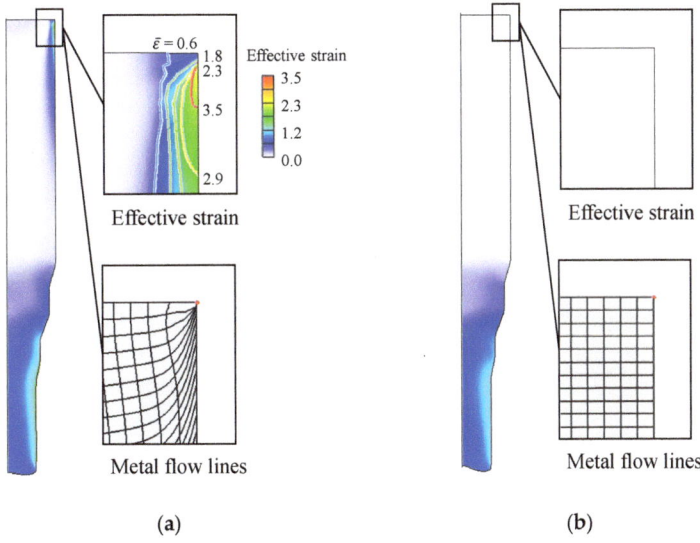

Figure 4. Effective strain of a multi-step cold forward extrusion process. (**a**) Fixed friction coefficient (Case (1)). (**b**) Variable friction coefficient (Case (2)).

It is empirically known that the lubricated film coated on the material does not suffer much damage within the container during the cold forward extrusion process. Experiments suggest that the material's surface at the third step should be severely damaged and scratched even though the upper material inside the container experiences no distinct plastic deformation. Because the contact surface with the undamaged lubricated film maintains low friction, the frictional stress does not cause a high enough pressure to cause plastic deformation at the inlet container. On the other hand, damage to the lubricated film on the exit side can be crucial due to a considerable frictional stress and excessive heat. This tribological phenomenon implies that the LRC occurs with a moderate area reduction during the multi-step cold forward extrusion. Therefore, the friction coefficient at the exit should be significant while that at the entry should be low, and the varying friction coefficient with the contact surface state is thus essential.

By simply expressing the damage to the lubricant as a function of surface strain, the method used in Case (2) solved the excessive friction problem of the Coulomb friction law. This tribological phenomenon can occur in the actual processes even though the material in the inlet side remains in the dead metal zone during the entire extrusion. This case emphasizes that the magnitude of the friction coefficient must change as the state of the contact surface changes to reflect the phenomenon of LRC. It is this example that demonstrates the LRC phenomenon step by step and highlights the importance of the friction model.

This example is very useful in explaining LRC because the effective strain occurring at all contact surfaces is directly related to the deterioration of the lubricant. In actual forging, significant strain may occur on the surface without contact with the die. In this case, it may be unrealistic to treat the friction coefficient as a function of the surface strain itself because the strain does not accelerate the deterioration of the lubricant. Quantification of the degree of lubricant degradation at the contact surface is thus important.

3.2. Case 2: Critical Surface Strain in Hot Forging of Aluminum Alloy Piston

Lee et al. [33] conducted a non-isothermal FE analysis of a hot forging process of an A4032 alloy piston. The surface of the material was well-lubricated by a solid graphite film. The flow function obtained from compression tests was evaluated in the previous study, revealing that the maximum error of this flow function was 1.43%. It is accurate enough for us to focus on friction.

Figure 5 compares the experiments and FE predictions of aluminum piston hot forging for an assumed constant friction coefficient of 0.2. As shown in Figure 5a, the experiments of aluminum piston hot forging showed that the bottom of the material was deformed to be convex. Still, the rigid-thermoviscoplastic FEM predicted a concave shape, as shown in Figure 5b. Lee et al. conducted FE analyses of the process using various commercial software and various traditional friction laws with various constant friction coefficients [33] and friction factors [49]. It was concluded that the FE predictions obtained by the continuous friction conditions cannot express the actual situation.

As can be seen in Figure 5a, the lubricant in the regions with large effective strains was significantly damaged. In contrast, the lubricant film on the lateral side does not appear significantly damaged, implying that the LRC from a thick film to boundary lubrication schemes occurred in the regions where the material surface's color became bright. The friction coefficient should thus change from position to position depending on the damage of the lubricant. A scientific approach to damage to the lubricant film is needed. The friction coefficient can thus be thought of as a function of state variables such as pressure at the contact surface, temperature [47], strain and strain rate, and relative speed of material to die.

However, this was assumed to be dependent on the strain of the material at the surface, which is called surface strain, since the surface strain of the material may represent the degree of damage to the lubricant. The effective strain of the material at each location is different, as can be seen in Figure 5b, where the pattern of effective strain distribution resembles the color of the scratched surface. Considering this, Lee et al. set the friction coefficient at $0.1\ W_E(\varepsilon)$.

Lee et al. used an optimization function of HyperStudy [50] to find the optimized weighting function $W_E(\varepsilon)$ in terms of deformed shape along with the initial guess shown in Figure 2. The results of the optimized friction coefficient function, $0.1 W_E(\varepsilon)$, are shown in Figure 6. These results highlight that the friction coefficient changes rapidly when the surface strain reaches around 1.5, called critical surface strain [33]. This implies that the changed color of the material's surface represents the degree of lubricant damage and the evidence of the LRC.

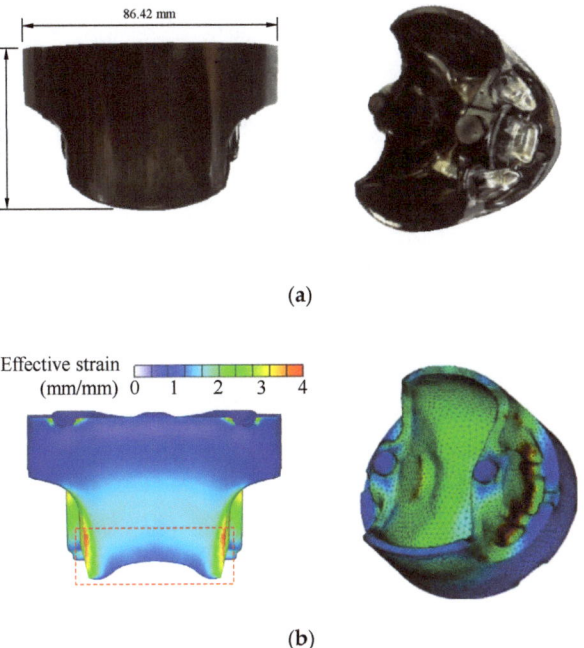

Figure 5. Comparison of the experiments and FE predictions of the aluminum piston hot forging process. (**a**) Experiments. (**b**) Predictions (Effective strain).

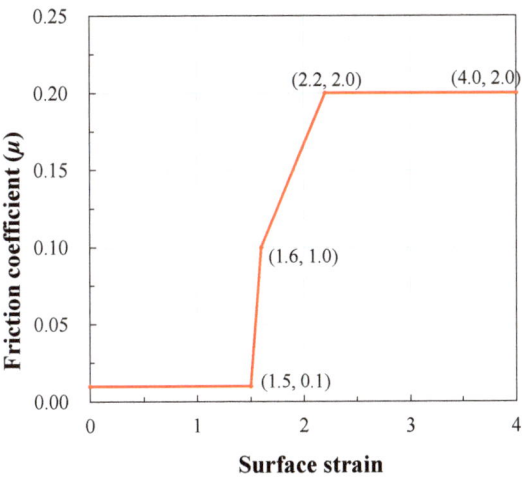

Figure 6. Optimized friction coefficient function.

With the variable friction coefficient expressed as this weighting function, a deformed shape similar to the experiment was predicted, as shown in Figure 7. This example is sufficient to dramatically demonstrate the LRC represented by the deterioration of lubricants and its effects on friction during hot forging. This can occur in the forging of materials with a low strain hardening capability when the material flows through narrow die gaps.

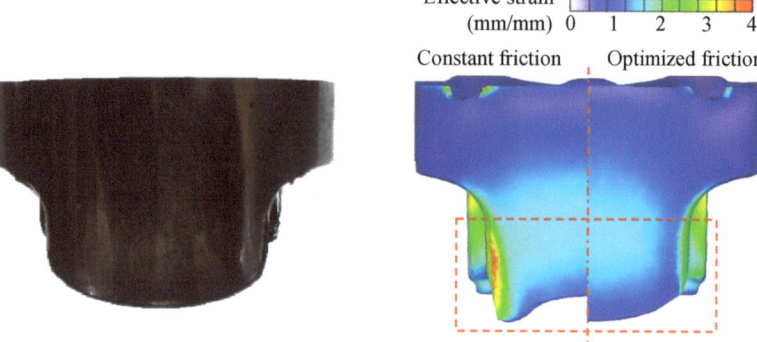

Figure 7. Comparison of the experiment and the prediction obtained using the variable friction coefficient.

It should be emphasized that the FE predictions obtained with the optimized friction condition show a smoother strain distribution than those of the constant friction, as shown in Figure 7. This phenomenon may mean that the lubricant itself has the tendency to spread deterioration to its neighbors to the extent possible.

3.3. Case 3: Sudden LRC in the Automatic Multistage Cold Forging of Aluminum Alloy Yoke

Figure 8 describes a part of the AMSCF [46] of an aluminum (Al6082-T6) steering yoke for passenger cars and its fifth stage's design. As shown in Figure 8a, the punch was offset approximately 0.2 mm from the center of the process. As a result, as shown in Figure 8b, it created an ear height difference of 2.2 mm. A hundred test specimens, which were sawn from an aluminum alloy rod, fabricated for forging but uncoated, were forged by the AMSCF machine in a forming oil environment. All the test forgings exhibited almost the same macroscopic configuration except the first 30 trials. Because the metal forming occurred in the very complex working conditions of the AMSCF, no information about friction could be scientifically obtained using conventional methods. Hamid et al. [46] numerically found the frictional conditions that caused this difference in the ear height.

Hamid et al. [46] characterized the material's flow behavior, and its validity was evaluated by comparing the FE predictions of the compression tests with their experiments in terms of the compression force–stroke curve. It was confirmed that the average error was within 1.9%, implying that the friction can be decoupled from the process-dominant flow behaviors of the material. However, using various constant friction conditions ($\mu = 0.03 \sim 0.5$), they could not obtain the results that satisfied the experimentally measured difference in the ear height of 2.2 mm, as shown in Figure 8b. With a constant coefficient of friction, the maximum predictable ear height difference was 0.6 mm, which is very small compared to the experimental value. Notably, other than LRC, there is no way to explain this result.

Meanwhile, Hamid et al. assumed the friction coefficient to be a function of the surface strain of the material at the contact surface, as shown in Figure 9. They conducted FE analyses for various friction coefficient functions using the rigid-thermoviscoplastic FEM. The number of tetrahedrons ranged from 99,000 to 101,000 and around 320 remeshings were conducted. As shown in Figure 9, a difference in ear height similar to the experimental result was predicted only in the case that a jump in the friction coefficient occurred at the critical surface strain of 1.75. Interestingly, the larger the slope of the function, the closer the predicted result was to the experimental value. It was confirmed that the optimal value of the critical surface strain was 1.75, at which the LRC [40] occurred.

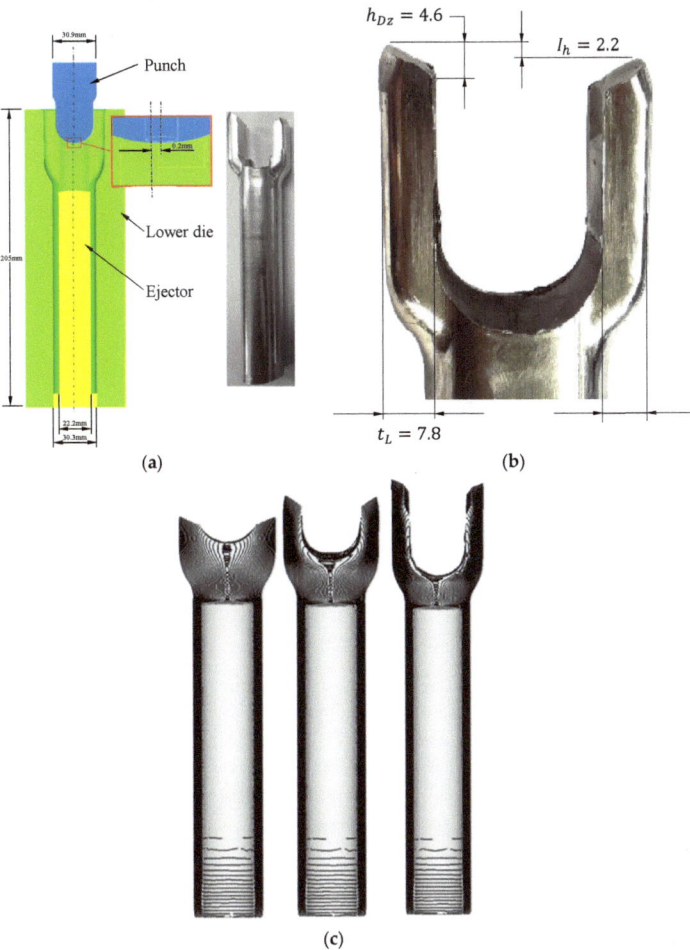

Figure 8. Unbalanced ears occurred during pilot production of the AMSCF of an aluminum yoke. (**a**) Process and product. (**b**) Detailed view of the head part. (**c**) FE-predicted evolution of the metal flow lines at Stage 5.

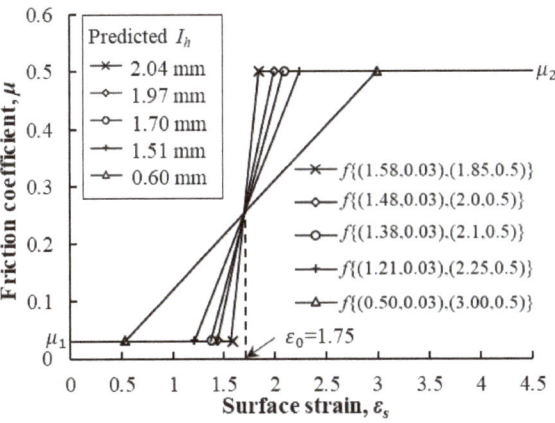

Figure 9. Relationship between surface strain and friction coefficient μ.

Hamid et al. explained this as a tribological shifting phenomenon. As the temperature rose, the viscosity of the forming oil for the AMSCF decreased. As the voids in the contact surface decreased due to high pressure and severe shear deformation, the lubricating forming oil suddenly lost its function or escaped from the contact surface. It caused the fluid lubrication-induced LRC, resulting in a rapid increase in friction. The ear height difference, owing mainly to the extraordinary lubrication phenomenon of tribological LRC, was believed to appear when the strain hardening of the material was slight. The temperature softening is extreme, and it occurs when the temperature, surface strain, and pressure conditions at the contact surface reach critical values. It is concluded that the tribological shifting phenomenon was owing to a drastic LRC, the reasons of which are mechanically and tribologically mixed by the deformation and temperature of the material and lubricant.

This example is also sufficient to highlight the dependence of friction on state variables, which can be explained as the effect of the LRC on friction even though the tribology in the AMSCF process is extremely complicated.

In this process, the final shape is determined by very complex macroscopic phenomena [46]. What is observed from the analysis results is that the greater the jump in friction coefficient at a specific strain, the closer the predicted results are to the experimental value. In addition, at the same time as LRC, all state variables such as hydrostatic pressure, effective stress, and effective strain rate and the increase rate of forming load change rapidly. Considering the influence of the friction coefficient especially at the critical surface strain, it can be argued that the LRC leads this change in macroscopic responses.

3.4. Case 4 Skin-Shearing Phenomenon in the Round-to-Half-Circle Drawing Process of SUS 304

Heo et al. [47] studied the skin-shearing phenomenon shown in Figure 10 that occurred during the round-to-half circle drawing process of SUS 304. The shape drawing process uses the grease containing the additives as a lubricant, and thus, considerable friction cannot occur at room temperature because its normal stress at the contact surface is smaller than that in the cold forward forging of Section 3.1, for example. However, the lubricating effect of the grease may rapidly be lost as the temperature of the contact surface rises due to friction heat.

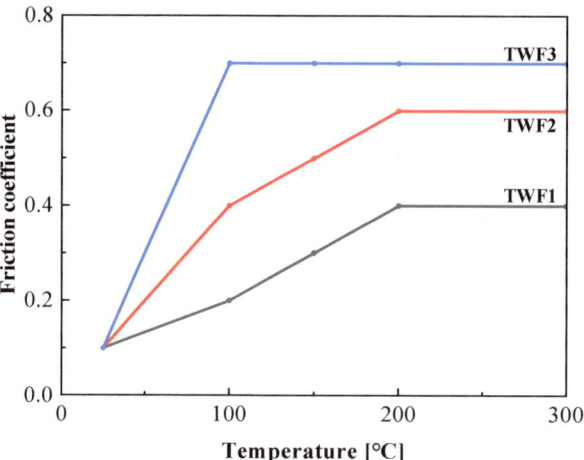

Figure 10. Friction coefficient–temperature curves, assumed.

This problem is a chronic feature of the drawing process in which the relative motion between the material and the die is extreme. When high pressure is applied in a specific part of the die, excessive generation of friction heat centered on this part can dominate the process and product quality of the drawing process. Therefore, temperature is important in terms of friction. Of course, an increase in temperature destroys the lubrication, which causes a change from the thick or thin film lubrication regime to the boundary lubrication regime and promotes die wear due to high heat and relative material–die movement, leading to increased friction. In any case, temperature is an important factor that affects friction and wear in drawing and can represent a state variable related to the lubrication phenomenon.

In the numerical study of Heo et al., a temperature-dependent friction coefficient was thus used, as shown in Figure 11. However, the hybrid friction model was used, following the Coulomb friction law. Still, the maximum shear stress was regulated not to exceed 95% of the pure shear yield stress. The flow stress of the material used is the result of previous research [51], and the maximum error was 7.7% in the range of test temperature and strain rate. It is of sufficient accuracy for macroscopic research purposes.

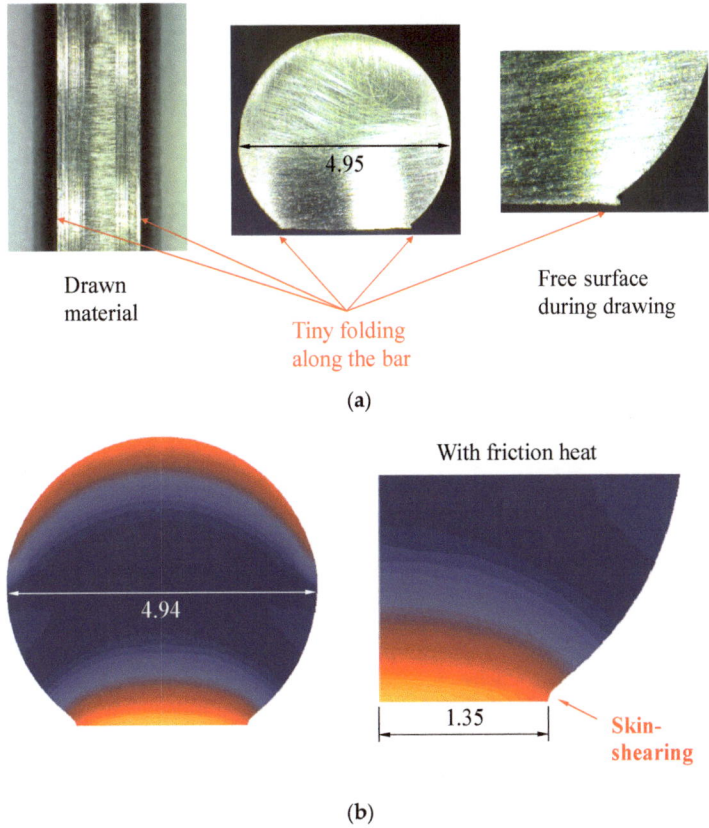

Figure 11. Experimental and FE-predicted folding defect owing to skin-shearing. (**a**) Experimental folding defect. (**b**) FE-predicted skin-shearing using TWF3 in Figure 10.

Heo et al. used the combined steady-state flow and unsteady-state temperature method to reveal the local heating generated by friction in the drawing process, that is, the friction heat ball phenomenon [47]. Due to this friction heat ball, a part of the die is heated in a vicious cycle of heat generation and increased friction. Eventually, the friction heat ball heats the material locally, resulting in a LRC and rapid temperature softening of the surface. It ultimately led to the phenomenon of skin-shearing during drawing. Note that before the skin-shearing phenomenon, the LRC continuously occurred during the shape drawing because of the vicious cycle, which brought out the steady increase in die temperature.

Heo et al. conducted FE analyses under various frictional conditions, including TWF1, TWF2, and TWF3, defined in Figure 10. Special mesh systems which were extremely refined in the contact interface were employed. The superfine tetrahedrons with edge lengths less than 0.02 mm were purposely generated near the burr-like defect and skin-shearing region. Figure 11 compares the experiments and the FE predictions of TWF3, showing an excellent qualitative agreement with each other. As shown in Figure 11, the TWF3 predicted the skin-shearing phenomenon, which is quite peculiar. The possibility of this phenomenon occurring was discovered in the FE predictions of TWF2, but TWF1 did not even predict the possibility of skin-shearing occurring. It means that high friction causes scratches and the wear of the die as the lubricant does not play its role due to the high temperature in the friction heat ball area, resulting in the deterioration of the boundary lubrication regime and an extreme increase in the coefficient of friction. The validity of the high friction coefficient was confirmed because the lubricant experimentally flowed like water at the outlet, and smoke was generated due to frictional heat.

Heo et al. showed that the size of the FE-predicted defect tended to increase as the tetrahedral element edge length at the edge and plate area where the defect was created became smaller. In other words, it was confirmed that the FE predictions converged with the experiments.

Although this case concerns a shape drawing process, which is an extreme metal forming process in terms of friction, it is sufficient to highlight the influence of LRCs that occur over time due to the temperature dependence of friction during drawing.

Skin-shearing that occurs during drawing is a very unusual phenomenon. Various attempts were made to predict this phenomenon but failed [47]. Not only does the loss of lubricant function due to heat energy accumulated in the die cause an extreme increase in friction, but the increased surface temperature of the die locally heats the surface of the material, causing a decrease in flow stress at the material's surface and finally increasing the skin-shearing of the material. Without the LRC by the drastically decayed function of the lubricant, it is impossible to explain the skin-shearing phenomenon.

4. Conclusions

The importance of friction during bulk metal forming was emphasized since the material and process during bulk metal forming experience extreme competition between flow and friction. It was revealed that the lubrication regime change is inevitable during bulk metal forming accompanying moderate plastic deformation. Multi-step forward extrusion and drawing processes are vulnerable to the lubrication regime change (LRC) owing to large sliding motion at the material–die interface. Aluminum forging is also vulnerable to the LRC because the aluminum alloys belong to low strain hardening materials, which may easily experience the exaggerated effect of the lubrication regime change. Strictly speaking, most forging processes accompany any LRC owing to the effect of material flow and heat transfer. This study showed the cases of typical LRC in the bulk metal forming processes, including the multi-step forward extrusion process, round-to-half circle drawing process, and cold and hot forging processes of aluminum alloys. A few new findings are summarized as follows:

(1) It was emphasized that a multi-step cold forward extrusion process with a moderate reduction of area is a typical example exposed to the problem related to the LRC since large sliding motion, high hydrostatic pressure, and large deformation of material at the exit exaggerate the continuous evolution of the LRC from the thick film to boundary lubrication regimes via thin film and mixed lubrication regimes;

(2) The flow characteristics of aluminum alloys for bulk metal forming purposes are expressed by low strain hardening, high temperature softening between 200 °C and 300 °C, and the narrow transition region between cold and hot forging. In this forming environment, friction can dominate the macroscopic plastic deformation during bulk metal forming, and LRC can thus easily occur, leading to macroscopic instability;

(3) When the lubrication regime changes, the friction coefficient greatly depends on the state of the contact surface and the state variables at the contact surface (strain rate, strain rate, pressure, temperature, relative speed of material to die, etc.). From a process analysis perspective, the major factors that govern the state of lubrication include the material's surface strain (a measure of the damage to the lubricant) and temperature, which are major factors affecting the LRC;

(4) An increase in surface strain promotes direct contact between the material and the die, which leads to the creation of debris and the acceleration of friction and wear. For this reason, when the strain at the contact surface reached the critical surface strain in the hot forging of an aluminum alloy material coated with a lubricant, the friction coefficient rapidly increased. A similar phenomenon occurred in the AMSCF of aluminum alloy operated in a forming oil environment, which resulted in a sudden change in the lubrication state at the specific surface strain during metal forming because of a drastic LRC.

In summary, as Wilson crucially criticized four and a half decades ago, the traditional Coulomb friction law, which assumes a constant value of friction coefficient, is not appropriate in most bulk metal forming where the state of the contact surface (pressure, lubrication state of the material surface) changes rapidly. Experiences say that most forging processes accompany a LRC during the process. However, most researchers rely on the traditional friction law based on a constant friction coefficient. In the case of lightweight materials, including aluminum and magnesium alloys, friction can have a significant impact on the process. Careful attention is thus required to solve sophisticated tribological problems in bulk metal forming.

Finally, it is recommended that a quantity to measure the damage of lubricant be developed to accelerate the study of friction in bulk metal forming.

Author Contributions: Conceptualization, M.-S.J.; Methodology, M.-S.J.; Validation, Y.H., N.-H.K. and N.-Y.K.; Investigation, Y.H., N.-H.K. and N.-Y.K.; Data curation, N.-H.K.; Writing—original draft, M.-S.J.; Writing—review & editing, Y.H.; Supervision, M.-S.J. All authors have read and agreed to the published version of the manuscript.

Funding: This work was partly supported by Korea Institute of Energy Technology Evaluation and Planning (KETEP) (20214000000520, Human Resource Development Project in Circular Remanufacturing Industry) grant funded by the Korea government (MOTIE) and supported by the Leaders in INdustry-university Cooperation 3.0 (LINC 3.0) (202406190001, Ministry of Education and National Research Foundation of Korea).

Data Availability Statement: Data are included in the article.

Conflicts of Interest: The author declares no conflict of interest.

References

1. Srikanth, A.; Zabaras, N. Shape optimization and preform design in metal forming processes. *Comput. Methods Appl. Mech. Eng.* **2000**, *190*, 1859–1901. [CrossRef]
2. António, C.; Castro, C.; Sousa, L. Optimization of metal forming processes. *Comput. Struct.* **2004**, *82*, 1425–1433. [CrossRef]
3. Bonte, M.H.A.; van den Boogaard, A.H.; Huétink, J. An optimization strategy for industrial metal forming processes. *Struct. Multidisc. Optim.* **2008**, *35*, 571–586. [CrossRef]
4. Ozturk, M.; Kocaoglan, S.; Sonmez, F.O. Concurrent design and process optimization of forging. *Comput. Struct.* **2016**, *167*, 24–36. [CrossRef]
5. Merklein, M.; Allwood, J.M.; Behrens, B.A.; Brosius, A.; Hagenah, H.; Kuzman, K.; Mori, K.; Tekkaya, A.E.; Weckenmann, A. Bulk forming of sheet metal. *CIRP Ann.-Manuf. Technol.* **2012**, *61*, 725–774. [CrossRef]
6. Mori, K.; Nakano, T. State-of-the-art of plate forging in Japan. *Prod. Eng.* **2016**, *10*, 81–91. [CrossRef]
7. Petrik, J.; Ali, S.I.; Feistle, M.; Bambach, M. CrystalMind: A surrogate model for predicting 3D models with recrystallization in open-die hot forging including an optimization framework. *Mech. Mater.* **2023**, *189*, 104875. [CrossRef]
8. Bambach, M. Fast simulation of incremental sheet metal forming by adaptive remeshing and subcycling. *Int. J. Mater. Form.* **2016**, *9*, 353–360. [CrossRef]

9. Chenot, J.L.; Bernacki, M.; Bouchard, P.O.; Fourment, L.; Hachem, E.; Perchat, E. Recent and future developments in finite element metal forming simulation. In Proceedings of the 11th International Conference on Technology of Plasticity, ICTP, Nagoya, Japan, 19–24 October 2014.
10. Joun, M.S. Recent advances in metal forming simulation technology for automobile parts by AFDEX. *IOP Conf. Ser. Mater. Sci. Eng.* **2020**, *834*, 012016. [CrossRef]
11. Lin, Y.C.; Chen, X.M. A critical review of experimental results and constitutive descriptions for metals and alloys in hot working. *Mater. Des.* **2011**, *32*, 1733–1759. [CrossRef]
12. Aliakbari Sani, S.; Ebrahimi, G.R.; Vafaeenezhad, H.; Kiani-Rashid, A.R. Modeling of hot deformation behavior and prediction of flow stress in a magnesium alloy using constitutive equation and artificial neural network (ANN) model. *J. Magnes. Alloys* **2018**, *6*, 134–144. [CrossRef]
13. Joun, M.S.; Razali, M.K.; Yoo, J.D.; Kim, M.C.; Choi, J.M. Novel extended C-m models of flow stress for accurate mechanical and metallurgical calculations and comparison with traditional flow models. *J. Magnes. Alloys* **2022**, *10*, 2516–2533. [CrossRef]
14. Şimşir, C.; Duran, D. A flow stress model for steel in cold forging process range and the associated method for parameter identification. *Int. J. Adv. Manuf. Technol.* **2018**, *94*, 3795–3808. [CrossRef]
15. Voyiadjis, G.Z.; Song, Y. A physically based constitutive model for dynamic strain aging in Inconel 718 alloy at a wide range of temperatures and strain rates. *Acta Mech.* **2020**, *231*, 19–34. [CrossRef]
16. Luan, J.; Sun, C.; Li, X.; Zhang, Q. Constitutive model for AZ31 magnesium alloy based on isothermal compression test. *Mater. Sci. Technol.* **2014**, *30*, 211–219. [CrossRef]
17. Mirone, G.; Barbagallo, R.; Bua, G.; Licignano, P.; Tedesco, M.M. An Enhanced Approach for High-Strain Plasticity in Flat Anisotropic Specimens with Progressively Distorting Neck Sections. *Metals* **2024**, *14*, 578. [CrossRef]
18. Engel, U. Tribology in microforming. *Wear* **2006**, *260*, 265–273. [CrossRef]
19. Joun, M.S.; Moon, H.G.; Choi, I.S.; Lee, M.C.; Jun, B.Y. Effects of friction laws on metal forming processes. *Tribol. Int.* **2009**, *42*, 311–319. [CrossRef]
20. Trzepiecinski, T.; Lemu, H.G. Recent developments and trends in the friction testing for conventional sheet metal forming and incremental sheet forming. *Metals* **2020**, *10*, 47. [CrossRef]
21. Wilson, W.R.D.; Marsault, N. Partial hydrodynamic lubrication with large fractional contact areas. *J. Tribol.* **1998**, *120*, 16–20. [CrossRef]
22. Shisodea, M.P.; Hazratia, J.; Mishrab, T.; de Rooijb, M.; van den Boogaarda, T. Modeling mixed lubrication friction for sheet metal forming applications. *Procedia Manuf.* **2020**, *47*, 586–590. [CrossRef]
23. Noder, J.; George, R.; Butcher, C.; Worswick, M.J. Friction characterization and application to warm forming of a high strength 7000-series aluminum sheet. *J. Mater. Process. Technol.* **2021**, *293*, 117066. [CrossRef]
24. Evin, E.; Tomáš, M. Influence of friction on the formability of Fe–Zn-coated IF steels for car body parts. *Lubricants* **2022**, *10*, 297. [CrossRef]
25. Trzepieciński, T. Experimental analysis of frictional performance of EN AW-2024-T3 Alclad aluminium alloy sheet metals in sheet metal forming. *Lubricants* **2023**, *11*, 28. [CrossRef]
26. Olsson, M.; Cinca, N. Mechanisms controlling friction and material transfer in sliding contacts between cemented carbide and aluminum during metal forming. *Int. J. Refract. Met. Hard Mater.* **2024**, *118*, 106481. [CrossRef]
27. Wang, H.; Chen, G.; Zhu, Q.; Zhang, P.; Wang, C. Frictional behavior of pure titanium thin sheet in stamping process: Experiments and modeling. *Tribol. Int.* **2024**, *191*, 109131. [CrossRef]
28. Svoboda, P.; Jopek, M. The Effect of Strain Rate on the Friction Coefficient. *Manuf. Technol.* **2024**, *24*, 289–293. [CrossRef]
29. Kchaou, M. New Framework for studying High Temperature Tribology (HTT) Using a Coupling Between Experimental Design and Machine Learning. *Tribol.-Finn. J. Tribol.* **2024**, *41*, 4–12. [CrossRef]
30. Hossen, M.S.; Westrum, J.J.; Shultz, M.; Tan, H.; Kim, D. Variable Friction Model Development and Implementation to the Pulling Force Prediction of the Split-Sleeve Cold Expansion Process for Aluminum 2024-T3. In Proceedings of the ASME 2024 19th International Manufacturing Science and Engineering Conference, Knoxville, TN, USA, 17–21 June 2024; ASME: New York, NY, USA, 2024; p. V002T06A005.
31. Kim, J.H.; Ko, B.H.; Kim, J.H.; Lee, K.H.; Moon, Y.H.; Ko, D.C. Evaluation of friction using double cup and spike forging test for dry-in-place coating and forming oils. *Tribol. Int.* **2020**, *150*, 106361. [CrossRef]
32. Kramer, P.; Groche, P. Friction Measurement under Consideration of Contact Conditions and Type of Lubricant in Bulk Metal Forming. *Lubricants* **2019**, *7*, 12. [CrossRef]
33. Lee, S.W.; Lee, J.M.; Joun, M.S. On critical surface strain during hot forging of lubricated aluminum alloy. *Tribol. Int.* **2020**, *141*, 05855. [CrossRef]
34. Kato, K. Wear in relation to friction—A review. *Wear* **2000**, *241*, 151–157. [CrossRef]
35. Holmberg, K.; Ronkainen, H.; Laukkanen, A.; Wallin, K. Friction and wear of coated surfaces—Scales, modelling and simulation of tribomechanisms. *Surf. Coat. Technol.* **2007**, *202*, 1034–1049. [CrossRef]
36. Sheng, S.; Zhou, H.; Wang, X.; Qiao, Y.; Yuan, H.; Chen, J.; Yang, L.; Wang, D.; Liu, Z.; Zou, J.; et al. Friction and wear behaviors of Fe-19Cr-15Mn-0.66 N steel at high temperature. *Coatings* **2021**, *11*, 1285. [CrossRef]
37. Zhong, W.; Liu, Y.; Hu, Y.; Li, S.; Lai, M. Research on the mechanism of flash line defect in coining. *Int. J. Adv. Manuf. Technol.* **2012**, *63*, 939–953. [CrossRef]

38. Groche, P.; Kramer, P.; Zang, S.; Rezanov, V. Prediction of the Evolution of the Surface Roughness in Dependence of the Lubrication System for Cold Forming Processes. *Tribol. Lett.* **2015**, *59*, 9. [CrossRef]
39. Hafis, S.M.; Ridzuan, M.J.M.; Mohamed, A.R.; Farahana, R.N.; Syahrullail, S. Minimum Quantity Lubrication in Cold Work Drawing Process: Effects on Forming Load and Surface Roughness. *Procedia Eng.* **2013**, *68*, 639–646. [CrossRef]
40. Wilson, W.R.D. Friction and lubrication in bulk metal-forming processes. *J. Appl. Metalwork.* **1978**, *1*, 7–19. [CrossRef]
41. Westeneng, J.D. *Modelling of Contact and Friction in Deep Drawing Processes*; University of Twente: Enschede, The Netherlands, 2001.
42. Berthier, Y. Experimental evidence for friction and wear modelling. *Wear* **1990**, *241*, 77–92. [CrossRef]
43. Hol, J.; Meinders, V.T.; de Rooij, M.B.; van den Boogaard, T. Multi-scale friction modeling for sheet metal forming: The boundary lubrication regime. *Tribol. Int.* **2015**, *81*, 112–128. [CrossRef]
44. Shisode, M.; Hazrati, J.; Mishra, T.; de Rooij, M.; ten Horn, C.; van Beeck, J.; van den Boogaard, T. Modeling boundary friction of coated sheets in sheet metal forming. *Tribol. Int.* **2021**, *153*, 106554. [CrossRef]
45. Fontalvo, G.A.; Humer, R.; Mitterer, C.; Sammt, K.; Schemmel, I. Microstructural aspects determining the adhesive wear of tool steels. *Wear* **2006**, *260*, 1028–1034. [CrossRef]
46. Hamid, N.A.; Kim, K.M.; Hwang, T.M.; Choi, J.M.; Joun, M.S. Tribological shifting phenomena during automatic multistage cold forging of an automotive Al6082-T6 steering yoke. *J. Manuf. Proc.* **2024**, *114*, 178–195. [CrossRef]
47. Heo, Y.; Kim, N.Y.; Nam, J.W.; Chung, I.G.; Joun, M.S. Friction heat ball in round-to-half circle drawing and its effect on the material's skin shearing. *Tribol. Int.* **2024**, *197*, 109755. [CrossRef]
48. Wilson, W.R.D.; Cazeault, P. Measurement of frictional conditions in lubricated strip drawing. In Proceedings of the Fourth NAMRC, Columbus, OH, USA, 17–19 May 1976; pp. 165–171.
49. Lee, S.W.; Jo, J.W.; Joun, M.S.; Lee, J.M. Effect of friction conditions on material flow in FE analysis of Al piston forging process. *Int. J. Precis. Eng. Manuf.* **2019**, *20*, 1643–1652. [CrossRef]
50. Available online: https://altair.co.kr/ (accessed on 10 September 2024).
51. Byun, J.B.; Jee, C.W.; Seo, I.D.; Joun, M.S. Characterization of double strain-hardening behavior using a new flow parameter of extremum curvature strain of Voce strain-hardening model. *J. Mech. Sci. Technol.* **2022**, *36*, 4115–4126. [CrossRef]

Disclaimer/Publisher's Note: The statements, opinions and data contained in all publications are solely those of the individual author(s) and contributor(s) and not of MDPI and/or the editor(s). MDPI and/or the editor(s) disclaim responsibility for any injury to people or property resulting from any ideas, methods, instructions or products referred to in the content.

Article

Development of a Machine Vision System for the Average Roughness Measurement of Shot- and Sand-Blasted Surfaces

Kyungmok Kim

School of Aerospace and Mechanical Engineering, Korea Aerospace University, 76 Hanggongdaehak-ro, Deogyang-gu, Goyang-si 412-791, Gyeonggi-do, Republic of Korea; kkim@kau.ac.kr; Tel.: +82-2-300-0288

Abstract: This article presents a machine vision system for measuring the arithmetic average roughness of shot- and sand-blasted surfaces. In the developed system, a digital microscope was used for capturing surface images after shot- and sand-blasting processes. The captured grayscale images were analyzed with the proposed algorithm using Otsu's global thresholding and a size bandpass filter. The algorithm detected white regions associated with the specular reflection of light on a binary image, and then calculated the size of selected regions. One-way ANOVA was used to identify the relation between the size of the regions and the arithmetic average roughness of blasted surfaces. It was noted that the average size of white regions showed a linear relation to the arithmetic average roughness of both shot- and sand-blasted surfaces. Different abrasives (shot or sand) were found to bring about differences in the rate of change of the average size within a chosen roughness range. When a surface image with unknown roughness is given, it is possible to predict the arithmetic average roughness on the basis of the relation. This machine vision system enables the fast and low-cost roughness measurement of shot- and sand-blasted surfaces. Thus, it could be useful in a quality inspection for shot- and sand-blasting.

Keywords: arithmetic average surface roughness; machine vision; specular reflection; blasting

1. Introduction

Machine vision and machine learning are known as promising technologies in tribology [1,2]. Particularly, advancements in machine vision for surface roughness characterization are attributed to improvements in computational power and in digital image sensing performance. Blasted surfaces prior to the deposition of lubricants are required to maintain precisely controlled roughness. Although conventional methods (e.g., visual or tactile comparison with a sample comparator) enable the roughness measurement of a blasted surface, there exist industrial demands for faster and more accurate measurement.

In order to satisfy these demands, various machine vision techniques have been developed for measuring surface roughness [3–6]. One approach involves analyzing light intensity on a grayscale level, which has been correlated with surface roughness [3]. Additionally, wavelet transform has been employed to examine surface images after grinding and milling, using frequency normalization to enhance precision [4]. Another method, subpixel edge detection, was introduced for the in-process measurement of surface roughness, providing accurate surface contours for roughness prediction [5]. Another reported machine vision approach is to characterize surface texture for evaluating the arithmetic average surface roughness of a grinded workpiece [6]. This approach uses grayscale co-occurrence matrix features for surface texture characterization. Correlation between surface roughness and the matrix feature-based principal component is modeled by using a multiple regression analysis.

Along with these developments in machine vision, machine learning models have been developed for the prediction of surface roughness and for the characterization of surface texture [7–9]. Particularly, neural networks has been used for predicting surface

Citation: Kim, K. Development of a Machine Vision System for the Average Roughness Measurement of Shot- and Sand-Blasted Surfaces. *Lubricants* **2024**, *12*, 339. https://doi.org/10.3390/lubricants12100339

Received: 11 September 2024
Revised: 27 September 2024
Accepted: 28 September 2024
Published: 30 September 2024

Copyright: © 2024 by the author. Licensee MDPI, Basel, Switzerland. This article is an open access article distributed under the terms and conditions of the Creative Commons Attribution (CC BY) license (https://creativecommons.org/licenses/by/4.0/).

roughness. One research work selected a convolutional neural network (CNN) for image-based roughness measurement [7]. The micro-scale images of a surface produced by die-sinking EDM were selected as training dataset for CNN. Other research has focused on a vision-based artificial neural network (ANN) for predicting surface roughness [8,9]. Surface images obtained at various milling parameters were selected for the training dataset in ANN [8]; the parameters included speed, feed rate, and the depth of cut. After the training and testing processes, the arithmetic average surface roughness was determined by the ANN model. A computer vision-based ANN method was developed for the surface roughness prediction of a turned workpiece [9]. The ANN model was built from datasets obtained with various process parameters (speed, depth of cut, feed rate, and grayscale). The arithmetic average surface roughness from the model was compared with those measured with a stylus-type profilometer. In order to increase the accuracy of these machine learning models, it is necessary to obtain as large of a training dataset as possible.

Despite these recent advancements in machine vision and machine learning, little has been found related to a machine vision system for the roughness measurement of blasted surfaces. In order to use a machine vision system for the quality inspection of a manufacturing process, the system needs to maintain fast execution and high accuracy.

In this article, a simple machine vision algorithm is proposed for predicting the arithmetic average roughness of shot- and sand-blasted surfaces. The proposed machine vision system using a digital microscope allows the capture of micro-images, analyzing captured images with Otsu's method and a proposed image filter, and predicting the arithmetic average roughness of blasted surfaces.

2. Materials and Methods

The proposed machine vision system includes a digital microscope with a lens hood, an external computer, and a display, as shown in Figure 1. A conical-shaped lens hood was installed to block out external lights that can affect a target surface. Micro-images captured with the microscope (charge-coupled device sensor with a resolution of 1.3 M pixels (1280 × 960), ring illumination method, 1200×) were transferred to an external computer for image processing (binarization of the submitted grayscale image and filtering). After the image processing, image and data analysis were employed on the computer: the area calculation of the selected specular reflection of light on a filtered image, and the prediction of the arithmetic average roughness (R_a) on predetermined data (e.g., data obtained with a surface roughness comparison sample). Finally, the predicted roughness was displayed.

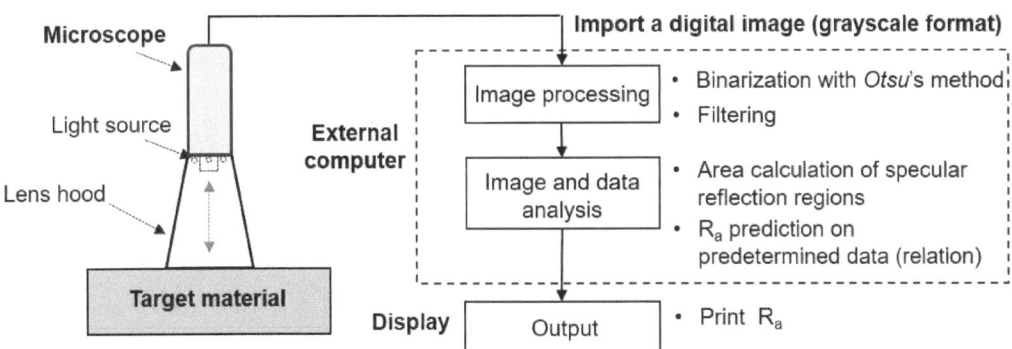

Figure 1. Schematic diagram of the proposed machine vision system.

A detailed procedure for determination of the arithmetic average roughness is described in Figure 2. In this machine vision algorithm, it is necessary to import a grayscale image. The imported grayscale image is binarized. In this algorithm, Otsu's method, a classic thresholding approach in image segmentation, was chosen because of its simplicity and speed [10,11]. The method is also known to be better than the original one in segmenting images corrupted by noise [12,13]. On a generated binary image, the specular reflection of light is presented as white regions. A single binary image contains various sizes and shapes of white regions (that is, a region is the group of white pixels).

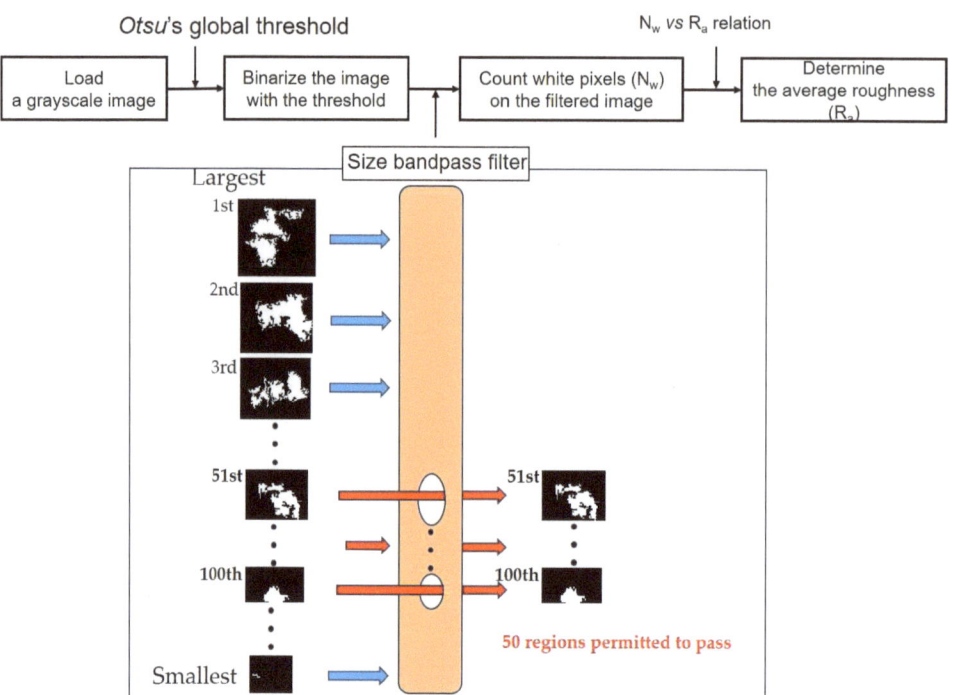

Figure 2. Flow chart for determining the arithmetic average roughness. The size bandpass filter passes specific white regions.

In this study, a specific number of regions were chosen with a size bandpass filter. The filter played the following roles. All regions were arranged in order from the largest. Among them, 50 regions (from the 51st- to 100th-largest ones) were allowed to pass through the filter. Here, '50' is the specific number of regions; the specific number, '50', allowed us to minimize image noise and to obtain the adequate correlation between the percentage of white pixels and the arithmetic average roughness.

On the filtered binary image, the number of white pixels was counted and the average size of 50 regions was calculated. Note that the white pixels correspond to the specular reflection of light. For the purpose of establishing the correlation between the number of white pixels and the arithmetic average roughness, this study used roughness comparison specimens, with blasted surfaces that were well controlled. Based on the relation between the size of the selected regions and the arithmetic average roughness, the arithmetic average surface roughness after a blasting process can be predicted. The proposed image-processing algorithm code is presented in Appendix A.

In this study, two types of blasted surfaces were prepared with shot and sand abrasives (commercial surface roughness comparison samples (nickel alloys) with fabrication standards GB6060.5, ISO 2632-2-1985, supplier: MC, China). The arithmetic average roughness (R_a) values of shot-blasted surfaces were 3.2 µm, 6.3 µm, and 12.5 µm. For shot blasting, high-carbon cast steel (53–60 HRC) shots ranging from 0.5 mm to 2 mm in size were selected. The continuous impact of shots was applied to the target surface at an air pressure of 2–7 bars. Stand-off distance between 15 cm and 45 cm was set based on the shot size.

The selected sand-blasted surfaces also maintained 3.2 μm, 6.3 μm, and 12.5 μm in the arithmetic average roughness. In sand blasting, silica (5 Mohs) particles ranging from 0.5 mm to 2 mm were propelled at an air pressure of 2−7 bars and at a stand-off distance of 15−45 cm.

For image processing and analysis, four micro-images (magnification level: 1200×) were obtained at random locations on a surface at each roughness level; the blasted plates were 20 cm × 25 cm in size. The surface of the plate was divided into four equal parts. Four images were captured in each quadrant.

3. Results

Figure 3 shows shot-blasted surface images with R_a of 12.5 μm. Figure 3a,d,g,j presents grayscale images. Four images captured at random locations on the surface were selected (5 mm × 5 mm in size). Brightness and darkness on the grayscale image were determined with the specular reflection and diffuse reflection of light. The grayscale images were binarized with Otsu's thresholding method, as shown in Figure 3b,e,h,k. Otsu's global thresholding method separates the submitted grayscale image into two classes, foreground and background. The foreground (white) corresponds to the regions of specular reflection, while the background (black) is associated with those of diffuse reflection. Sections B and C present the histograms of grayscale images, representing the distribution of pixel intensities. The global thresholds for each image were automatically determined on the histogram, as shown in Table 1. The thresholds were observed to vary with respect to the arithmetic average roughness of shot-blasted surfaces, while significant difference was not found among the four values at each roughness level. For sand blasting, the automatically determined thresholds were found to remain lower than those for shot blasting, ranging from 0.18 to 0.27. The values were close to each other without regard to the roughness level.

Table 1. Automatically determined Otsu's global thresholds.

Figures No	Type of Abrasive	R_a [μm]	Image a	Image d	Image g	Image j
3	Shot	12.5	0.4118	0.4078	0.4000	0.4118
4		6.3	0.2549	0.2471	0.2431	0.2510
5		3.2	0.2980	0.2784	0.2745	0.2980
6	Sand	12.5	0.1765	0.2235	0.2255	0.1882
7		6.3	0.2431	0.2471	0.2392	0.2353
8		3.2	0.2549	0.2549	0.2627	0.2667

Figure 3c,f,i,l has the images to which a proposed size bandpass filter was applied. In the filtered images, fifty white regions were presented. The total number of white pixels was counted on each image and averaged, as shown in Table 2.

Table 2. Number of white pixels in filtered images. Note that the total number of pixels in a binary image was 360,000.

| Figures No | Type of Abrasive | R_a [μm] | Number of White Pixels | | | | |
			Image a	Image d	Image g	Image j	Average
3	Shot	12.5	25,229	25,397	23,370	25,817	24,953
4		6.3	13,762	13,071	11,481	13,077	12,847
5		3.2	8462	7549	7241	9479	8182
6	Sand	12.5	19,501	21,640	20,242	19,345	20,182
7		6.3	15,750	14,092	16,152	16,448	15,611
8		3.2	12,297	11,708	13,667	10,191	11,966

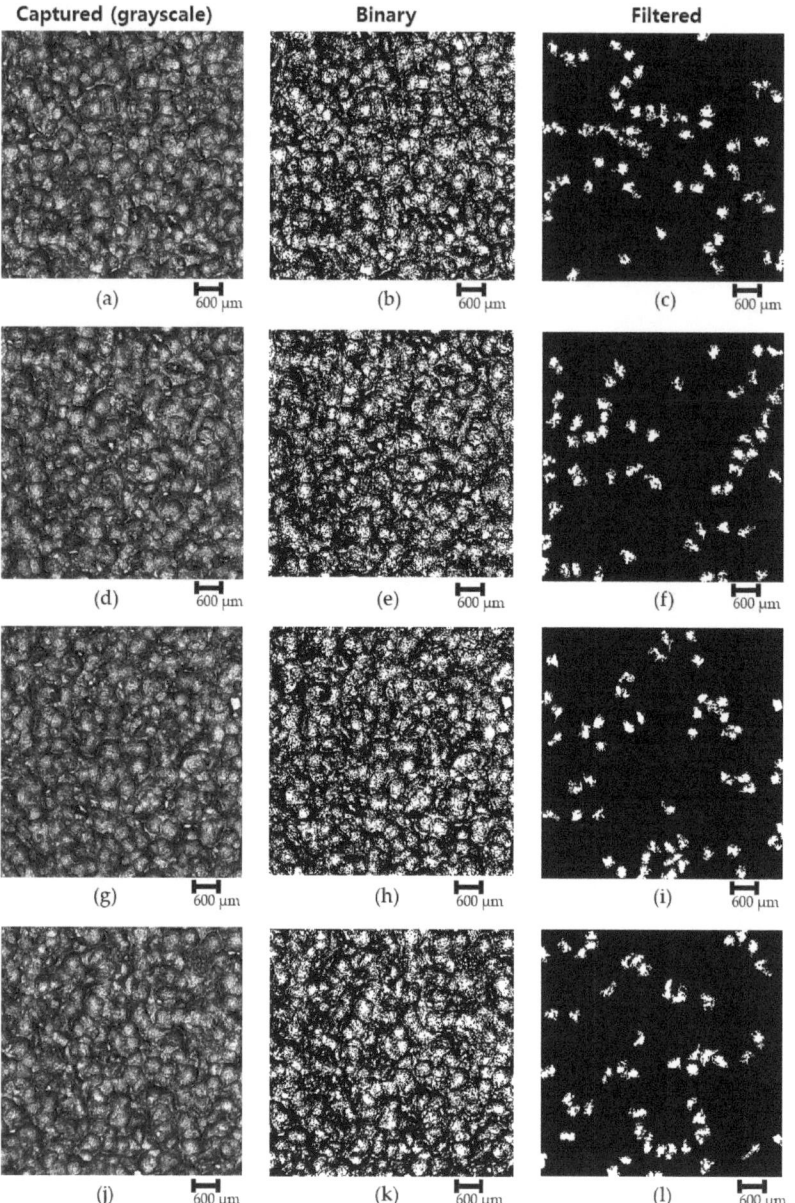

Figure 3. Shot-blasted surface images with R_a of 12.5 μm: (**a,d,g,j**) present grayscale images; (**b,e,h,k**) show binary ones obtained with Otsu's method; (**c,f,i,l**) are filtered images.

Figure 4 shows shot-blasted surface images with R_a of 6.3 μm. Figure 4a,d,g,j presents captured grayscale images. The surfaces were found to be smoother than those in Figure 3 since the arithmetic average roughness for Figure 4 was half lower that in Figure 3. Figure 4b,e,h,k shows binary ones obtained with Otsu's method. Figure 4c,f,i,l presents filtered binary images. Figure 5 shows shot-blasted surface images with an R_a of 3.2 μm. It is clear that the grayscale images showed the smoothest texture among the three shot-blasted surfaces, as shown in Figure 5a,d,g,j. Binary images are presented in Figure 5b,e,h,k. The binary images after the application of the size bandpass filter are shown in Figure 5c,f,i,l.

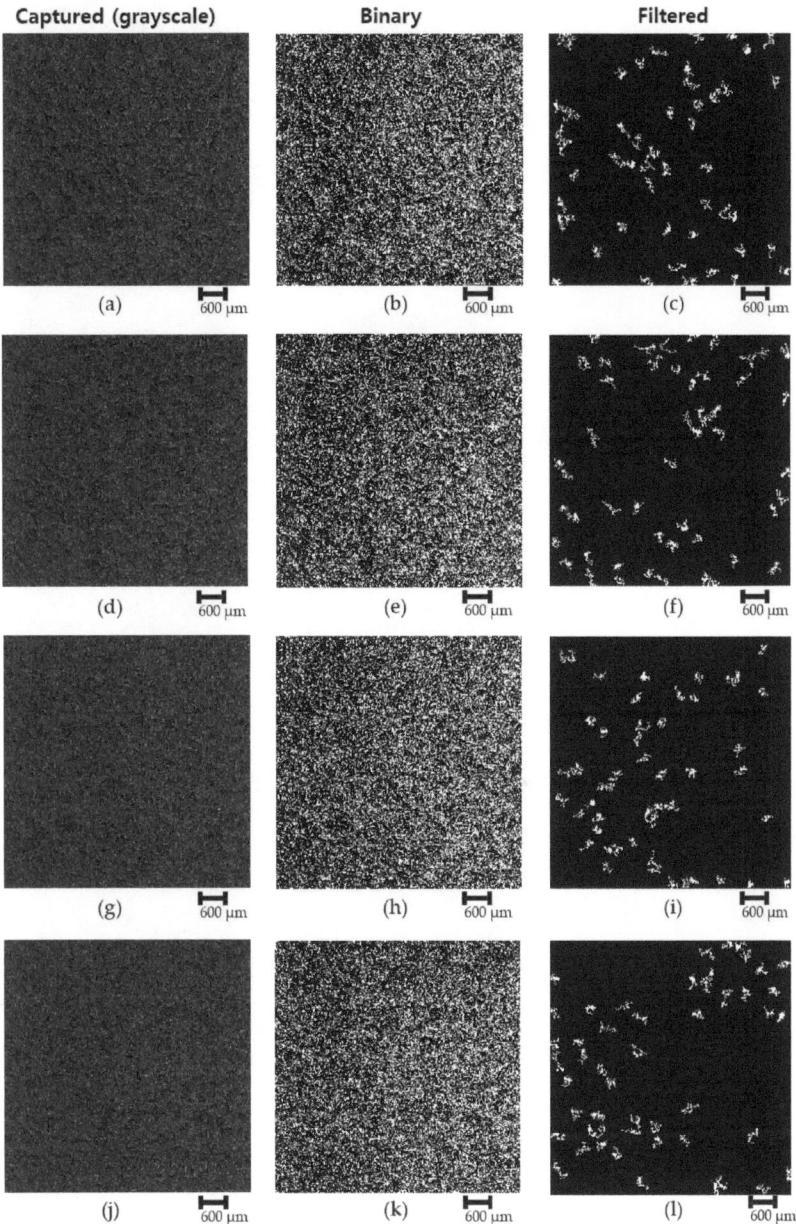

Figure 4. Shot-blasted surface images with R_a of 6.3 µm: (**a,d,g,j**) present grayscale images; (**b,e,h,k**) show binary ones obtained with Otsu's method; (**c,f,i,l**) are filtered images.

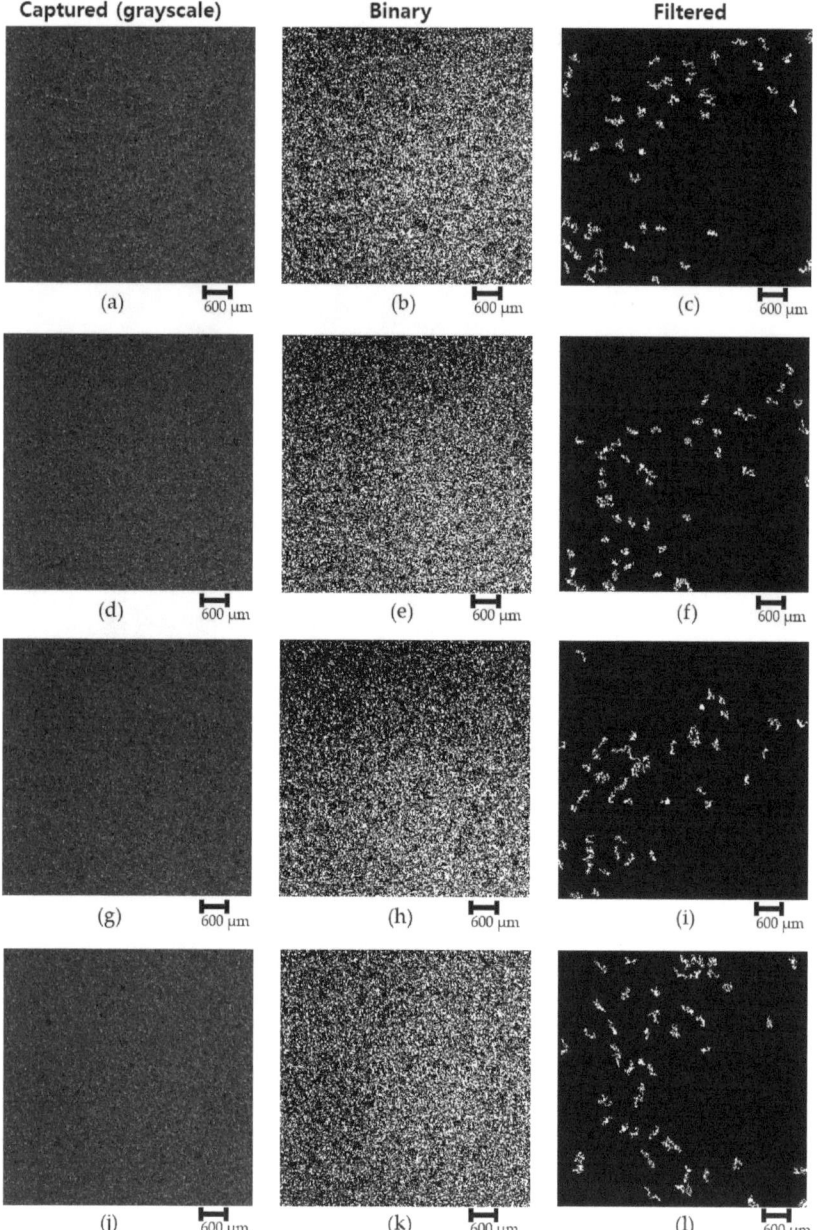

Figure 5. Shot-blasted surface images with R_a of 3.2 μm: (**a,d,g,j**) present grayscale images; (**b,e,h,k**) show binary ones obtained with Otsu's method; (**c,f,i,l**) are filtered images.

Figures 6–8 show images of sand-blasted surfaces with an R_a of 12.5 μm, 6.3 μm, and 3.2 μm, respectively. Grayscale images for sand-blasted surfaces (Figure 6a,d,g,j) present that surface texture differed greatly from those observed on the shot-blasted surfaces, despite the same R_a level. Shot-blasted surfaces maintained more rounded small peaks, compared with sand-blasted surfaces. While capturing the surface images for sand blasting, the induced light intensity in the microscope was maintained as constant. That is,

Figures 6–8 were captured under the same light intensity condition. The binary images presented in Figures 7 and 8 were also obtained with Otsu's threshold method. Figures 9 and 10 present the size distributions of white regions in a binary image before filtering. The size was defined as the number of white pixels. Note that tiny regions, having fewer than ten pixels, were not taken into account. There existed more than 400 regions with a range from 10 to 20 pixels in size (this corresponded to about 40 percent of the total). In the figures, 50 white regions in a gray area were selected.

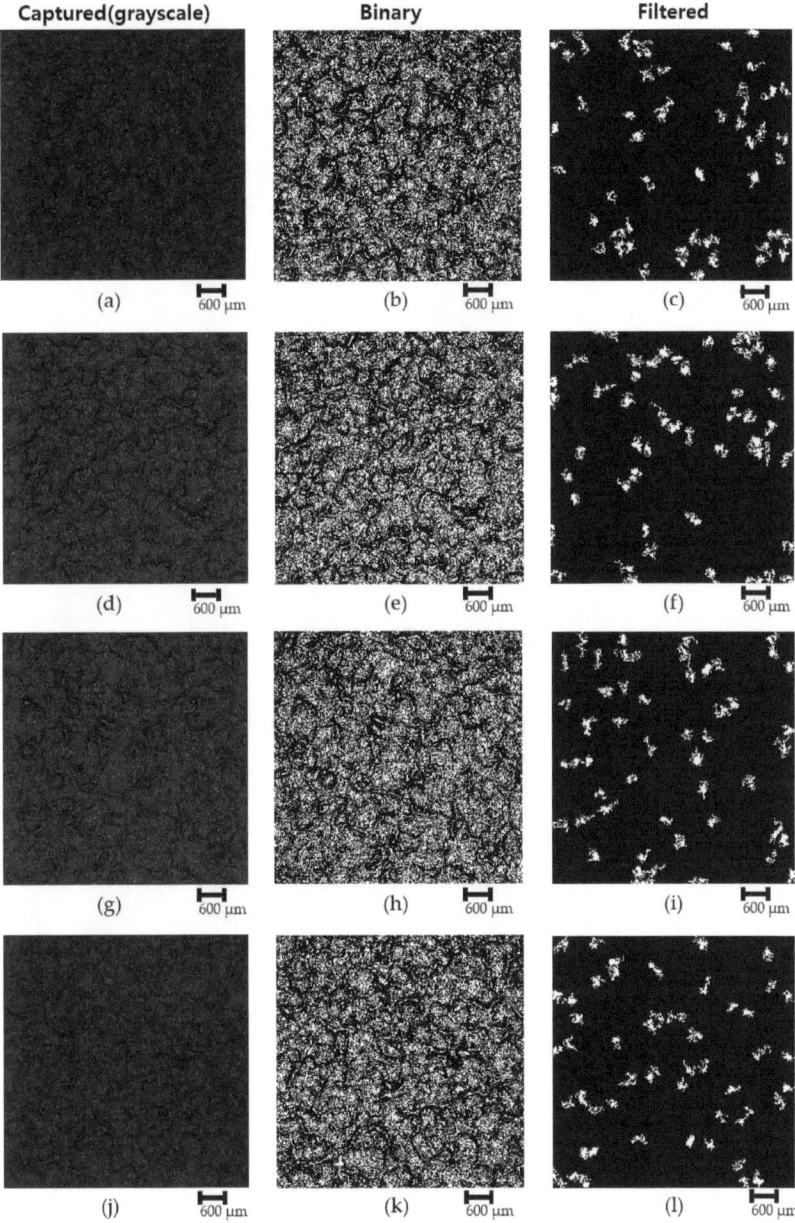

Figure 6. Sand-blasted surface images with an R_a of 12.5 μm: (**a,d,g,j**) present grayscale images; (**b,e,h,k**) show binary ones obtained with Otsu's method; (**c,f,i,l**) are filtered images.

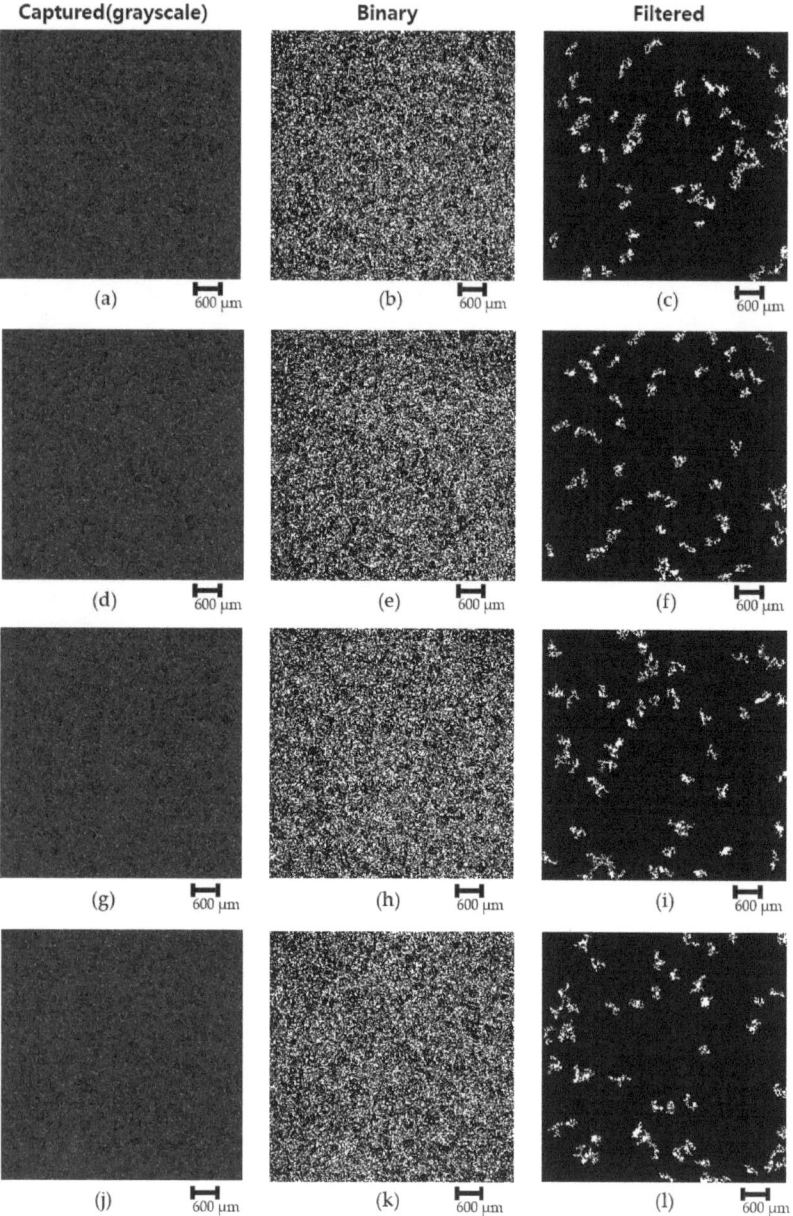

Figure 7. Sand-blasted surface images with an R_a of 6.3 µm: (**a**,**d**,**g**,**j**) present grayscale images; (**b**,**e**,**h**,**k**) show binary ones obtained with Otsu's method; (**c**,**f**,**i**,**l**) are filtered images.

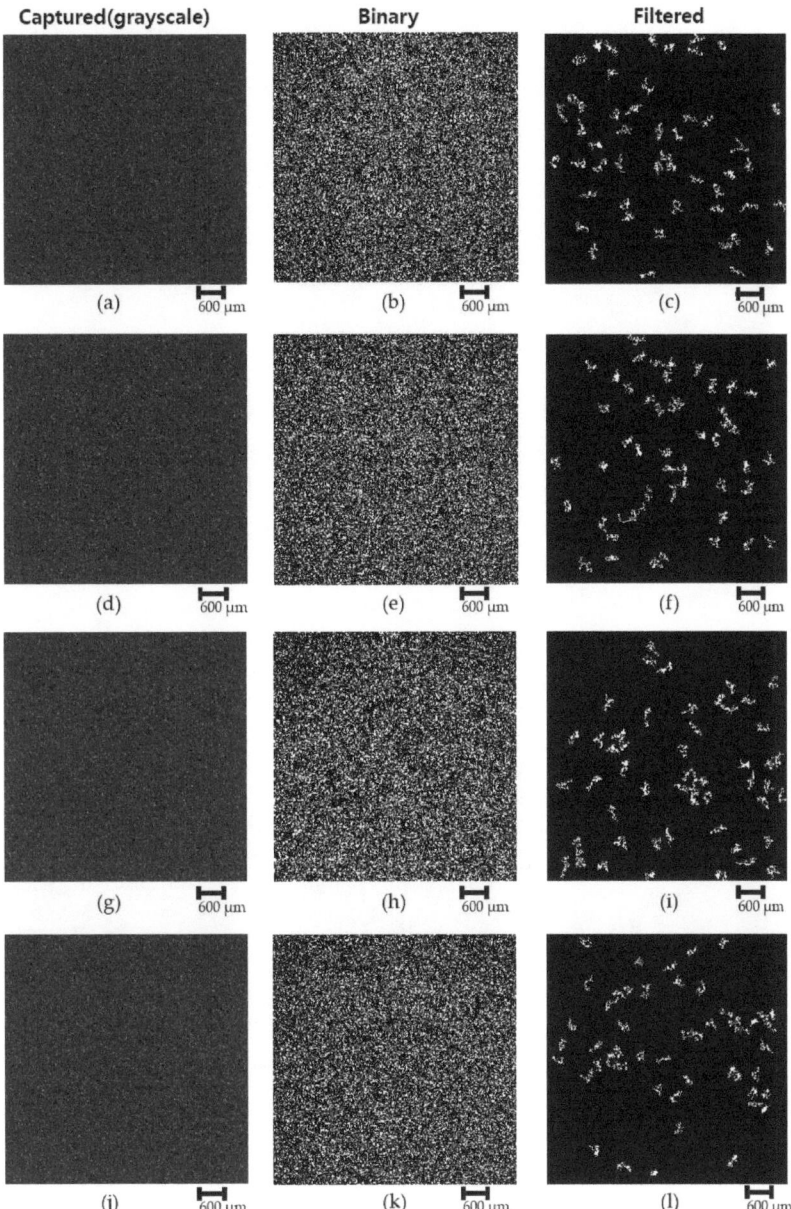

Figure 8. Sand-blasted surface images with an R_a of 3.2 μm: (**a**,**d**,**g**,**j**) present gray-scale images; (**b**,**e**,**h**,**k**) show binary ones obtained with Otsu's method; (**c**,**f**,**i**,**l**) are filtered images.

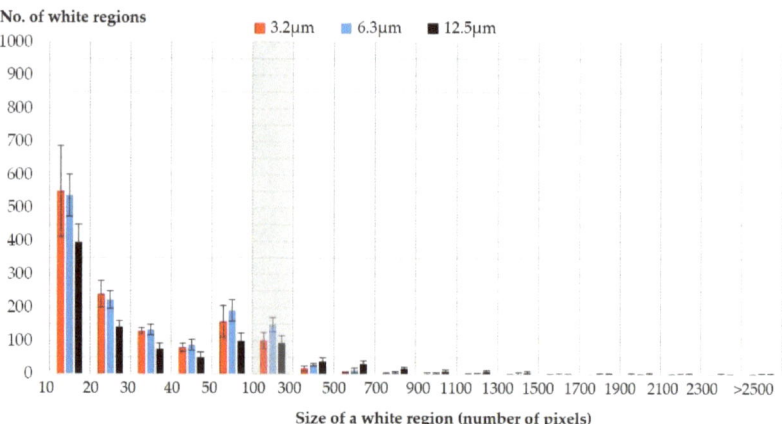

Figure 9. Size distribution of white regions in binary images (shot-blasted surface). Error bars denote 95 percent confidence intervals. Fifty white regions within the gray area were selected.

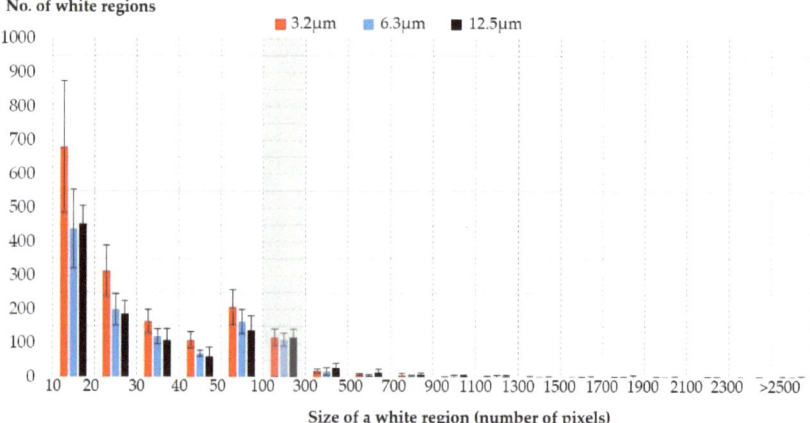

Figure 10. Size distribution of white regions on binary images (sand-blasted surface). Error bars denote 95 percent confidence intervals. Fifty white regions within the gray area were selected.

The same filter was applied to the binary images, and the filtered images are found in Figures 7c,f,i,l and 8c,f,i,l. The white pixels of the filtered images were counted and the number of the pixels is presented in Table 2. The number of white pixels on the filtered images tends to increase with increased arithmetic average roughness for both shot- and sand-blasted surfaces.

4. Discussion

In order to analyze a relation between the average size of white regions and the arithmetic average roughness, the numbers of white pixels presented in Table 2 were interpreted with a one-way analysis of variance (ANOVA). In the one-way ANOVA, null hypothesis and alternative hypothesis were defined as.

Null Hypothesis (H_0): *There is no significant difference among three groups (numbers of white pixels) measured at each arithmetic average roughness (R_a) of blasted surfaces.*

Alternative Hypothesis (H_1): *Statistically significant difference exists among the mean levels of the groups.*

Tables 3 and 4 show the summary and the result of one-way ANOVA for shot- and sand-blasted surfaces, respectively. The summaries of the standard errors and the confidence intervals for each roughness level are shown (Table 3(1) and Table 4(1)). The standard error for shot-blasted surfaces was lower than that for sand-blasted surfaces. It is possible to compare the variability of data between shot- and sand-blasted surfaces: at the R_a values of 3.2 µm and 12.5 µm, the variances of shot blasting were greater than those of sand blasting, indicating that shot-blasting data maintained greater variability than sand-blasting ones. For shot blasting, there was no significant difference in variances among the three groups. Meanwhile, for sand blasting, the variance for Group 3.2 µm was almost at one-fifth the level compared with those of other two groups.

Table 3. One-way ANOVA summary and results for shot-blasted surfaces.

(1) Summary

Groups (R_a)	Count	Sum	Average	Variance	Standard Error	Lower Bound	Upper Bound
3.2 µm	4	32,731	8182	1,015,592	510.4	7028.1	9337.4
6.3 µm	4	51,391	12,847	935,418	510.4	11,693.1	14,002.4
12.5 µm	4	99,813	24,953	1,175,232	510.4	23,798.6	26,107.4

(2) Result

Cause of Variation	Sum of Squares	Deg. of Freedom	Mean Square	F	p Value	F-Critical
Between groups	5.99×10^8	2	3.0×10^8	287.6	6.99×10^{-9}	4.26
Within groups	9.38×10^6	9	1.04×10^6			
Total	6.09×10^8	11				

Table 4. One-way ANOVA summary and results for sand-blasted surfaces.

(1) Summary

Groups (R_a)	Count	Sum	Average	Variance	Standard Error	Lower Bound	Upper Bound
3.2 µm	4	47,863	11,966	207,338	597.1	10,615.1	13,316.4
6.3 µm	4	62,442	15,611	1,106,643	597.1	14,259.8	16,961.1
12.5 µm	4	80,728	20,182	1,097,898	597.1	18,831.3	21,532.7

(2) Result

Cause of Variation	Sum of Squares	Deg. of Freedom	Mean Square	F	p Value	F-Critical
Between groups	1.36×10^8	2	6.7×10^7	47.5	1.64×10^{-5}	4.26
Within groups	1.28×10^7	9	1.4×10^6			
Total	1.48×10^8	11				

During the ANOVA test procedure, an F-statistic was produced, which was used to calculate a p-value. The p-value indicates whether there exists a difference in measured data groups or not. If the p-value is lower than 0.05, the null hypothesis is rejected and the alternative hypothesis is accepted. As shown in Table 3(2)) and Table 4(2), the calculated p-values were found to be 6.99×10^{-9} for shot blasting and 1.64×10^{-5} for sand blasting. This means that there existed statistically significant differences among the mean levels of the three measured groups (the numbers of white pixels measured at different roughness levels).

For the purpose of establishing the relation between the average size of selected white regions and the arithmetic average roughness, a curve fitting was employed. Figure 11 shows the relation between the percentage of white pixels (P_w) in the filtered image and the arithmetic average roughness (R_a). Figure 11 shows that it is possible to fit the data with a linear function; note that R-squared values for shot- and sand-blasted surfaces were found to be 0.9963 and 0.9844, respectively. The error bars represent 95 percent confidence intervals. There exists a difference between the two fitted curves for shot- and sand-blasted surfaces. Particularly, the slopes of the curves were found to be different. It could be identified that the type of abrasive gave rise to the difference. The fitted curves are informative for predicting the arithmetic average roughness of blasted surfaces in a given manufacturing process within a chosen range (from 3.2 μm to 12.5 μm); note that the curve should be re-calculated when light intensity and abrasive material are changed.

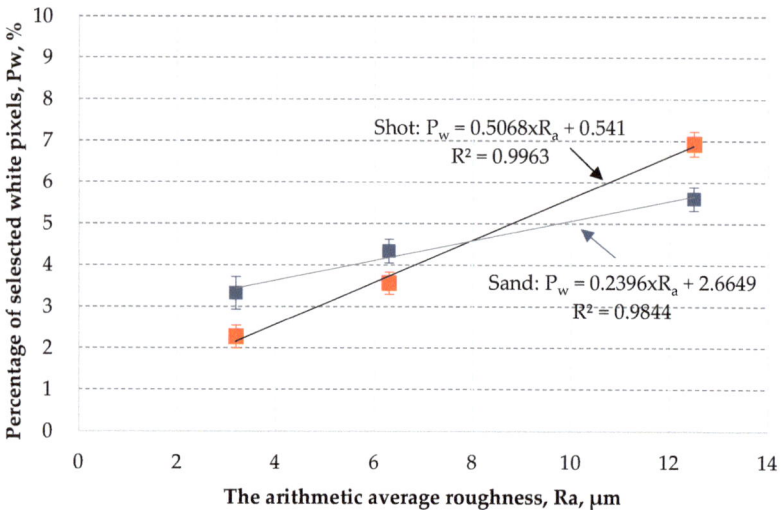

Figure 11. Relation between the percentage of white pixels (P_w) and the arithmetic average roughness (R_a). Here, markers show the average values and error bars denote 95 percent confidence levels.

This proposed machine vision system can operate quickly without human error, compared with conventional methods such as a stylus profilometer [14,15] or manual comparison [16]. Although a stylus profilometer provides accurate roughness measurement, a physical contact between a stylus probe and a target surface must be made. Continuous measurement brings about the wear of the stylus probe. Conventional laser or optical profilometers provide non-contact roughness measurement. However, it is difficult to use the profilometers in an actual blasting process due to cost. Thus, in practice, manual comparison is widely used, including tactile or visual perception with a sample comparator. However, this manual comparison gives rise to human error that can decrease the accuracy of surface roughness.

Further work needs to include the investigation of possible factors that affect the slope of the linear function. The proposed system was developed for shot- and sand-blasting processes at fixed magnification and light intensity conditions. For identifying actual feasibility in shot- and sand-blasting processes, it is necessary to perform a sensitivity analysis by changing capture parameters such as magnification and light intensity. In addition, other raw materials should be taken into account.

Other dry-type blasting cases, including blower, suction, and dry ice blasting, need to be taken into account in the future system.

Dry lubricants are often applied onto shot-peened surfaces in aerospace engineering. Thus, the future application of the system should include shot-peened surfaces having an R_a value lower than 3.2 μm.

5. Conclusions

In this article, a machine vision system was developed for measuring the arithmetic average roughness of shot- and sand-blasted surfaces. In the system developed in-house, grayscale images were captured at the microscale, and then the images were analyzed with a proposed algorithm using Otsu's method. A binary image determined with Otsu's threshold was filtered; that is, only white regions satisfying specific size conditions were extracted from the binary image. The sizes of extracted regions associated with the specular reflection of light were determined by counting the number of white pixels in the filtered image.

Blasted nickel alloy plates with R_a values of 3.2 μm, 6.3 μm, and 12.5 μm were prepared with two types of abrasives: shot and sand. The system captured four surface images for each R_a level, so twelve surface images were analyzed for each blasting.

A one-way analysis of variance (ANOVA) was used to identify the relation between the size of the selected white regions (number of selected white pixels) and the arithmetic average roughness of blasted surfaces. It was found that the size of the regions was related to the arithmetic average roughness of shot- and sand-blasted surfaces within the chosen R_a range.

For the purpose of discovering a mathematical relationship between the average size of selected white regions and the arithmetic average roughness, curve fitting was employed. It was identified that the percentage of selected white pixels can be expressed as a linear function of the average roughness of both shot- and sand-blasted surfaces. There exists a difference between the slopes of the linear function of the shot- and sand-blasted surfaces. The type of abrasive was one factor bringing about such difference. Further work needs to include an investigation of possible factors affecting the slope of the linear function.

The proposed system was developed for the quality inspection of near-white metal blasting with shot and sand abrasives. Before applying the actual quality inspection process, the system was evaluated at a fixed magnification and under constant light intensity conditions. For identifying actual feasibility in shot- and sand-blasting processes, it is necessary to perform a sensitivity analysis by changing capture parameters such as magnification and light intensity. In addition, other raw materials and dry-type blasting, including blower, suction, and dry ice blasting, need to be taken into account in the future system. Dry lubricants are often applied onto shot-peened surfaces in aerospace engineering. Thus, the future application of the system should include shot-peened surfaces having a lower R_a value than 3.2 μm.

When a blasted surface with unknown roughness is given, it is possible to predict the arithmetic average roughness on the basis of the relationship within the range from 3.2 μm to 12.5 μm in R_a. Considering that this machine vision system can provide faster surface roughness measurement and can be built at lower cost than conventional methods, it could be used in the quality inspection process of shot and sand blasting.

Funding: This research was conducted with the support of the Korea Institute for Advancement of Technology funded by the government (Ministry of Trade, Industry and Energy) in 2023 (Project No. P0023691, Specialized Workforce Development Program for Space Materials, Parts, and Equipment).

Data Availability Statement: The data used to support the findings of this study are available from the corresponding author upon request.

Conflicts of Interest: The author declares no conflicts of interest.

Appendix A. The Proposed Algorithm Code for Image Processing (Written in *Matlab*)

```
Image_gray = imread(image);              % import a grayscale image
level = graythresh(Image_gray);          % Otsu's threshold
BW = imbinarize(Image_gray, level);      % Binary image generation with Otsu's threshold
maskfilter = imfill(BW, 'holes');

maskfil = bwareafilt(maskfilter, 100);   % Selection of top 100 largest white regions
Filtered_img = bwareafilt(maskfil, 50,"smallest");   % Selection of bottom 50 regions among 100.

Nc=0;
for i=1:1:width                          % width of the image (No. of pixels)
for j=1:1:height                         % height of the image (No. of pixels)
if Filtered_img (i,j) > 0.5              % Unity corresponds to white
Nc=Nc+1;
end
end
end
```

Appendix B. Otsu's Thresholding Histogram for Shot-Blasted Surfaces

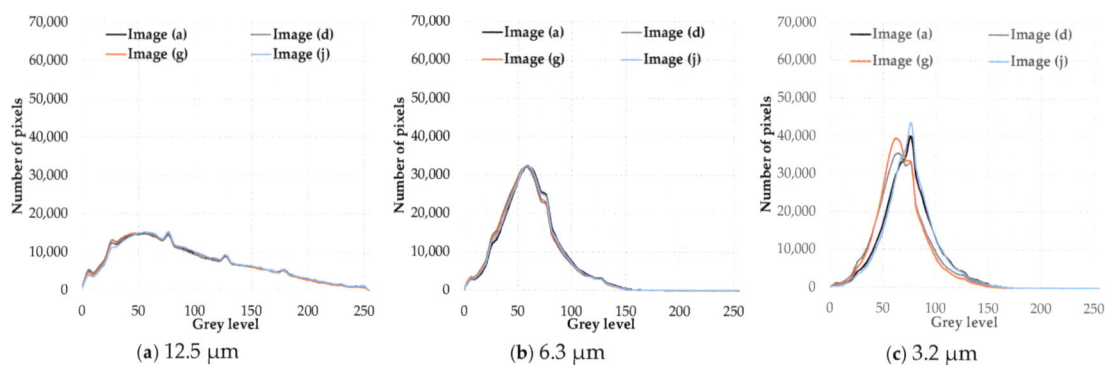

(**a**) 12.5 μm (**b**) 6.3 μm (**c**) 3.2 μm

Appendix C. Otsu's Thresholding Histogram for Sand-Blasted Surfaces

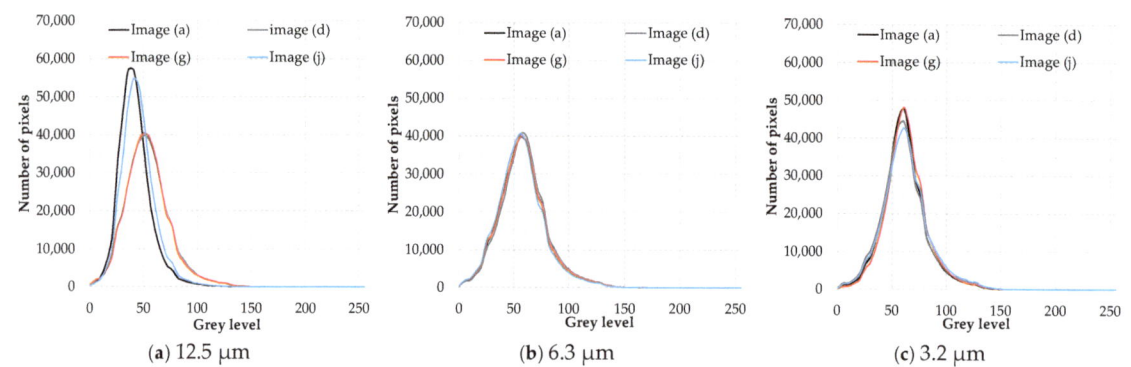

(**a**) 12.5 μm (**b**) 6.3 μm (**c**) 3.2 μm

References

1. Marian, M.; Tremmel, S. Current treads and applications of machine learning in tribology—A review. *Lubricants* **2021**, *9*, 86. [CrossRef]
2. Paturi, U.M.R.; Palakurthy, S.T.; Reddy, N.S. The Role of Machine Learning in Tribology: A Systematic Review. *Arch. Computat Methods Eng.* **2023**, *30*, 1345–1397. [CrossRef]

3. Luk, F.; North, W. Measurement of surface roughness by a machine vision system. *J. Phys. E Sci. Instrum.* **1989**, *22*, 977–980. [CrossRef]
4. Josso, B.; Burton, D.R.; Lalor, M.J. Frequency normalised wavelet transform for surface roughness analysis and characterization. *Wear* **2002**, *252*, 491–500. [CrossRef]
5. Balasundaram, M.K.; Ratnam, M.M. In-process measurement of surface roughness using machine vision with sub-pixel edge detection in finish turning. *Int. J. Precis. Eng. Manuf.* **2014**, *15*, 2239–2249. [CrossRef]
6. Joshi, K.; Patil, B. Prediction of surface roughness by machine vision using principal components based regression analysis. *Procedia Comput. Sci.* **2020**, *167*, 382–391. [CrossRef]
7. Giustia, A.; Dotta, M.; Maradia, U.; Boccadoro, M.; Gambardella, L.M. Image-based measurement of material roughness using machine learning techniques. *Procedia CIRP* **2020**, *95*, 377–382. [CrossRef]
8. Sanjeevi, R.; Nafaraja, R.; Radha Krishnan, B. Vision-based surface roughness accuracy prediction in the CNC milling process (Al6061) using ANN. *Mater. Sci.* **2020**, *2214*, 7853. [CrossRef]
9. Karthikeyan, S.; Subbarayan, M.R.; Beemaraj, R.K.; Sivakandhan, C. Computer vision-based surface roughness measurement using artificial neural network. *Mater. Today Proc.* **2022**, *60*, 1325–1328.
10. Otsu, N. A threshold selection method from gray level histograms. *IEEE Trans. Syst. Man. Cybern.* **1979**, *9*, 62–66. [CrossRef]
11. Sezgin, M.; Sankur, B. Survey over image thresholding techniques and quantitative performance evaluation. *J. Electron. Imaging* **2004**, *13*, 146–165.
12. Sha, C.; Hou, J.; Cu, H. A robust 2D Otsu's thresholding method in image segmentation. *J. Vis. Commun. Image R.* **2016**, *41*, 339–351. [CrossRef]
13. Kim, K. Evaluation of fretting wear damage on coated system using computer vision technique. *Int. J. Surf. Sci. Eng.* **2022**, *16*, 150–162. [CrossRef]
14. Bhushan, B. *Introduction to Tribology*; John Wiley & Sons: Hoboken, NJ, USA, 2022; pp. 55–84.
15. Poon, C.Y.; Bhushan, B. Comparison of surface roughness measurements by stylus profiler, AFM and non-contact optical profiler. *Wear* **1995**, *190*, 76–88. [CrossRef]
16. *ASTM D4417*; Standard Test Methods for Field Measurement of Surface Profile of Blast Cleaned Steel. ASTM International: West Conshohocken, PA, USA, 2021. [CrossRef]

Disclaimer/Publisher's Note: The statements, opinions and data contained in all publications are solely those of the individual author(s) and contributor(s) and not of MDPI and/or the editor(s). MDPI and/or the editor(s) disclaim responsibility for any injury to people or property resulting from any ideas, methods, instructions or products referred to in the content.

Article

Numerical Optimization Analysis of Floating Ring Seal Performance Based on Surface Texture

Zhenpeng He [1,2], Yuhang Guo [3], Jiaxin Si [4,*], Ning Li [4,5], Lanhao Jia [3], Yuchen Zou [1] and Hongyu Wang [3]

1 Aeronautical Engineering Institute, Civil Aviation University of China, Tianjin 300300, China; hezhenpeng@tju.edu.cn (Z.H.); zyc6346@163.com (Y.Z.)
2 State Key Laboratory of Mechanical Transmission, Chongqing University, Chongqing 400044, China
3 Sino-European Institute of Aviation Engineering, Civil Aviation University of China, Tianjin 300300, China; 2022122012@cauc.edu.cn (Y.G.); 2022122013@cauc.edu.cn (L.J.); 2023122021@cauc.edu.cn (H.W.)
4 AECC Hunan Aviation Powerplant Research Institute, Zhuzhou 412002, China; lining-608@163.com
5 Science and Technology on Helicopter Tranmission Laboratory, Zhuzhou 412002, China
* Correspondence: 19158706887@163.com

Abstract: Much research and practical experience have shown that the utilization of textures has an enhancing effect on the performance of dynamic seals and the dynamic pressure lubrication of gas bearings. In order to optimize the performance of floating ring seals, this study systematically analyzes the effects of different texture shapes and their parameters. The Reynolds equation of the gas is solved by the successive over-relaxation (SOR) iteration method. The pressure and thickness distributions of the seal gas film are solved to derive the floating force, end leakage, friction, and the ratio of buoyancy to leakage within the seal. The effects of various texture shapes, including square, 2:1 rectangle, triangle, hexagon, and circle, as well as their parameters, such as texture depth, angle, and area share, on the sealing performance are discussed. Results show that the texture can increase the air film buoyancy and reduce friction, but it also increases the leakage by a small amount. Square textures and rectangular textures are relatively effective. The deeper the depth of the texture within a certain range, the better the overall performance of the floating ring seal. As the texture area percentage increases, leakage tends to increase and friction tends to decrease. A fractal roughness model is developed, the effect of surface roughness on sealing performance is briefly discussed, and finally the effect of surface texture with roughness is analyzed. Some texture parameters that can significantly optimize the sealing performance are obtained. Rectangular textures with certain parameters enhance the buoyancy of the air film by 81.2%, which is the most significant enhancement effect. This rectangular texture reduces friction by 25.8% but increases leakage by 79.5%. The triangular textures increase buoyancy by 28.02% and leakage increases by only 10.08% when the rotation speed is 15,000 r/min. The results show that texture with appropriate roughness significantly optimizes the performance of the floating ring seal.

Keywords: floating ring seal; surface texture; surface roughness; Reynolds equation; successive over-relaxation (SOR)

Citation: He, Z.; Guo, Y.; Si, J.; Li, N.; Jia, L.; Zou, Y.; Wang, H. Numerical Optimization Analysis of Floating Ring Seal Performance Based on Surface Texture. *Lubricants* **2024**, *12*, 241. https://doi.org/10.3390/lubricants12070241

Received: 27 May 2024
Revised: 21 June 2024
Accepted: 1 July 2024
Published: 3 July 2024

Copyright: © 2024 by the authors. Licensee MDPI, Basel, Switzerland. This article is an open access article distributed under the terms and conditions of the Creative Commons Attribution (CC BY) license (https://creativecommons.org/licenses/by/4.0/).

1. Introduction

A floating ring seal is a non-contact fluid seal installed on a rotor, which separates two chambers with a high-pressure drop by a high-pressure fluid film formed by high-speed rotation of the rotor and eccentric operation of the floating ring [1,2]. Its excellent reliability and high sealing efficiency are of great significance for the safe operation of rotating machinery. Floating ring seals are typically installed within steel or titanium supporting rings to minimize the sealing clearance and avoid rubbing with the rotor. However, the non-contact seals that are installed on the rotor are constrained by the balance between achieving optimal sealing clearance and minimizing the risk of rotor rubbing.

With the operating environment of rotating machinery becoming increasingly complex, the stability and service life of the sealing system are subject to higher demands. In response, researchers have undertaken extensive studies on the mechanisms of floating ring seals and the diverse factors that influence their sealing performance. Lee et al. [3] proposed a numerical method for calculating the static and dynamic characteristics of floating ring seals. By using the finite difference method, they solved the bulk-flow model under eccentric conditions and researched the influences of rotating speed and pressure drop on the performance of the floating ring seal. Bae et al. [4] used numerical analysis to evaluate the friction on the front end of the floating ring of a turbopump oxidation pump, and researched the effects of rotor vibration on the radial force acting on the floating ring and the flow characteristics around the floating ring seal. Furthermore, they researched the effects of rotational speeds and clearance dimensions on the leakage of the floating ring seal, and utilized the findings to optimize the performance. The experimental results are in good agreement with the numerical analysis. Shi et al. [5] established a calculation method for the working clearance of gas floating ring seals. By selecting appropriate insert ring materials, the sealing clearance of the floating ring can be kept stable. By using the "micro-variable clearance" design method to calculate the ideal floating ring material, better results can be achieved. Anbarsooz et al. [6] conducted computational fluid dynamics simulations on the floating ring seal of a carbon dioxide centrifugal compressor. They studied the effect of the annular clearance on the sealing performance and found that the leakage rate significantly increased with increase in the clearance. The leakage function remained almost unchanged with increase in the inlet total pressure and temperature. The results also showed that the minimum flow velocity required to block the leakage significantly increased with increase in the clearance. Their works provide guidance for improving the performance of the floating ring seal.

Surface texture technology is an emerging and effective means of achieving lubrication and friction reduction. By changing the micro-surface morphology of the shaft sleeve in fluid sealing systems, significant achievements have been realized in improving the fluid dynamic pressure effect of the lubricating surface and avoiding surface adhesion. It has now become a research hotspot in the field of advanced sealing technology [7–9]. Zhang et al. [10] have established an analysis model of oil film force for floating ring seals considering the influence of the axial pressure gradient. They studied the influence of groove geometry size on dynamic response and analyzed the stability of the floating ring. The analysis model is based on the Reynolds equation for oil lubrication and the assumption of short bearings [11], with consideration of the Lomakin effect of the fluid. Adjemout et al. [12] studied the influence of surface textures on the mechanical sealing performance using numerical analysis. They investigated the local and global effects of various texture shapes, directions, area ratios, and depth ratios on the fluid mechanics performance of mechanical seals, providing valuable guidance for enhancing the performance of floating ring seals through textures. Pei et al. [13] used a deterministic lumped-parameter thermal model to estimate the performance of floating ring bearings (FRBs), and systematically studied the effects of nine types of textures on the performance of the bearing. The results showed that textures can significantly increase side leakage. Shi et al. [14] compared the effects of micro-grooves and micro-dimples on the performance of the mechanical gas seal. Both micro-grooves and micro-dimples can generate a positive effect on the fluid dynamic performance of the seal under different conditions. In engineering practice, micro-grooves with a depth of 3 μm may be the best choice for enhancing the performance of the mechanical gas seal. Wang et al. [15] studied the directional effect of rectangular textures on the performance of floating ring seals. By selecting the texture direction angle judiciously, the performance of the floating ring gas film seal can be effectively optimized. He et al. [16] studied the effects of six different shapes of textures as well as their area ratio, depth ratio, and angle on the dynamic lubrication performance of rigid bearing rotors. Their study of textures is very detailed and has significance for the study of texture sealing systems. In addition, they also considered the influence of surface roughness on bearing performance.

They modeled a Gaussian rough surface for the bearing surface, and the results showed that the distribution of pressure peaks in gas bearings with rough surfaces was more dispersed, with significantly higher pressures at the bearing ends. Zhang et al. [17] proposed a new fractal roughness model to predict the wear performance of the end face of floating ring seals. The model's predictions of wear rate are consistent with experimental results. The above research has found that factors such as rotational speed, pressure, and floating ring size can affect the performance of floating ring seals. However, in operating conditions, when a floating ring seal is in stable operation, most of the working conditions are fixed and cannot be adjusted to optimize the performance of the floating ring seal. Reducing the sealing clearance or increasing the eccentricity can improve the sealing performance to a certain extent, but it also increases the contact and wear, making the system tend to be unstable and shortening the life of the sealing system. What is more, the above literature and numerous studies [18–20] have shown that suitable surface textures have a positive effect on improving the performance of sealing and rotor dynamic systems.

In this paper, a Reynolds equation model for the floating ring seal is constructed. A floating-leakage ratio is used to judge the performance of the floating ring seal, and the effects of various shapes of textures, texture area ratio, texture depth, and angles on the performance of the floating ring seal are analyzed. Finally, optimal texture parameters for optimizing the performance of the floating ring seal are obtained. In addition, a fractal surface roughness model is developed, and the optimization of the floating ring seal by texture under surface roughness is discussed.

2. Theory Model

2.1. Working Principle

A floating ring seal, as illustrated in Figure 1a, primarily consists of a check ring, a gasket, a wave spring, a floating ring, and a shell. Initially, the floating ring is in touch with the outer wall of the shaft, with the maximum eccentricity between the ring and the shaft. As the seal operates, the dynamic pressure due to the wedge effect at high rotational speeds causes the floating ring to float upwards, creating a rigid fluid film between the ring and the shaft, which serves as the main sealing surface. Additionally, the high-pressure gas and the wave spring generate a force making the floating ring's end face against the shell, forming a sub-sealing surface.

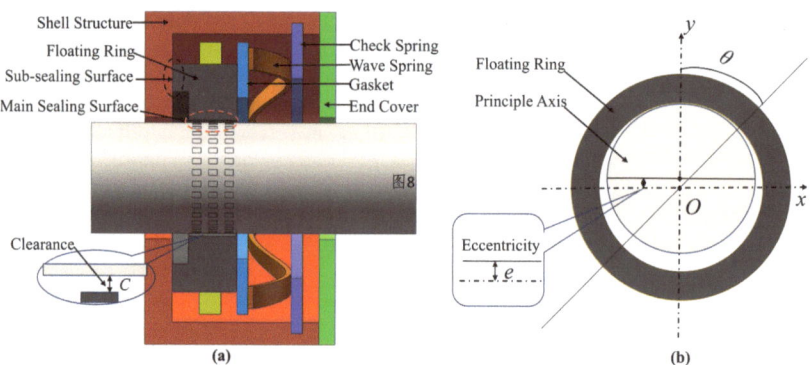

Figure 1. Schematic diagram of floating ring seal structure: (**a**) Three-dimensional structure of floating ring seal; (**b**) Diagram of eccentricity.

Adding texture holes on the outer surface of the shaft is an effective way to enhance the dynamic pressure effect of the floating ring seal. When the sealing equipment is working normally, the gas on the high-pressure side is brought into the texture hole by the pump suction effect. The medium gas is continuously compressed at the edge of the micro-hole, and a significant dynamic pressure effect is generated in the surrounding area, forming a thin and stiff air film between the dynamic ring and the static ring, thereby improving the performance of the floating ring seals.

This paper simplifies the floating ring seal as a rigid radial gas bearing, and uses the following Reynolds Equation (1) to solve the performance of the floating ring seal:

$$\frac{1}{R^2}\frac{\partial}{\partial \theta}\left(ph^3\frac{\partial p}{\partial \theta}\right) + \frac{\partial}{\partial z}\left(ph^3\frac{\partial p}{\partial z}\right) = \frac{6\mu U}{R}\frac{\partial}{\partial \theta}(ph) \quad (1)$$

The sealing cross-section of the floating ring forming an eccentricity with the rotor is shown in Figure 1b. The film thickness of the floating ring seal with surface texture can be expressed as:

$$h = c + e\cos(\theta) + h_1 + h_2 \quad (2)$$

where c is the average gas film clearance, e is the eccentricity, and θ is the coordinate of the angle centered at the origin from the positive direction of the y axis. h_1 represents the height of surface roughness. h_2 is the film thickness caused by textures.

2.2. Surface Texture Model

In this paper, the effects of many different types of textures on the performance of floating ring gas seals are discussed. The shapes of texture shapes include rectangle, triangle, ortho-hexagon, and circle. Rectangular textures involve rectangular types with different ratios of length to width. Circular textures involve different ratios of long axis to short axis. For these different shapes of textures, the effects of texture depth, texture angle, and texture area ratio on the performance of the floating ring seal were analyzed.

The fundamental design scheme for different types of textures is shown below.

① Set the number of texture distribution in the circumferential and longitudinal directions;
② Set the ratio of textured area on the surface of the rotor axis;
③ Set a defined shape of the texture, e.g., define its aspect ratio for a rectangle;
④ Set the remaining parameters, including texture depth and angle.

With the steps ① ②, the area of a single texture can be calculated. For textures with a defined shape by steps ③, the defined edge length or radius can be calculated. The region in which each of the textures are distributed on the surface is determined by step ①. A new coordinate system is established in each region and the constraint equations for the shape of the texture are constructed from the known parameters. The direction angle of the texture is set in the new coordinate system. Under this design scheme, the shape and size of the texture can be easily modified. When the area ratio of the surface texture increases, the corresponding edge length or radius of a single texture will be larger, and the gap between the surface textures will be smaller. At the end of this subsection, the coordinates, shape constraint functions, and shape illustrations of the texture are shown in Figure 2.

Texture Shape	Size and Constraint Equation	Film Thickness / μm
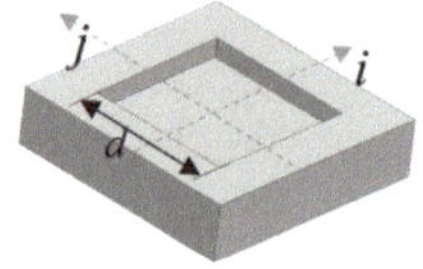 (square pit)	Size: $d = \sqrt{\dfrac{s}{(N_\theta \times N_z)}}$ Constraint Equation: $h(i,j) = \begin{cases} h_2, & \text{if } |i| \le \dfrac{d}{2}, |j| \le \dfrac{d}{2} \\ 0, & \text{else} \end{cases}$	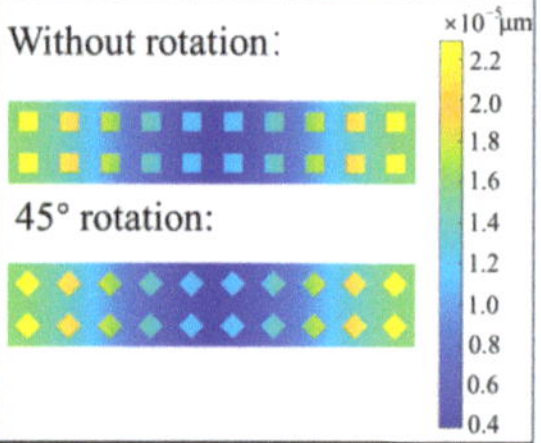 Without rotation: 45° rotation:
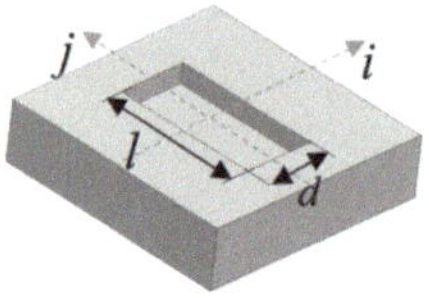 (rectangular pit)	Size: $d = \sqrt{\dfrac{s}{2 \times (N_\theta \times N_z)}}$ $l = 2 \times d$ Constraint Equation: $h(i,j) = \begin{cases} h_2, & \text{if } |i| \le \dfrac{d}{2}, |j| \le \dfrac{l}{2} \\ 0, & \text{else} \end{cases}$	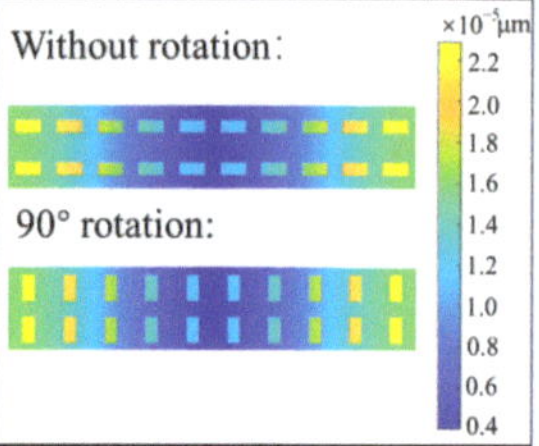 Without rotation: 90° rotation:
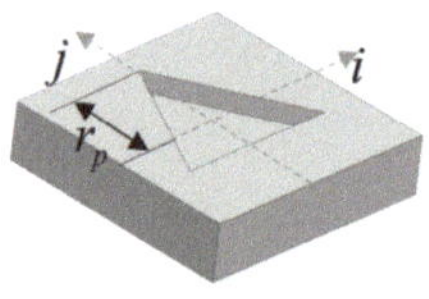 (triangular pit)	Size: $r_p = \dfrac{\sqrt{\dfrac{4 \times s}{\sqrt{3} \times N_\theta \times N}}}{\sqrt{3}}$ Constraint Equation: $h(i,j) = \begin{cases} h_2, & \text{if } -\dfrac{1}{2}r_p \le j \le -\sqrt{3}|i| + r_p \\ 0, & \text{else} \end{cases}$	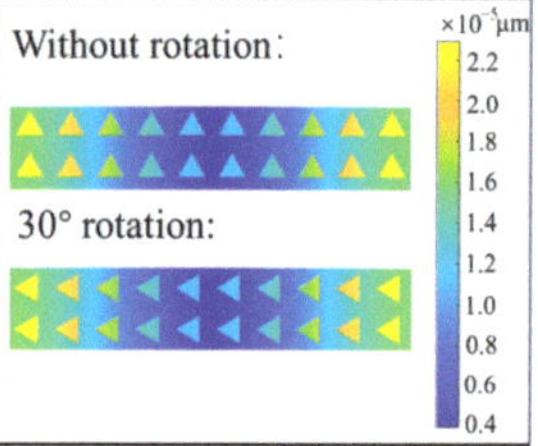 Without rotation: 30° rotation:
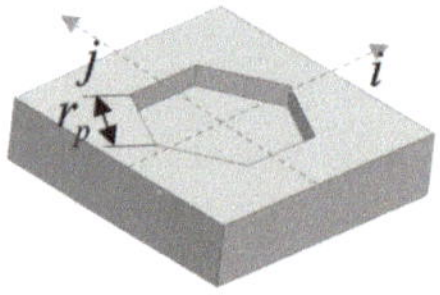 (hexagonal pit)	Size: $r_p = \sqrt{\dfrac{2 \times s}{3\sqrt{3} \times N_\theta \times N}}$ Constraint Equation: $h(i,j) = \begin{cases} h_2, & \text{if } |j| \le \min\left\{\dfrac{\sqrt{3}}{2}r_p, \sqrt{3}(r_p - |i|)\right\} \\ 0, & \text{else} \end{cases}$	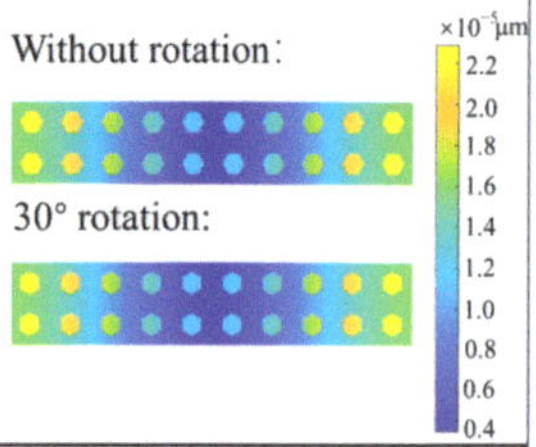 Without rotation: 30° rotation:
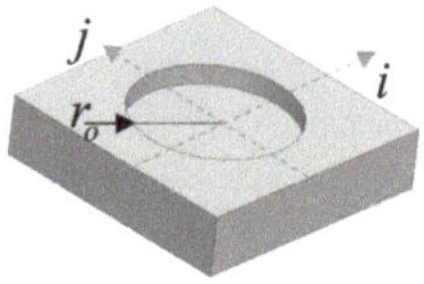 (circular pit)	Size: $r_o = \sqrt{\dfrac{s}{N_\theta \times N_z \times \pi}}$ Constraint Equation: $h(i,j) = \begin{cases} h_2, & \text{if } \sqrt{i^2 + j^2} \le r_o \\ 0, & \text{if } \sqrt{i^2 + j^2} > r_o \end{cases}$	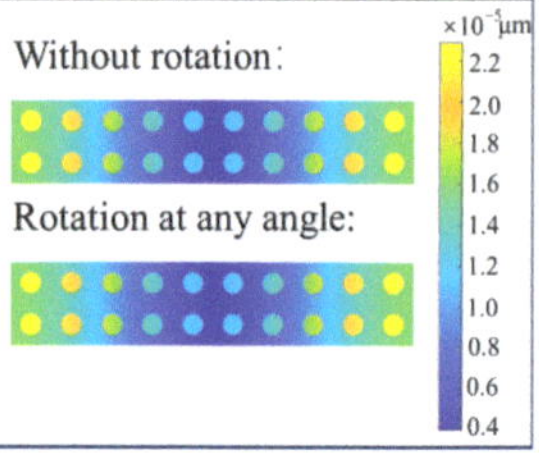 Without rotation: Rotation at any angle:

Figure 2. Schematic of the surface texture in different shapes.

2.3. Surface Roughness Model

During the machining process of the floating ring seals, the sealing surface and journal surface will produce small spacing and tiny peaks and valleys, forming a surface roughness. Different machining methods will produce rough surfaces with different characterization properties. A closer look at a rough surface reveals that it is also statistically similar, while having scale-free properties under scale transformations. Fractal theory can capture the different sizes and attendant properties of solid surfaces and can be used to model the micro-scale structure of rough surfaces.

The main idea of adding the surface roughness is as follows: We refer to the equations of the fractal roughness model and use the computer to randomly generate the height matrix of the fractal roughness surface, and then superimpose the height value of the fractal roughness surface to the thickness of the sealing film, and finally, put the thickness of the sealing film into the iteration of the main program to obtain the sealing performance parameters.

It is feasible to use the theory of fractals to study rough surfaces. Its 3D functional model [21] is expressed as follows:

$$Z(x,y) = \sum_{n=1}^{\infty} C_n \gamma^{-(3-D_s)n} \sin[\gamma^n(x\cos B_n + y\sin B_n) + A_n] \tag{3}$$

where C_n is a random value from a normal distribution with a mean of 0 and a variance of 1; A_n and B_n are mutually independent and uniformly distributed random numbers in the interval $[0, 2\pi]$; D_s is the theoretical fractal dimension $(2 < D_s < 3)$; the characteristic parameter γ is a constant greater than 1, which usually takes the value of 1.5; n is the number of natural sequences.

A typical function commonly used to generate random fractal surfaces is given by the following equation [22–25]:

$$z(x,y) = C \sum_{m=1}^{M} \sum_{n=0}^{n_{\max}} \gamma^{(D-3)n} \times \left\{ \cos \phi_{mn} - \cos\left[\frac{2\pi\gamma^n(x^2+y^2)^{1/2}}{L} \times \cos(\tan^{-1}(y/x) - \frac{\pi m}{M}) + \phi_{mn} \right] \right\} \tag{4}$$

$$C = L\left(\frac{G}{L}\right)^{(D-2)} \left(\frac{\ln \gamma}{M}\right)^{1/2} \tag{5}$$

where C is the height coefficient of the surface, D is the fractal dimension, and a larger fractal dimension D represents a denser rough surface profile. ϕ_{mn} is a series of random phases between 0 and π, γ controls the frequency density in the surface contour, M is the number of spatial undulations, L is the size of the simulated image, and G is the feature scale parameter.

In the actual simulation, n_{max} cannot be taken to infinity, and its value is taken as Equation (6):

$$n_{max} = \left\lfloor \frac{\log(L_{\max}/L_{\min})}{\log \gamma} \right\rfloor \tag{6}$$

where $L_{\max}$ is the sampling length, $L_{\min}$ is determined by the instrument resolution, and $\lfloor \cdot \rfloor$ means rounding down the nearest integer.

Fractal roughness surface characteristics depend mainly on the parameters G and D. The fractal roughness G is a frequency-independent height scale parameter that determines the specific dimensions of the rough surface—the rougher the surface the higher the value of G [26]. The fractal dimension D determines the proportion of high-frequency components and low-frequency components in the surface profile. Thus, a smaller value of D indicates that low-frequency components are more dominant in the surface profile. When the fractal dimension D is a fixed value, the higher the value of fractal roughness G, the smoother the topology [27]. Taking the fractal roughness $G = 5.5 \times 10^{-11}$, the fractal rough surfaces for different values of fractal dimension are shown in Figure 3. It can be seen that the fractal dimension affects the roughness of the fractal surface. The smaller the fractal dimension D, the more complex the variation in the surface contour and the rougher the surface will be [28]. From the comparison in Figure 3, we can see that the height of the surface roughness gradually increases as the fractal dimension decreases. From Table 1, we can see that both the arithmetic mean deviation and the maximum height of the profile increase as the fractal dimension decreases.

In fact, different values of G and different values of fractal dimension D can be taken one-by-one to analyze the results. However, the focus of this research is on the optimization of the texture on the floating ring seal, and the study of roughness is an extension and exploration based on this. It has to be admitted that the rough surface generated is probably not the real topography of the surface of the floating ring seal material. When $G = 5.5 \times 10^{-11}$ and the fractal dimension is taken as 2.5–2.7, the magnitude of the generated surface roughness is just at the micrometer level. These values are used and it is found that the generated surface topography is able to optimize the performance of the floating ring seal, and such a surface topography is selected and combined with the surface texture for further analysis of the floating ring seal.

Table 1. Roughness parameters for different fractal dimensions.

Fractal Dimension	$D = 2.4$	$D = 2.5$	$D = 2.6$
Arithmetic Mean Deviation/m	1.5137×10^{-6}	2.7300×10^{-7}	4.9646×10^{-8}
Maximum Height of Profile/m	1.0964×10^{-5}	2.0488×10^{-6}	3.9024×10^{-7}

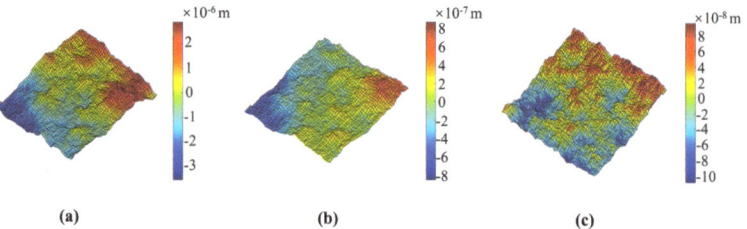

(a) (b) (c)

Figure 3. Fractal rough surfaces with different fractal dimensions D: (**a**) $D = 2.4$; (**b**) $D = 2.5$; (**c**) $D = 2.6$.

2.4. Differential Grids and Iteration

The numerical method used in this paper is the finite difference method, whose basic idea is to divide the solution field into orthogonal grids and use central difference formulas to substitute into the differential equations. The differential equations obtained are then solved by a computer using a successive over-relaxation (SOR) iteration method [29]. This approach is particularly useful for solving partial differential equations in two or three dimensions, and it has been widely applied in many scientific and engineering fields, such as heat transfer, fluid dynamics, and electromagnetic analysis. The approach for solving the problem is as follows: firstly, the Reynolds equation is simplified to a dimensionless form. Secondly, the dimensionless Reynolds equation is discretized and differentiated to obtain the difference equation. Finally, the SOR method is used to solve the difference equation.

The dimensionless Reynolds equation is presented as follows:

$$\frac{R^2}{L^2}\frac{3}{H}\frac{\partial H}{\partial \bar{z}}\frac{\partial \bar{p}}{\partial \bar{z}} + \frac{R^2}{L^2}\frac{\partial^2 \bar{p}}{\partial \bar{z}^2} + \frac{3}{H}\frac{\partial H}{\partial \theta}\frac{\partial \bar{p}}{\partial \theta} + \frac{\partial^2 \bar{p}}{\partial \theta^2} = \frac{2\Lambda p_a}{pH^3}\left(\frac{H}{2}\frac{\partial \bar{p}}{\partial \theta} + \bar{p}\frac{\partial H}{\partial \theta}\right) \quad (7)$$

where the symbols (θ, z) are coordinates in the circumferential and axial directions, $\Lambda = \frac{6\mu\omega}{p_a}\left(\frac{R}{c}\right)^2$ are the seal coefficients, and the dimensionless quantities are expressed with $\bar{z} = \frac{z}{L}, \bar{p} = \frac{p^2}{p_a^2}, H = \frac{h}{c}$.

The grid of the 2D flow field in the floating ring seal is shown in Figure 4. Δz is the step size in the differential grid axis direction, where $\Delta z = L/(n-1)$. $\Delta \theta$ is the step size in the differential grid circumferential direction, where $\Delta \theta = 2\pi/(m-1)$. i and j represent the circumferential and axial grid positions of the floating ring seal, respectively.

By using the following central difference scheme, the discrete difference of the continuous 2D Reynolds equation is differentiated and solved iteratively by computer. The difference schemes are shown as follows:

$$\frac{\partial^2 \bar{p}}{\partial \bar{z}^2} = \frac{\bar{p}_{i,j+1} - 2\bar{p}_{i,j} + \bar{p}_{i,j-1}}{\Delta \bar{z}^2}, \frac{\partial^2 \bar{p}}{\partial \theta^2} = \frac{\bar{p}_{i+1,j} - 2\bar{p}_{i,j} + \bar{p}_{i-1,j}}{\Delta \theta^2}, \frac{\partial \bar{p}}{\partial \bar{z}} = \frac{\bar{p}_{i,j+1} - \bar{p}_{i,j-1}}{2\Delta \bar{z}},$$

$$\frac{\partial \bar{p}}{\partial \theta} = \frac{\bar{p}_{i+1,j} - \bar{p}_{i-1,j}}{2\Delta \theta}, \frac{\partial H}{\partial \bar{z}} = \frac{H_{i,j+1} - H_{i,j-1}}{2\Delta \bar{z}}, \frac{\partial H}{\partial \theta} = \frac{H_{i+1,j} - H_{i-1,j}}{2\Delta \theta}.$$

By substituting the central difference scheme into the dimensionless Reynolds Equation (7), the differential Reynolds equation is obtained. The simplified difference equation for solving the dimensionless pressure of the floating ring seal is obtained as follows:

$$A_{i,j}\bar{p}_{i-1,j} + B_{i,j}\bar{p}_{i+1,j} + C_{i,j}\bar{p}_{i,j} + D_{i,j}\bar{p}_{i,j-1} + E_{i,j}\bar{p}_{i,j+1} = F_{i,j} \tag{8}$$

For the dimensionless pressure $\bar{p}$, the value of its $n+1$ times iteration can be calculated from the value of the n times iteration. The equation is presented as follows:

$$\bar{p}_{i,j}^{(n+1)} = \frac{A_{i,j}\bar{p}_{i-1,j}^{(n+1)} + B_{i,j}\bar{p}_{i+1,j}^{(n)} + D_{i,j}\bar{p}_{i,j-1}^{(n+1)} + E_{i,j}\bar{p}_{i,j+1}^{(n)} - F_{i,j}}{-C_{i,j}} \tag{9}$$

The iteration stops when the dimensionless pressure $\bar{p}$ satisfies the following convergence condition:

$$\sqrt{\sum_{i=1}^{m}\sum_{j=1}^{n}\left(\frac{\Delta \bar{p}_{i,j}}{\bar{p}_{i,j}}\right)^2} \leq 10^{-6} \tag{10}$$

The boundary conditions are presented as follows:

$$\begin{cases} p(z=0,\theta) = p_{\text{high}} \\ p(z=L,\theta) = p_{\text{low}} \\ p(z,\theta=0) = p(z,\theta=2\pi) \end{cases} \tag{11}$$

where p_{low} is the pressure of the low-pressure end of the floating ring seal, and p_{high} is the pressure of the high-pressure end of the floating ring seal. The total procedure for solving the sealing performance is shown in Figure 5.

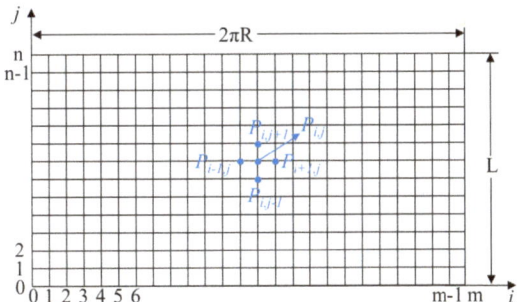

Figure 4. Schematic of the grid in the 2D flow field.

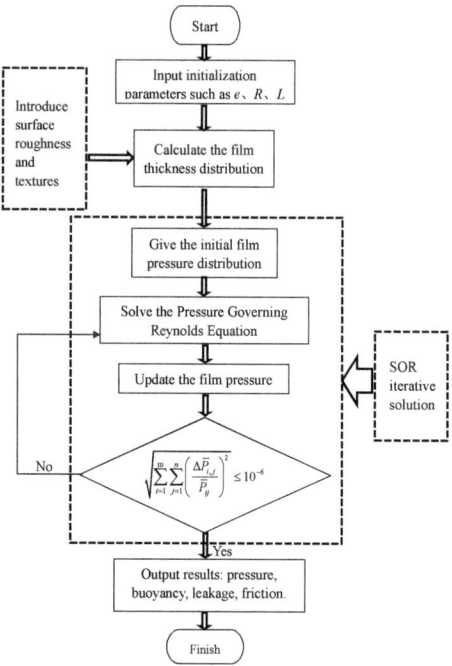

Figure 5. Flowchart of the numerical method.

2.5. Performance Evaluation

There is no contact between the floating ring and the rotating shaft in normal operation, resulting in low frictional wear. If the buoyancy force is insufficient, the floating ring will struggle to rise completely, resulting in contact with the rotor shaft, which can cause severe wear and significantly impact the performance and service life of the floating ring seal. Gas leakage is an important indicator of the performance of a floating ring gas seal. The smaller the gas leakage the better, which means a better seal and more efficient operation of the equipment. In practice, no leakage at all is unattainable, and there will always be a certain amount of gas leakage. If the leakage is too large, it will not only reduce the operating efficiency of the equipment, but may also cause pollution to the environment.

Through the coupled solution of the Reynolds equation and the film thickness equation, the pressure distribution of the gas film in the floating ring seal is obtained, and the seal gas film performance parameters are calculated.

The gas film dimensionless buoyancy force $\bar{w}$ for a floating ring seal is calculated as:

$$\bar{w} = \sqrt{\bar{w}_x^2 + \bar{w}_y^2} \tag{12}$$

where $\bar{w}_x$ and $\bar{w}_y$ are, respectively, the dimensionless buoyancy force in the x- and y-directions with the formula:

$$\begin{cases} \bar{w}_x = \int_0^1 \int_0^{2\pi} \left(\sqrt{\bar{p}(\theta, \bar{z})} - 1 \right) \sin\theta \, d\theta \, d\bar{z} \\ \bar{w}_y = \int_0^1 \int_0^{2\pi} \left(\sqrt{\bar{p}(\theta, \bar{z})} - 1 \right) \cos\theta \, d\theta \, d\bar{z} \end{cases} \tag{13}$$

The air film buoyancy force is $w = p_a RL\tilde{w}$. The gas leakage Q of a floating ring seal is calculated by the formula:

$$Q = \int_0^{2\pi} R \frac{\rho h^3}{12\mu} \frac{\partial p}{\partial z}\bigg|_{z=L} d\theta \tag{14}$$

The ratio of the air film buoyancy to the leakage of the floating ring seal is introduced to better measure the performance of the floating ring seal, and the formula for the buoyancy leakage ratio is $k = \frac{Q}{w}$.

In addition, this study considers calculating the friction to provide a comprehensive measure of the improved performance of the floating ring seal. The components of friction in the horizontal and vertical directions are shown in Equation (15):

$$\begin{cases} T_x = \int_0^1 \int_0^{2\pi} (\frac{\Lambda}{6}\frac{1}{h} - \frac{h}{2}\frac{\partial p}{\partial \theta}) \sin\theta d\theta dz \\ T_y = \int_0^1 \int_0^{2\pi} (\frac{\Lambda}{6}\frac{1}{h} - \frac{h}{2}\frac{\partial p}{\partial \theta}) \cos\theta d\theta dz \end{cases} \tag{15}$$

3. Validation

This paper compares and validates the results of Ma et al. [30] on floating ring cylinder surface air film seals. Their results were subsequently verified in the paper of Chen [31] and the paper of Xu [28]. The parameters of the floating ring cylinder surface air film are shown in Table 2.

A comparison between the results of this paper and Ma et al. [30] is shown in Figure 6. The curves in the figure show the dimensionless pressure of the cross-section in the seal at pressures of 0.11, 0.20, 0.32, and 0.45 MPa at the high-pressure end, respectively. From Figure 6, it can be seen that the trend of the dimensionless pressure in the mid-section obtained in this study is basically consistent with the literature for different conditions. The differences in the values are also very small, with the maximum difference not exceeding 5%. The above results and analyses show that the solution methodology of this study for floating ring seals is well supported and the results are relatively accurate.

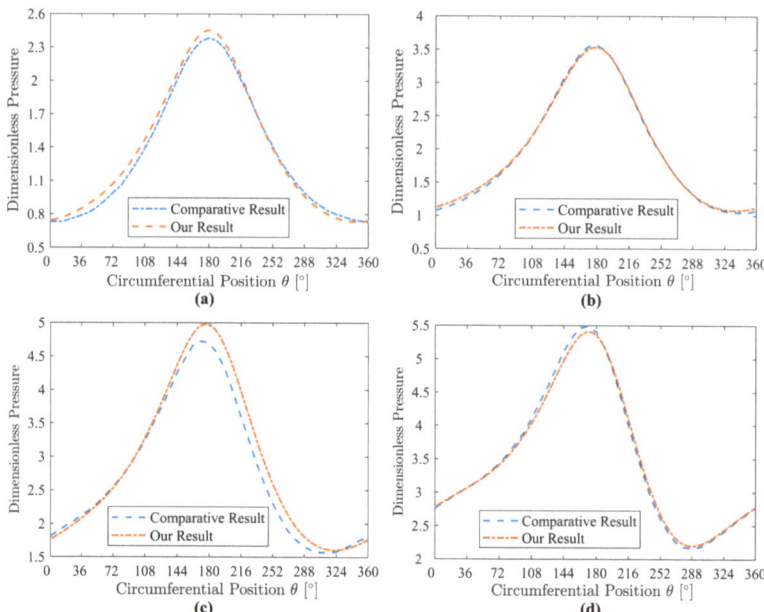

Figure 6. Dimensionless pressure of mid-section with different high pressure: (**a**) 0.11 MPa in the high-pressure side; (**b**) 0.20 MPa in the high-pressure side; (**c**) 0.32 MPa in the high-pressure side; (**d**) 0.45 MPa in the high-pressure side. The comparative results refer to the results of Ma et al. [30].

Table 2. Calculation parameters of seal.

Parameter	Value	Parameter	Value
Eccentricity	0.5	Viscosity [10^{-5} Pa· s]	1.932
Clearance [10^{-6} m]	10	Density [kg/m^3]	1.1614
Width of seal [10^{-3} m]	40	Pressure of low side [MPa]	0.11
Diameter of rotor [10^{-3} m]	160	Pressure of high side [MPa]	{0.11 0.20
Speed of rotor [r/min]	25,000		0.32 0.45}

4. Result and Discussion

4.1. Seal Mechanism

Figure 7 shows the effect of parameters like rotor speed, average clearance, and eccentricity on the seal's performance in terms of buoyancy, leakage, and friction. The vertical axis is the magnitude of the value of the floating ring seal performance parameters after the reduction to a dimensionless form and normalization. From Figure 7a, it can be seen that as the average gas film clearance of the floating ring seal increases, the gas film buoyancy and friction decrease and the gas leakage increases rapidly. When the clearance is large, the tendency of the gas film buoyancy force and friction force to decrease is relatively small. When the gas film clearance increases, the hydrodynamic effect becomes weaker leading to a reduction in the pressure of the gas in the sealing gap [32], which results in a corresponding reduction in the gas film buoyancy and friction. Due to the increase in the gas film clearance, the leakage channel of the gas in the sealing gap becomes looser [6], which leads to a rapid increase in the gas leakage.

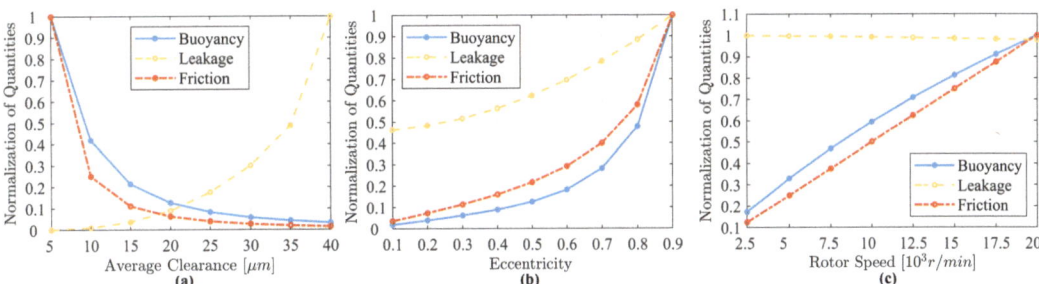

Figure 7. Normalized seal buoyancy, leakage, and friction: (**a**) at different clearances; (**b**) at different eccentricities; (**c**) at different speeds.

From Figure 7b, it can be seen that as the eccentricity of the floating ring increases, the gas film buoyancy and friction increase dramatically and the gas leakage increases consequently. The increase in the eccentricity results in a stronger converging wedge between the rotor and the floating ring. At the same time, the minimum air film thickness decreases, and the dynamic fluid pressure in the wedge region increases dramatically, resulting in a sharp increase in the air film buoyancy and friction. The increase in eccentricity results in an uneven distribution of the sealing gap, which leads to larger leakage channels in the low-pressure region and a corresponding increase in leakage.

Figure 7c illustrates the variation in seal performance parameters with rotor speed, where both the gas film buoyancy and friction increase accordingly with increasing speed, while the gas leakage decreases slightly. It is easy to understand that as the rotational speed increases, the hydrodynamic effect increases [32], resulting in an increase in the pressure of the air film inside the seal ring, which correspondingly results in an increase in the air film buoyancy [4]. At the same time, due to the viscous effect, the friction between the air film and the surface increases accordingly. As can be seen from the variation in leakage, the leakage is mainly related to the dimension of the leakage channel. The pressure gradient at the end face does not change much, making it stable for the leakage.

4.2. Effect of Texture

In this subsection, we calculate the effect of five different shapes of textures as well as their depth, rotation angle, and area ratio on the performance of the floating ring seal. The parameters of the textures are shown in Tables 2 and 3.

Table 3. Calculation parameters of seal and textures.

Parameters of Seal	Value	Parameters of Textures	Value
Eccentricity	0.5	Texture depth [10^{-6} m]	4
Clearance [10^{-6} m]	10	Texture area ratio	0.2
Width of seal [10^{-3} m]	40	Texture distribution	25×2
Diameter of rotor [10^{-3} m]	160	Texture angle	0
Speed of rotor [r/min]	12,500		
Pressure of low side [MPa]	0.101325		
Pressure of high side [MPa]	0.5		
Viscosity [10^{-5} Pa·s]	1.932		
Density [kg/m^3]	1.1614		

Figure 8 shows the effect of five texture shapes and their depths on the performance of the floating ring seal, where the horizontal coordinate is the variation in the depth of the surface texture and the vertical coordinate is the variation in the performance parameters of the floating ring seal, which include the gas film buoyancy force, gas leakage, friction force, and float-leakage ratio.

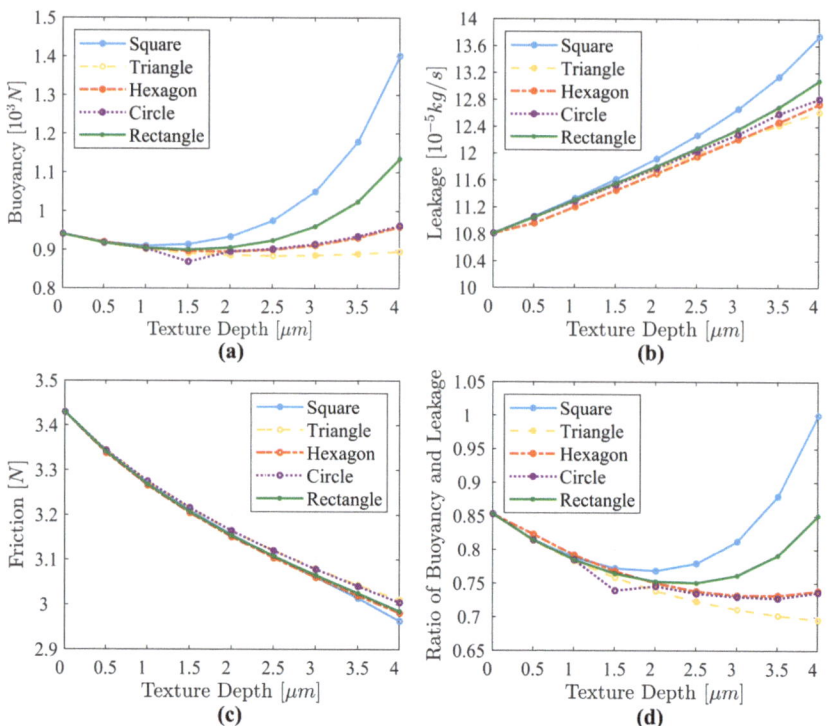

Figure 8. Effect of texture depth variation on seal performance: (**a**) effect on buoyancy; (**b**) effect on leakage; (**c**) effect on friction; (**d**) effect on ratio of buoyancy and leakage.

From Figure 8a, it can be seen that the air film buoyancy force decreases and then increases as the depth of the texture increases. The increasing trend is more drastic for squares and rectangles, the trend is almost the same for circles and square hexagons, and triangles bring the weakest effect. At a texture depth of 4 μm, compared to the non-textured surface, all other shapes of textures, except for the triangular textures, can enhance the gas film buoyancy, and the square texture improved the buoyancy by up to 91.5%. As can be seen in Figure 8b, the five shapes of textures lead to an increase in the gas leakage of the floating ring seal as the depth of the surface texture increases, with the square bringing the best increase and the rectangle the second best. The square texture increases the leakage by up to 34.66%. Figure 8c demonstrates the trend of the gas viscous friction, with the friction decreasing accordingly as the depth of the texture increases. Figure 8d shows the variation in the ratio of buoyancy and leakage. From Figure 8d, it can be seen that the trend is similar to that of the buoyancy, since the change in leakage is linear overall. Compared to the floating ring seal without texture, the square texture with a depth of 4 μm generates a significant improvement in the float-leakage ratio.

A comprehensive comparison of the four graphs shows that when the depth of the texture is more than 2 μm, the difference in the effect of the texture shape on the performance of the floating ring seals gradually increases. Especially, the gas film buoyancy and gas leakage of the square texture increase substantially. The increase in those two quantities is smaller for the rectangular texture. The change in the ratio of the buoyancy and leakage then illustrates that the enhancement of buoyancy is greater for square and rectangular textures compared with the leakage.

The texture can form tiny concave–convex structures [33] on the surface of the material, which leads to strong dynamic pressure effects [34] and reduces the actual contact area [35]. As the depth of the texture increases, the average thickness of the air film increases, which causes the sealing gap fluid channel to widen, enhancing the circulation of the lubricating air film and increasing leakage.

Figure 9 illustrates the effect of the rotation angle of the texture on the sealing performance of the floating ring seal, where the horizontal coordinate is the angle of rotation of the texture from 0° to 180°, since the shapes of the textures are all centrosymmetric. The discussion in this figure does not involve circular textures as circles have rotational invariance.

As can be seen from Figure 9a, for the square texture, the increase in the buoyancy force generated by the texture is greatest when no rotation occurs or when the angle of rotation is an integer multiple of 90°. Meanwhile, it can be seen from Figure 9b that the leakage in this case is also the largest for the square textures. However, we can find that the change in leakage is relatively small compared to the buoyancy force. For the square texture, the maximum change in the floating force is 52.01% while the maximum change in the leakage is 7.46%. This is also illustrated in Figure 9d, where the buoyancy dominates the trend of the buoyancy-leakage ratio. From Figure 9c, it can be found that the effect of the rotation angle of the texture on the friction is weak. Except for the case where the rectangular texture is rotated by 90°, the changes in friction caused by rotation angle are all within 1.35%. The reason for this special case for rectangles is probably because it is relatively less symmetrical. On the whole, the texture rotation angle has a relatively large effect on the gas film buoyancy force of the floating ring seal. The overall performance of the floating ring seal is optimized when the square texture is not rotated, the triangular texture is rotated by 30°, or the rectangular texture is rotated by 90°.

The effects of different texture area percentages on the performance of the floating ring seal are shown in Figure 10, which involves the textures in area percentages from 0 to 50%. From Figure 10b,c, it can be found that with increase in the percentage of texture area, the leakage of the floating ring seal is increased accordingly, and the friction is reduced accordingly. The reason for this phenomenon is similar to the situation of Figure 8, where the increase in the area of the texture increases the lubrication film [35] on the seal surface, and also increases the leakage channel of the fluid [34], thus reducing the friction and increasing the leakage. For the proposed shapes of textures, the effect of the area ratio of the texture on the buoyancy force is not consistent. From Figure 10a, it can be seen that both square and rectangular textures generate more buoyancy as the area ratio of the texture increases. The circular texture also provides the effect of buoyancy enhancement in general, but the trend of the change is tortuous. Changes in the area ratio of the triangular and hexagonal textures have a weak effect on the buoyancy force, which gradually decreases when the area ratio is larger. In addition, when the area ratio of the texture reaches 40%, the buoyancy force generated by the rectangular texture falls in a faulty manner. By calculation, the gap between the circumferential textures in this case is only 12.41% of the length of a single texture. The texture distribution with depth is too dense, making the effect equivalent to increasing the sealing clearance.

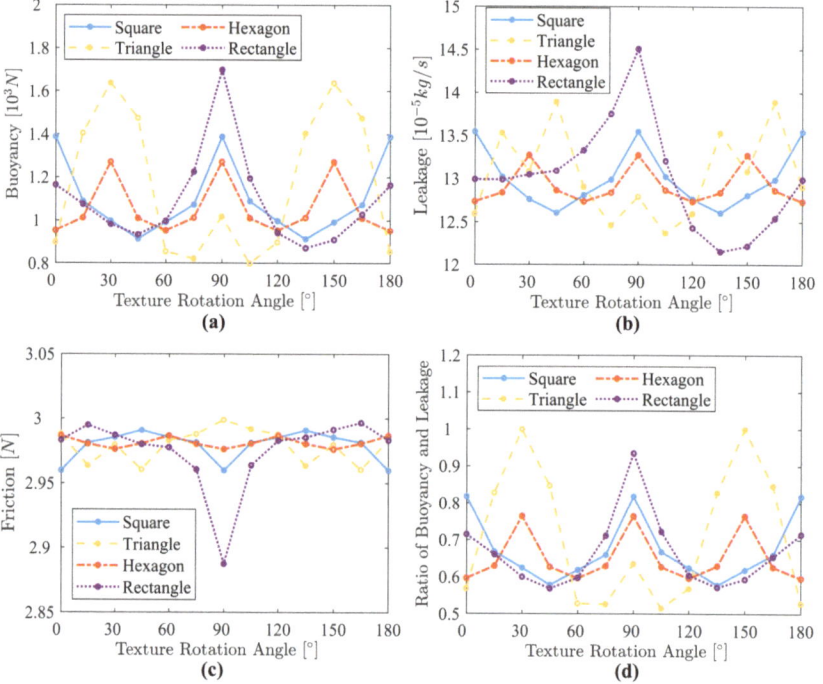

Figure 9. Effect of texture angle variation on seal performance: (**a**) effect on buoyancy; (**b**) effect on leakage; (**c**) effect on friction; (**d**) effect on ratio of buoyancy and leakage.

Overall, the effect of square and rectangular textures on the buoyancy of the floating ring seal is more significant, with an increase in either the depth or area ratio of the texture resulting in an increase in its buoyancy. Other shapes have a weaker effect on the buoyancy without angular rotation. For all shapes of textures, an increase in the depth or area ratio of the textures leads to a decrease in friction and an increase in leakage. The degree of friction reduction is close for different shapes of textures. Comparatively speaking, the effect of variation in the percentage of texture area on the floating ring seal is more significant than that generated by variation in the depth of the texture. Different shapes of textures have their optimal rotation angles: 0° for square textures, 90° for rectangular textures, and 30° for triangular textures. In general, the square texture with a rotation angle of 0° and an area ratio of 0.25, the rectangular texture with a rotation angle of 90° and an area ratio of 0.4, and the triangular texture with a rotation angle of 30° and an area ratio of 0.05 give the best results. In the next subsection, the folded three types of textures are further analyzed and discussed.

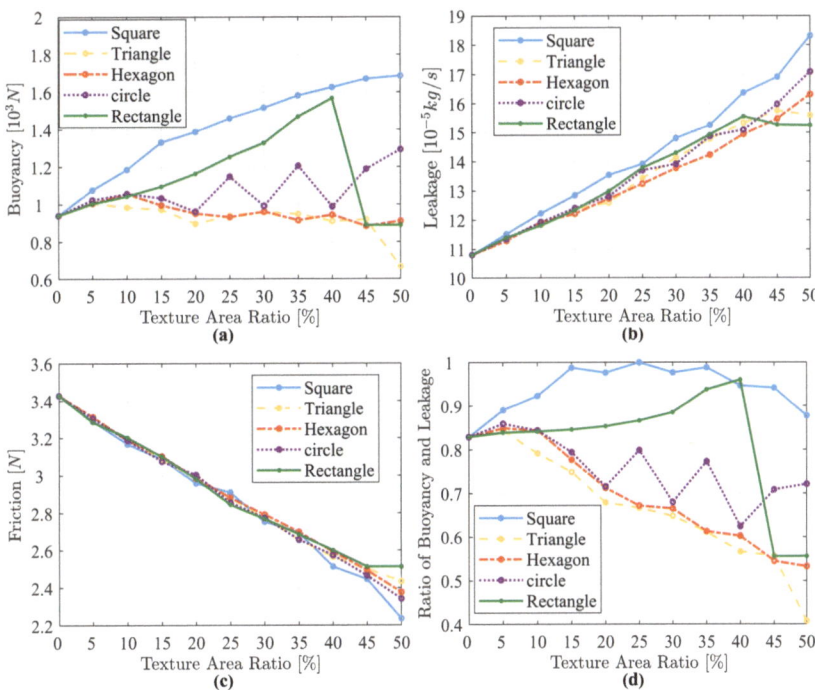

Figure 10. Effect of texture area ratio variation on seal performance: (**a**) effect on buoyancy; (**b**) effect on leakage; (**c**) effect on friction; (**d**) effect on ratio of buoyancy and leakage.

The results of the effect of no texture and three types of textures, as shown in Table 4, on the performance of the floating ring seal at different rotational speeds are shown in Figure 11. It can be seen from Figure 11a that all three types of textures can contribute to increasing the buoyancy force of the floating ring seal, and the increase is greater at higher rotational speeds. The specific rectangular textures generate the largest increase in buoyancy force, which reaches 81.2% at a rotation speed of 20,000 r/min. However, it can be noticed from Figure 11b that this particular rectangular texture increases the amount of leakage while enhancing the buoyancy. The textures increase the leakage with increasing speed compared to the no texture. But the specific triangular texture keeps the increase in leakage within a small range. This specific triangular texture increases the buoyancy by 28.02% when the rotation speed is 15,000 r/min and the leakage increases by only 10.08%. Figure 11c illustrates the variation in friction with rotational speed for four cases. It is obvious that the friction increases as the rotational speed increases and all three types of textures reduce the friction to a certain extent as compared to no texture. Taking into account the variation in buoyancy and leakage, it can be noticed from Figure 11d that all three specific textures have a good effect. Among them, the specific square texture and the specific triangular texture are slightly better than the rectangular texture.

Table 4. Three types of textures that work better.

Texture Shape	Texture Depth	Rotation Angle	Area Ratio
Square	4 μm	0°	25%
Rectangle	4 μm	90°	40%
Triangle	4 μm	30°	5%

Figure 12 shows the dimensionless pressure distribution for the case of a smooth surface and three types of surfaces with the aforementioned textures. From the figure, it can be found that all three textures enhance the peak air film pressure compared to the smooth surface. Among them, the peak air film pressure with rectangular texture is the largest. This is in accordance with the results of Figure 11. In addition, it can be noticed that the pressure in the texture region is lower than that in the non-texture region, but the texture makes the gas film pressure overall higher than that in the case of a smooth surface. And because of the increased pressure peaks, it leads to an increase in the gas pressure gradient making the gas leakage increase.

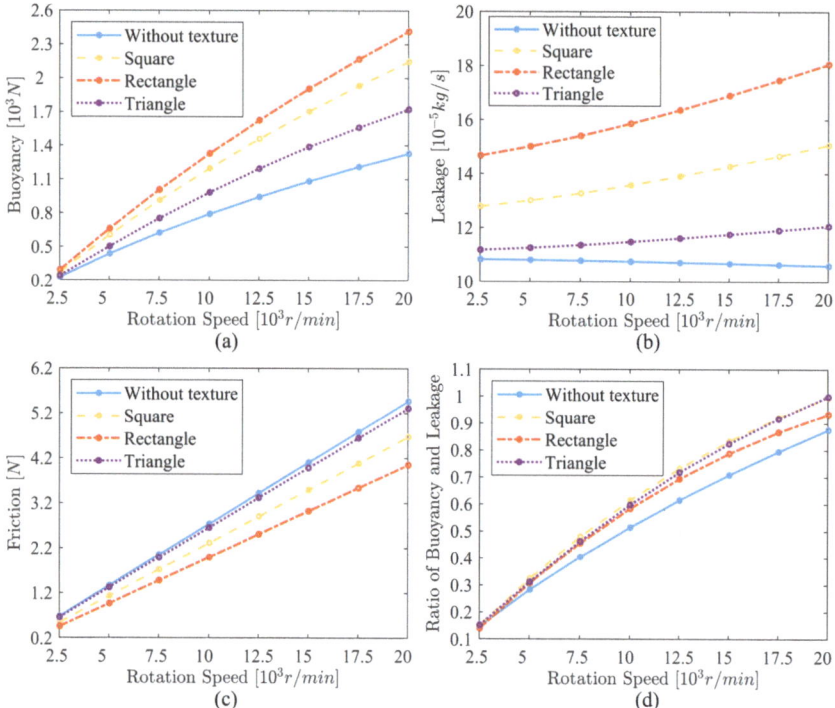

Figure 11. Effect of surface textures on sealing performance at different rotation speeds: (**a**) effect on buoyancy; (**b**) effect on leakage; (**c**) effect on friction; (**d**) effect on ratio of buoyancy and leakage.

Overall, the specific rectangular texture can be prioritized when buoyancy enhancement is most important. The specific triangular texture can be preferred when buoyancy enhancement is of high importance while at the same time strictly preventing an increase in leakage. In addition, this rectangular texture is the most effective in terms of reducing friction. The textures can be selected according to the requirements of the operation.

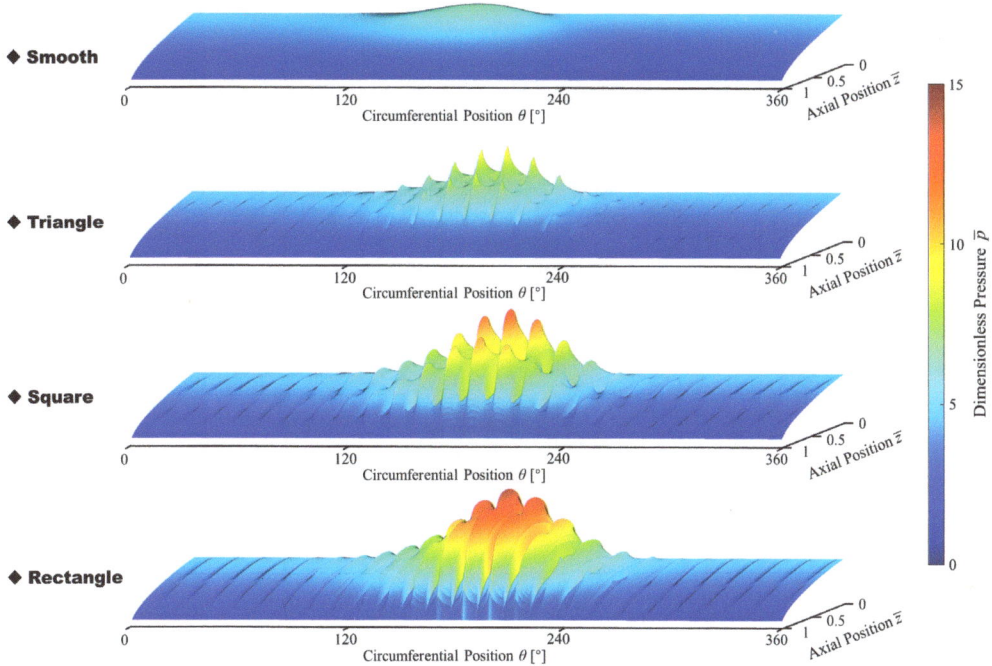

Figure 12. Dimensionless pressure in floating ring gas seals with different textures.

4.3. Effect of Texture with Roughness

Figure 13 shows the thickness and dimensionless pressure of the floating ring seal air film for a smooth surface and fractal rough surfaces with fractal dimensions $D = 2.5$ and $D = 2.6$, respectively. It can be seen that the surface roughness is presented in the air film thickness, which is more pronounced when the fractal dimension $D = 2.5$. This is because the arithmetic mean deviation is greater for fractal dimension $D = 2.5$ relative to fractal dimension $D = 2.6$. At the same time, we can see a significant increase in the pressure peak. The pressure comes from the dynamic pressure effect generated by the eccentricity of the floating ring, and the air film thickness is smaller at the position of the eccentricity of the floating ring, so the effect of roughness becomes more significant. The air film pressure of the floating ring seal with surface roughness is enhanced. When the fractal dimension $D = 2.6$, the fractal roughness surface has less effect, because the order of magnitude of its height is negligible compared to the air film thickness.

It can also be seen from Figure 14 that when the fractal dimension of the fractal rough surface $D = 2.6$, the parameters such as air film buoyancy, leakage, and friction of the floating ring seal are close to those when the surface is smooth. When the fractal dimension $D = 2.5$, the buoyancy of the floating ring seal is somewhat enhanced at different speeds, while the gas leakage is slightly increased. This is consistent with the air film pressure being boosted in Figure 13. The air film buoyancy force increases as the rotational speed increases. The buoyancy-leakage ratio is enhanced overall, and the greater the rotational speed the greater the enhancement. It can be seen that a certain surface roughness has an improving effect on the performance of the floating ring seal. This is consistent with the conclusions of [36]. It can be found that the surface roughness has a similar effect as the surface texture. However, as can be seen from Figure 14c, the surface roughness increases the friction of the floating ring seal, in contrast to the surface texture.

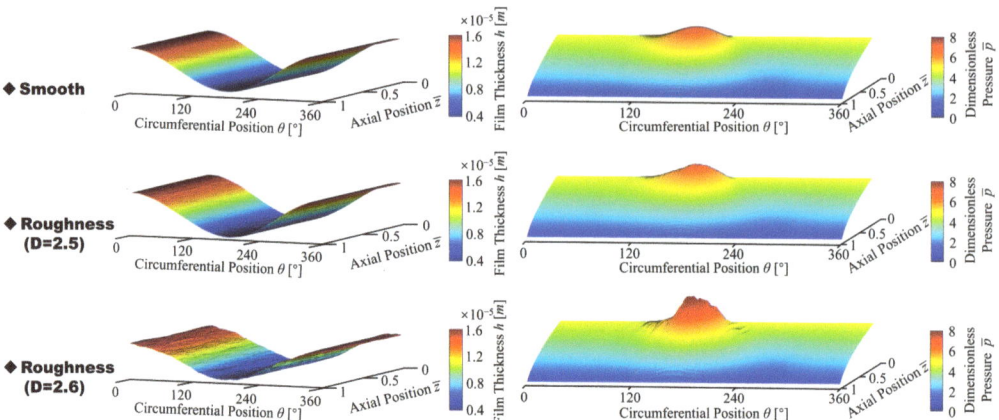

Figure 13. Air film thickness and dimensionless pressure formed on the smooth surface and rough surfaces with different fractal dimensions.

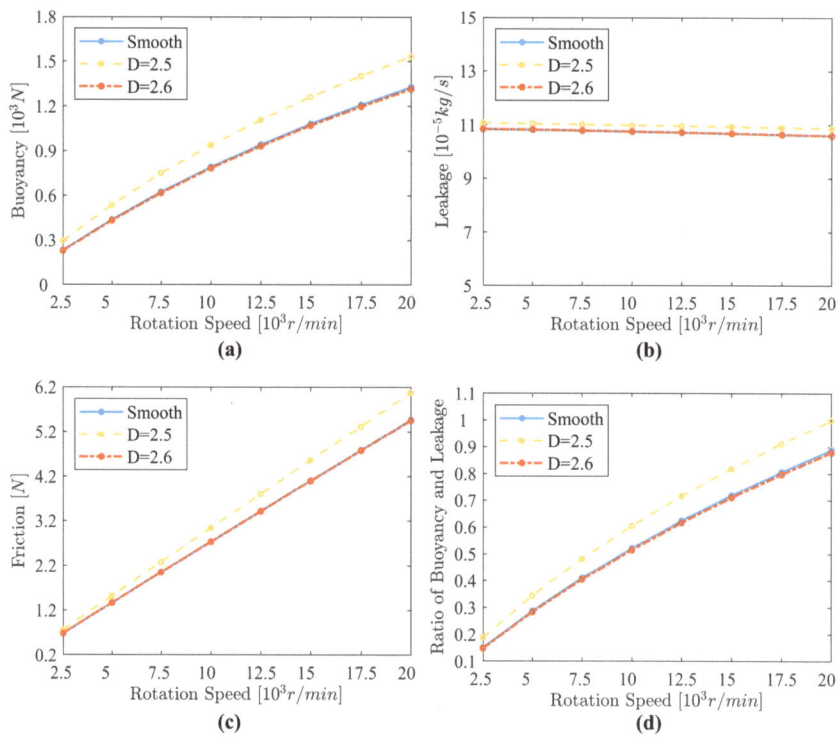

Figure 14. Effect of surface roughness on sealing performance at different rotation speeds: (**a**) effect on buoyancy; (**b**) effect on leakage; (**c**) effect on friction; (**d**) effect on ratio of buoyancy and leakage.

Therefore, the effect of certain textures with surface roughness on the performance of the floating ring seal is discussed and the results are shown in Figure 15. The texture parameters are shown in Table 4 and the fractal dimension $D = 2.5$. Comparison of Figures 11 and 14 shows that the textures with a certain roughness boost more buoyancy and only slightly increase leakage. Compared to a smooth surface, when the rotation speed is 15,000 r/min, the triangular textures increase buoyancy by 28.02% and increase leakage by only 10.08%. As a comparison, the triangular textures with roughness increase buoyancy by 67.39% and leakage increases by 14.92%. The rectangular surface textures improve buoyancy by 183.29% at 17,500 rpm when the surface of the floating ring seal is rough. Figure 11c illustrates the variation in friction with rotational speed for four cases. From comparison of the buoyancy-leakage ratio, the textures with a certain roughness generate a better enhancement effect on the floating ring seal. In addition, the surface roughness does not change the mechanism of the effect of textures, but affects the degree of the effect of the textures. The design of textures combined with surface roughness can be one of the means to enhance the performance of a floating ring seal. However, the process and cost of mechanically creating surface roughness need to be considered.

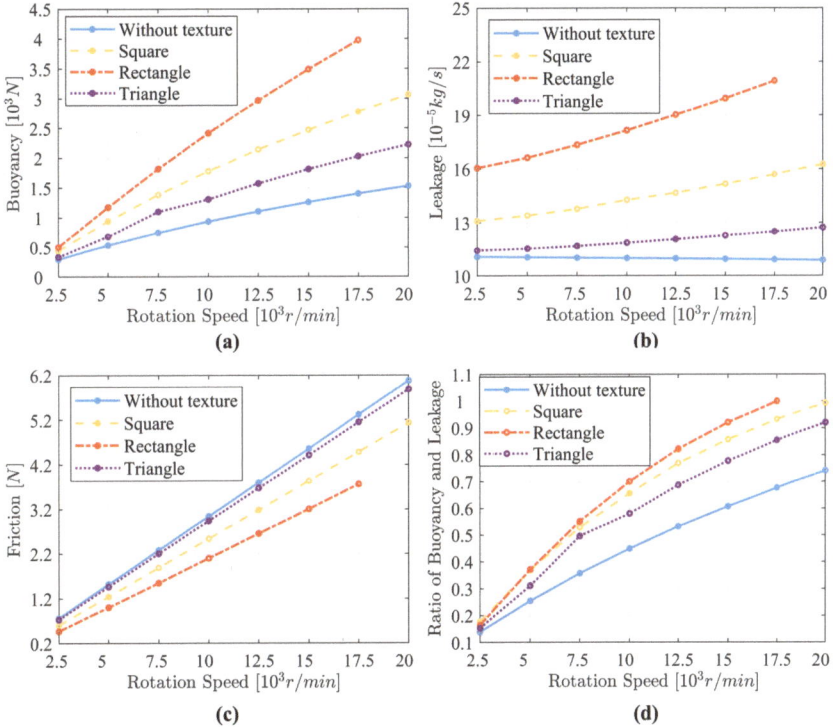

Figure 15. Effect of surface textures with roughness on sealing performance at different rotation speeds: (**a**) effect on buoyancy; (**b**) effect on leakage; (**c**) effect on friction; (**d**) effect on ratio of buoyancy and leakage.

5. Conclusions

In summary, this work employed the finite difference method to iteratively solve the Reynolds equation for the sealing fluid, and then analyzed the effects of surface texture and surface roughness on the performance of floating ring seals. The effects of different surface textures and their parameters on the performance of floating ring seals are systematically presented and discussed. The key conclusions are summarized as follows:

① Increasing the clearance can reduce friction, but it will compromise buoyancy and increase leakage. Conversely, increasing the eccentricity can enhance buoyancy, but it will significantly increase friction and leakage. Moreover, the increase in rotational speed significantly boosts buoyancy and friction, with a negligible impact on leakage.

② For an arbitrarily shaped texture in this paper, the depth of the texture has a significant influence on the performance of the floating ring seal. The greater the depth of the texture, the greater the leakage and the lower the friction. The change in buoyancy caused by the depth is related to the shape. The greater the depth of the texture within a certain range, the better the overall performance of the floating ring seal.

③ The best angles for making the textures most effective are 30° for triangles and hexagons, 90° for rectangles, and 0° for squares.

④ An increase in the texture area leads to increased leakage and reduced friction. The square and rectangular textures exhibit a stronger effect, and as the texture area increases, the buoyancy force also increases significantly.

⑤ The textures mentioned in Table 4 can significantly enhance the performance of the floating ring seals. The specific rectangular texture shows the largest increase in buoyancy force, which reaches 81.2% at a rotation speed of 20,000 r/min. This specific triangular texture increases buoyancy by 28.02% when the rotation speed is 15,000 r/min and leakage increases by only 10.08%.

⑥ Appropriate surface roughness can serve to enhance the performance of the floating ring seal; the effect of texture is substantially enhanced when the floating ring seal surface is rough. The triangular textures mentioned in Table 4 with roughness increase buoyancy by 67.39% and leakage increases only by 14.92%.

Both optimized surface texture and surface roughness can contribute to enhancing the overall performance of floating ring seals, particularly in terms of buoyancy. The combination of surface texture and surface roughness is a promising direction that yields positive results. However, the rough surface simulated by the computer model is generated by randomization, and the feasibility of the process still requires further verification and experimental confirmation.

Author Contributions: Conceptualization, Z.H., Y.G. and J.S.; Methodology, Z.H., Y.G., L.J., Y.Z. and H.W.; Validation, Y.G.; Data curation, Y.G.; Writing – original draft, Y.G.; Writing – review & editing, Z.H., Y.G., J.S., Ning Li, L.J., Y.Z. and H.W.; Visualization, Y.G.; Supervision, Z.H., J.S. and N.L.; Funding acquisition, Z.H. All authors have read and agreed to the published version of the manuscript.

Funding: This research was funded by the Fundamental Research Funds for the Central Universities grant number 3122019178, the Tianjin Municipal Science and Technology Bureau Science and Technology Plan Project grant number 23JCYBJC00110 and the Civil Aircraft Special Scientific Research Project initiated by the Ministry of Industry and Information Technology.

Data Availability Statement: The original contributions presented in the study are included in the article, further inquiries can be directed to the corresponding author.

Acknowledgments: The authors would like to thank the anonymous referees for their careful reading and valuable comments to this paper.

Conflicts of Interest: Author Jiaxin Si and author Ning Li are employed by the company AECC Hunan Aviation Powerplant Research Institute. The remaining authors declare that the research was conducted in the absence of any commercial or financial relationships that could be construed as a potential conflict of interest.

Nomenclature

A	textured area
A_n	random numbers in the interval $[0, 2\pi]$
b	position of the shaft surface
B_n	random numbers in the interval $[0, 2\pi]$
c	radial clearance, m
C	height coefficient of the surface
C_n	a normally distributed random variable with a mean of 0 and variance of 1
d	lengths of square textures; the width of the rectangular texture, m
D	fractal dimension
D_s	theoretical fractal dimension
e	eccentricity, m
G	feature scale parameter

h	gas film thickness, m
h_1	distribution of depth of texture
h_2	distribution of depth of surface roughness
h_d	depth of texture, m
H	dimensionless film thickness
k	ratio of buoyancy and leakage, N·kg/s
l	length of the rectangular texture, m
L	size of the simulated image, m
L_{max}	sampling length, m
L_{min}	a number which is determined by the instrument resolution
m	frequency coefficient
M	number of spatial undulations
n	frequency coefficient
n_{max}	maximum frequency coefficient
p	gas film pressure, Pa
$\bar{p}$	dimensionless film pressure
p_a	air pressure, Pa
p_{low}	pressure in the low-pressure side, Pa
p_{high}	pressure in the high-pressure side, Pa
Q	leakage, kg/s
R	radius of the rotating shaft, m
r_p	radius of the outer circle of a triangular or hexagonal texture, m
r_o	radius of the outer circle of a triangular or hexagonal texture, m
T_x	dimensionless friction component in the x-direction
T_y	dimensionless friction component in the y-direction
U	linear velocity of rotating shaft, m/s
w	buoyancy of gas film, N
$\bar{w}$	dimensionless buoyancy of gas film, N
$\bar{w}_x$	dimensionless buoyancy component in the x-direction
$\bar{w}_y$	dimensionless buoyancy component in the y-direction
γ	characteristic parameter
Λ	seal coefficient
μ	aerodynamic viscosity, Pa·m
ϕ_{mn}	a series of random phases between 0 and π
ρ	gas density, kg/m^3

References

1. Xia, P.; Chen, H.; Liu, Z.; Ma, W.; Yang, B. Analysis of whirling motion for the dynamic system of floating ring seal and rotor. *Proc. Inst. Mech. Eng. Part J J. Eng. Tribol.* **2019**, *233*, 1221–1235. [CrossRef]
2. Duan, W.; Chu, F.; Kim, C.H.; Lee, Y.B. A bulk-flow analysis of static and dynamic characteristics of floating ring seals. *Tribol. Int.* **2007**, *40*, 470–478. [CrossRef]
3. Lee, Y.B.; Kim, C.H.; Duan, W.; Chu, F. Analysis of Leakage Flow and Dynamic Characteristics in Floating Ring Seals for High Pressure Turbopump. In Proceedings of the Turbo Expo: Power for Land, Sea, and Air, Barcelona, Spain 8–11 May 2006; Volume 42401, pp. 1341–1349.
4. Bae, J.H.; Kwak, H.D.; Heo, S.J.; Choi, C.H.; Choi, J.S. Numerical and experimental study of nose for Lox floating ring seal in turbopump. *Aerospace* **2022**, *9*, 667. [CrossRef]
5. Shi, R.J.; Li, S.X.; Zhen, R.; Ma, L.j.; Zhang, J.B. Design method and research of high temperature gas floating ring seal with slightly variable gap. *IOP Conf. Ser. Mater. Sci. Eng.* **2021**, *1081*, 012005. [CrossRef]
6. Anbarsooz, M.; Amiri, M.; Erfanian, A.; Benini, E. Effects of the ring clearance on the aerodynamic performance of a CO_2 centrifugal compressors annular seal: A numerical study. *Tribol. Int.* **2022**, *170*, 107501. [CrossRef]
7. Li, G.; Zhang, Q.; Huang, E.; Lei, Z.; Wu, H.; Xu, G. Leakage performance of floating ring seal in cold/hot state for aero-engine. *Chin. J. Aeronaut.* **2019**, *32*, 2085–2094. [CrossRef]
8. Nagai, K.; Kaneko, S.; Taura, H.; Watanabe, Y. Numerical and experimental analyses of dynamic characteristics for liquid annular seals with helical grooves in seal stator. *J. Tribol.* **2018**, *140*, 052201. [CrossRef]
9. Lu, X.; Andrés, L.S. Leakage and Rotordynamic Force Coefficients of a Three-Wave (Air in Oil) Wet Annular Seal: Measurements and Predictions. *J. Eng. Gas Turbines Power* **2019**, *141*, 032503. [CrossRef]

10. Zhang, G.H.; Wang, G.L.; Liu, Z.S.; Ma, R.X. Stability Characteristics of Steam Turbine Rotor Seal System with Analytical Floating Ring Seal Force Model. In Proceedings of the Turbo Expo: Power for Land, Sea, and Air. American Society of Mechanical Engineers, San Antonio, TX, USA, 3–7 June 2013; Volume 55270, p. V07BT30A004.
11. Kirk, R.; Miller, W. The influence of high pressure oil seals on turbo-rotor stability. *ASLE Trans.* **1979**, *22*, 14–24. [CrossRef]
12. Adjemout, M.; Brunetiere, N.; Bouyer, J. Numerical analysis of the texture effect on the hydrodynamic performance of a mechanical seal. *Surf. Topogr. Metrol. Prop.* **2015**, *4*, 014002. [CrossRef]
13. Pei, S.; Xu, H.; Yun, M.; Shi, F.; Hong, J. Effects of surface texture on the lubrication performance of the floating ring bearing. *Tribol. Int.* **2016**, *102*, 143–153. [CrossRef]
14. Shi, L.; Zhang, Y.; Chen, S.; Zhu, W. Comparative research on gas seal performance textured with microgrooves and microdimples. *J. Braz. Soc. Mech. Sci. Eng.* **2019**, *41*, 280. [CrossRef]
15. Wang, S.; Ding, X.; Li, N.; Ding, J.; Zhang, L. Orientation effect on sealing characteristics of rectangular micro-textured floating ring gas film seal. *J. Beijing Univ. Aeronaut. Astronaut.* **2023**, 1–18. Available online: https://kns.cnki.net/kcms2/detail/11.2625.V.20230710.1709.004.html (accessed on 30 June 2024).
16. He, Z.; Song, Q.; Liu, Q.; Xin, J.; Yang, C.; Liu, M.; Li, B.; Yan, F. Analysis of the effect of texturing parameters on the static characteristics of radial rigid bore aerodynamic journal bearings. *Surf. Topogr. Metrol. Prop.* **2022**, *10*, 035025. [CrossRef]
17. Zhang, Z.; Ding, X.; Xu, J.; Jiang, H.; Li, N.; Si, J. A study on building and testing fractal model for predicting end face wear of Aeroengine's floating ring seal. *Wear* **2023**, *532*, 205079. [CrossRef]
18. Pattnayak, M.R.; Ganai, P.; Pandey, R.; Dutt, J.; Fillon, M. An overview and assessment on aerodynamic journal bearings with important findings and scope for explorations. *Tribol. Int.* **2022**, *174*, 107778. [CrossRef]
19. Zhang, G.; Xu, K.; Han, J.; Huang, Y.; Gong, W.; Guo, Y.; Huang, Z.; Luo, X.; Liang, B. Performance of textured foil journal bearing considering the influence of relative texture depth. *Proc. Inst. Mech. Eng. Part J J. Eng. Tribol.* **2022**, *236*, 2105–2117. [CrossRef]
20. Tewelde, F.B.; Allen, Q.; Zhou, T. Multiscale Texture Features to Enhance Lubricant Film Thickness for Prosthetic Hip Implant Bearing Surfaces. *Lubricants* **2024**, *12*, 187. [CrossRef]
21. Yan, W.; Komvopoulos, K. Contact analysis of elastic-plastic fractal surfaces. *J. Appl. Phys.* **1998**, *84*, 3617–3624. [CrossRef]
22. Ausloos, M.; Berman, D. A multivariate Weierstrass–Mandelbrot function. *Proc. R. Soc. London. A Math. Phys. Sci.* **1985**, *400*, 331–350.
23. Zhang, X.; Xu, Y.; Jackson, R.L. An analysis of generated fractal and measured rough surfaces in regards to their multi-scale structure and fractal dimension. *Tribol. Int.* **2017**, *105*, 94–101. [CrossRef]
24. Xiao, H.; Sun, Y.; Chen, Z. Fractal modeling of normal contact stiffness for rough surface contact considering the elastic–plastic deformation. *J. Braz. Soc. Mech. Sci. Eng.* **2019**, *41*, 11. [CrossRef]
25. Flores Alarcón, J.L.; Figueroa, C.G.; Jacobo, V.H.; Velázquez Villegas, F.; Schouwenaars, R. Statistical Study of the Bias and Precision for Six Estimation Methods for the Fractal Dimension of Randomly Rough Surfaces. *Fractal Fract.* **2024**, *8*, 152. [CrossRef]
26. Liu, Y. Digital Modeling for Functional Surface Structure Based on Fractal Representation. Master's Thesis, South China University of Technology, Guangdong, China, 2011.
27. Jiang, S.; Zheng, Y. An analytical model of thermal contact resistance based on the Weierstrass—Mandelbrot fractal function. *Proc. Inst. Mech. Eng. Part C J. Mech. Eng. Sci.* **2010**, *224*, 959–967. [CrossRef]
28. Xu, J. Analysis and Experimental Research on Dynamic Pressure Lubrication Characteristics of Compliant Foil Gas Film Seal. Ph.D. Thesis, Lanzhou University of Technology, Lanzhou, China, 2022.
29. Youssef, I.K.; Taha, A. On the modified successive overrelaxation method. *Appl. Math. Comput.* **2013**, *219*, 4601–4613. [CrossRef]
30. Ma, G.; Ping, X.; Shen, X.; Hu, G. Analysis of quasi-dynamic characteristics of compliant floating ring gas cylinder seal. *J. Aerosp. Power* **2010**, *25*, 1190–1196.
31. Chen, T. The Groove Parameters Optimization and Numerical Simulation on Cylindrical Gas Seal. Master's Thesis, Kunming University of Science and Technology, Kunming, China, 2015.
32. Shipeng, W.; Xuexing, D.; Junhua, D.; Jingmo, W. Numerical Analysis on Improving the Rectangular Texture Floating Ring Gas-film Seal Characteristics with Different Bottom Shapes. *J. Appl. Fluid Mech.* **2023**, *16*, 1371–1385.
33. Ye, S.; Tang, H.; Ren, Y.; Xiang, J. Study on the load-carrying capacity of surface textured slipper bearing of axial piston pump. *Appl. Math. Model.* **2020**, *77*, 554–584. [CrossRef]
34. Ding, X.; Wu, J.; Wang, Y.; Cui, B.; An, S.; Su, B.; Wang, Y. Influence of surface texture on sealing performance of PTFE materials. *Macromol* **2022**, *2*, 225–235. [CrossRef]
35. Luz, F.K.C.; Profito, F.J.; dos Santos, M.B.; Silva, S.A.; Costa, H.L. Deterministic Simulation of Surface Textures for the Piston Ring/Cylinder Liner System in a Free Piston Linear Engine. *Lubricants* **2024**, *12*, 12. [CrossRef]
36. Yuan, Z.; Chen, J. Influences of Floating-ring Seal Parameters on Clearance Leakage. *IOP Conf. Ser. Mater. Sci. Eng.* **2020**, *790*, 012077. [CrossRef]

Disclaimer/Publisher's Note: The statements, opinions and data contained in all publications are solely those of the individual author(s) and contributor(s) and not of MDPI and/or the editor(s). MDPI and/or the editor(s) disclaim responsibility for any injury to people or property resulting from any ideas, methods, instructions or products referred to in the content.

Article

Thermo-Fluid–Structural Coupled Analysis of a Mechanical Seal in Extended Loss of AC Power of a Reactor Coolant Pump

Youngjun Park [1], Gwanghee Hong [1], Sanghyun Jun [2], Jeongmook Choi [2], Taegyu Kim [2], Minsoo Kang [2] and Gunhee Jang [1,*]

[1] Department of Mechanical Convergence Engineering, Hanyang University, Seoul 04763, Republic of Korea; lkjk0418@hanyang.ac.kr (Y.P.); palla12@hanyang.ac.kr (G.H.)
[2] Department of Engineering, Flowserve KSM, Gimpo 10040, Republic of Korea; shjun@flowserveksm.co.kr (S.J.); jmchoi@flowserveksm.co.kr (J.C.); tgkim@flowserveksm.co.kr (T.K.); mskang@flowserveksm.co.kr (M.K.)
* Correspondence: ghjang@hanyang.ac.kr

Abstract: We proposed a numerical method to investigate the thermo-fluid–structural coupled characteristics of a mechanical seal of a reactor coolant pump (RCP), especially during extended loss of AC power (ELAP) operation. We developed a finite element program for the general Reynolds equation, including the turbulence effect to calculate the pressure, opening force, and leakage rate of fluid lubricant and the two-dimensional energy equation to calculate the temperature distribution of the fluid lubricant. We verified the accuracy of the developed program by comparing the simulated temperature distribution and leakage rate of this study with those of previous research. Heat conduction and elastic deformation due to pressure and temperature changes at the seal structure were analyzed using an ANSYS program. The results showed that temperature more significantly affected the elastic deformation of the seal structure near clearance than pressure both under normal and ELAP operating conditions. High temperature and pressure of the coolant under ELAP operating conditions deform the seal structure, resulting in a much smaller clearance of the fluid film than normal operating condition. However, even with a small clearance under ELAP operation, the leakage rate slightly increases due to the high internal pressure of the coolant. This research will contribute to the development of robust mechanical seals for RCPs by accurately predicting the characteristics of mechanical seals, especially when the RCP is operating under ELAP.

Keywords: ELAP; mechanical seal; thermo-fluid–structural coupled analysis; numerical analysis

Citation: Park, Y.; Hong, G.; Jun, S.; Choi, J.; Kim, T.; Kang, M.; Jang, G. Thermo-Fluid–Structural Coupled Analysis of a Mechanical Seal in Extended Loss of AC Power of a Reactor Coolant Pump. *Lubricants* **2024**, *12*, 212. https://doi.org/10.3390/lubricants12060212

Received: 3 May 2024
Revised: 3 June 2024
Accepted: 6 June 2024
Published: 10 June 2024

Copyright: © 2024 by the authors. Licensee MDPI, Basel, Switzerland. This article is an open access article distributed under the terms and conditions of the Creative Commons Attribution (CC BY) license (https://creativecommons.org/licenses/by/4.0/).

1. Introduction

A reactor coolant pump (RCP) used in a pressurized water reactor of a nuclear power plant is a device that circulates pressurized water coolant between the reactor and the steam generator. To prevent leakage of the coolant, a three-stage cartridge-type mechanical seal is applied to the RCP. Under normal operating conditions of the RCP, the coolant temperature is around 40 °C, and the total pressure reduction through the three-stage mechanical was around 150 bar, where the high pressure of the coolant was roughly distributed to three mechanical seals with a distribution of 42% (63 bar reduction), 42% (63 bar reduction), and 16% (24 bar reduction). However, with extended loss of AC power (ELAP), the coolant temperature is over 300 °C, and a single stage of the mechanical seal has to be safely operated to withstand a high-pressure difference of 150 bar. It is important to predict the leakage and the seal deformation of the mechanical seal in the design stage to prevent leakage of coolant under the high-pressure and high-temperature conditions of ELAP [1]. Figure 1 illustrates the structure of the mechanical seal of an RCP, consisting of a seal that rotates with the shaft and a stationary seal attached to the housing. Either a stationary or a rotating seal has grooves to generate pressure. During operation, the mechanical seal balances the closing force exerted by external pressure and spring force with the opening force generated by the pressure in the fluid film. The behavior of the fluid film can be determined by the Reynolds equation, where the viscosity of the lubricant is mainly dependent on temperature. Pressure and leakage vary depending on the state of the fluid, the thickness of the fluid film, and surrounding pressure. The thermal energy generated in the fluid film affects the temperature of the fluid film and the surrounding structures through conduction and convection, and the structural deformation of the seal due to temperature and pressure influences the behavior of the fluid film. Therefore, an accurate calculation of the leakage from the mechanical seal requires a coupled analysis that considers the pressure and temperature distribution of the fluid film and the elastic deformation of the seal structure.

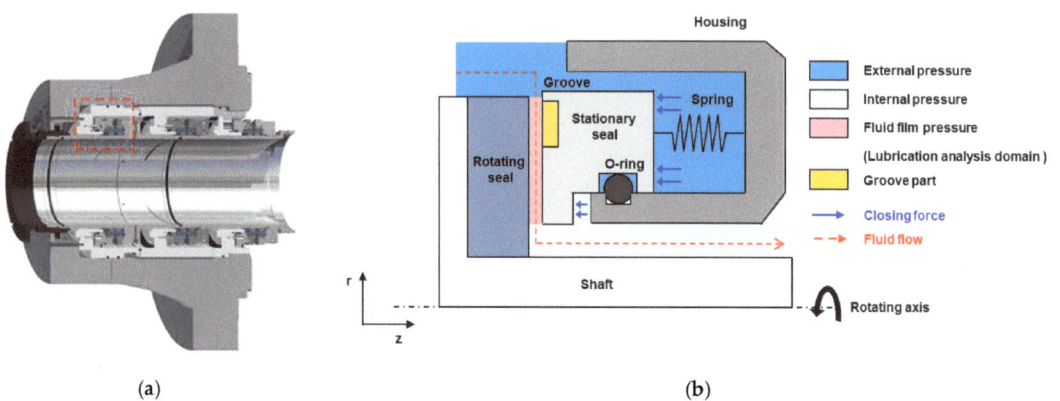

Figure 1. (**a**) Structure of a mechanical seal in an RCP and (**b**) a magnified drawing of the mechanical seal in the dotted box of (**a**).

Several researchers have analyzed the performance of mechanical seals through thermo-hydrodynamic analysis. Thomas et al. analyzed the pressure, temperature, and leakage rate by solving the Reynolds equation and the energy equation to study the performance of mechanical face seals operating under high pressure. They suggested that this calculation required consideration of heat transfer of the fluid film for a stable design of seals under high pressure [2,3]. Other researchers also predicted the performance of seals by analyzing the pressure and temperature of the fluid film through thermo-hydrodynamic analysis [4–8]. However, they did not consider the changes in clearance caused by elastic deformation of the seal structure occurring at high a temperature and pressure. Brunetiere et al. proposed a method to analyze a mechanical face seal, including the structural thermal deformation under normal operating conditions of 40 °C and a pressure difference of 20 bar, in addition to solving the two-dimensional Reynolds equation and the three-dimensional energy equation. They showed that the calculation of leakage without considering structural thermal deformation can lead to significant error [9–11]. Other researchers also investigated the performance of a mechanical seal considering structural thermal deformation to accurately calculate the leakage rate of the seals in mechanical devices [12–14].

Galenne et al. proposed an elasto-hydrodynamic model to analyze the failure of the seals for an RCP under normal steady operating conditions at 50 °C and a total pressure difference of 150 bar [15]. They compared the simulated leakage rates from the elastic deformation of seals by solving the two-dimensional Reynolds equation and the three-dimensional energy equation and comparing the structural analysis with the experimental values. Moreover, they investigated lubrication characteristics by including the effects of structural heat transfer and elastic deformation. However, they did not consider harsh operating conditions. Su et al. proposed a coupled thermo-fluid–structural model to analyze the lubrication characteristics of wavy pattern grooved seals for an RCP under three temperatures (20 °C, 40 °C, 75 °C) and pressures (53 bar, 106 bar, 159 bar) [16]. They showed that the elastic deformation of the seals due to pressure and temperature affected the leakage rates of mechanical seals, with leakage rates increasing as the sealing pressure increased. Other researchers have also conducted a thermo-elasto-hydrodynamic analysis of the lubrication characteristics of seals for RCPs [17–20]. However, prior researchers did not investigate the effects of the high temperature and pressure of ELAP operating conditions on the performance of RCP seals.

In this study, we proposed a thermo-fluid–structural coupled analysis method to predict the pressure and temperature distributions, the change in the clearance of fluid film due to elastic deformation of the mechanical seal, and the resulting leakage rate under normal and ELAP operating conditions of an RCP. We developed the finite element program of the general Reynolds equation, including the turbulence effect to calculate the pressure, opening force, and leakage rate of fluid lubricant and the two-dimensional energy equation to calculate the temperature distribution of the fluid lubricant. The accuracy of the developed program was verified by comparing the simulated temperature distribution and leakage rate with those of the previous research. Heat conduction and elastic deformation of the seal structure due to pressure and temperature change were analyzed using the ANSYS program. The proposed method was used to investigate the lubricating characteristics of the mechanical seal under normal and ELAP conditions of the RCP.

2. Method of Analysis

2.1. Lubrication Analysis in Fluid Film

The two-dimensional Reynolds equation can be derived from the three-dimensional Navier–Stokes equation and the continuity equation with the assumption of incompressible laminar flows. By neglecting body force, inertia force, and the pressure gradient along the direction of film thickness, the two-dimensional Reynolds equation is derived to calculate the pressure of thin film flow. In fluid between two plates, the fluid state changes from laminar to turbulent condition when the Reynolds number is greater than about 1000, and the turbulent effect becomes significant as the Reynolds number increases [21,22]. Hirs proposed the general Reynolds equation in Equation (1) to model the turbulence occurring in thin-film flow [21,22].

$$\frac{\partial}{\partial r}\left(G_r \frac{h^3}{\mu}\frac{\partial p}{\partial r}\right) + \frac{\partial}{r\partial \theta}\left(G_\theta \frac{h^3}{\mu}\frac{\partial p}{r\partial \theta}\right) = \frac{U}{2}\frac{\partial h}{r\partial \theta} \quad (1)$$

Here, h, μ, p, U, G_r, and G_θ represent the film thickness, viscosity coefficient, pressure, velocity of the fluid, and the flow rate coefficients in the radial and circumferential directions, respectively, and they are the functions of r and θ. The flow rate coefficients vary depending on the state of the fluid. For laminar flow, these coefficients are 12 in both radial and circumferential directions; for turbulent flow, they are $0.0392Re^{0.75}$ and $0.0687Re^{0.75}$, where Re is the Reynolds number. The film thickness varies depending on the deformation of the seal structure and can be expressed as shown in Equation (2).

$$h = h_c + h_g + h_d \quad (2)$$

where h_c, h_g, and h_d represent the minimum film thickness, groove depth, and elastic deformation of the seal face, respectively. The internal and external boundary conditions of the Reynolds equation are described in Equation (3). The external boundary conditions are applied to the inner radius and the outer radius. The periodic boundary condition implies that the pressure is continuous along the circumferential direction.

$$\begin{array}{l} p(r = r_i, \theta) = p_i \\ p(r = r_o, \theta) = p_o \\ p(r, \theta) = p(r, \theta + 2\pi) \end{array} \quad (3)$$

Here, p_i and p_o represent the internal and external pressures, respectively.

To develop the finite element formulation of the Reynolds equation, a weighting function is multiplied; partial integration is performed, and Green's theorem is employed to yield the following expression.

$$\int_\Omega \left(G_r \frac{h^3}{\mu} \frac{\partial w}{\partial r} \frac{\partial p}{\partial r} + G_\theta \frac{h^3}{\mu} \frac{\partial w}{r\partial \theta} \frac{\partial p}{r\partial \theta} \right) d\Omega - \int_\Omega \frac{Uh}{2} \frac{\partial w}{r\partial \theta} d\Omega = 0 \tag{4}$$

where Ω and w represent the interested domain and weighting function, respectively. The pressure of a four-node element can be defined through the nodal pressure $\mathbf{P}_e$, arbitrary vector $\mathbf{\eta}_e$, and shape function $\mathbf{N}$ and is expressed as shown in Equations (5) and (6).

$$p = \mathbf{N}^T \mathbf{P}_e \tag{5}$$

$$w = \mathbf{\eta}_e^T \mathbf{N} \tag{6}$$

By substituting Equations (5) and (6) into Equation (4), the following local finite-element equation is produced:

$$\mathbf{\eta}_e^T \int_\Omega \left(G_r \frac{h^3}{\mu} \frac{\partial \mathbf{N}}{\partial r} \frac{\partial \mathbf{N}^T \mathbf{P}_e}{\partial r} + G_\theta \frac{h^3}{\mu} \frac{\partial \mathbf{N}}{r\partial \theta} \frac{\partial \mathbf{N}^T \mathbf{P}_e}{r\partial \theta} \right) d\Omega - \mathbf{\eta}_e^T \int_\Omega \frac{Uh}{2} \frac{\partial \mathbf{N}}{r\partial \theta} d\Omega = 0 \tag{7}$$

Equation (7) can be rewritten by introducing the local element stiffness matrices and load vectors, as shown in Equations (8)–(10):

$$\mathbf{K}_e \mathbf{P}_e = \mathbf{F}_e \tag{8}$$

$$\mathbf{K}_e = \int_\Omega \left(G_r \frac{h^3}{\mu} \frac{\partial \mathbf{N}}{\partial r} \frac{\partial \mathbf{N}^T}{\partial r} + G_\theta \frac{h^3}{\mu} \frac{\partial \mathbf{N}}{r\partial \theta} \frac{\partial \mathbf{N}^T}{r\partial \theta} \right) d\Omega \tag{9}$$

$$\mathbf{F}_e = \int_\Omega \frac{Uh}{2} \frac{\partial \mathbf{N}}{r\partial \theta} d\Omega \tag{10}$$

Assembling the local finite element equations to construct the global finite element equation, applying the boundary condition, and solving the global finite element equation, we can determine the pressure of the fluid film. Then, the opening force, leakage rate, and heat generation can be calculated, as shown in Equations (11)–(13) [21,22].

$$F_{open} = \int_0^{2\pi} \int_{r_i}^{r_o} p r \, dr \, d\theta \tag{11}$$

$$Q = \int_0^{2\pi} G_r \frac{h^3}{\mu} \frac{\partial p}{\partial r} r \, d\theta \tag{12}$$

$$H = \int_{r_i}^{r_o} \int_0^{2\pi} \left(\frac{h}{2} \frac{\partial p}{r\partial \theta} + \frac{\mu U}{h} \right) r \, d\theta \, dr \times r\omega \tag{13}$$

2.2. Thermal Analysis in Fluid Film

Assuming convection at the boundary between the fluid film and the solid (based on the principle of energy conservation), the two-dimensional energy equation in the fluid film can be established as follows [23,24]:

$$\rho C_p \left[\left(\frac{\omega r h}{2} - \frac{G_\theta h^3}{\mu} \frac{\partial p}{r\partial \theta} \right) \frac{\partial T}{r\partial \theta} + \left(-\frac{G_r h^3}{\mu} \frac{\partial p}{\partial r} \right) \frac{\partial T}{\partial r} \right]$$
$$= \left[\frac{G_\theta h^3}{\mu} \left(\frac{\partial p}{r\partial \theta} \right)^2 + \frac{G_r h^3}{\mu} \left(\frac{\partial p}{\partial r} \right)^2 \right] + \frac{\mu (\omega r)^2}{h} + h_{fr}(T_r - T) + h_{fs}(T_s - T) \tag{14}$$

where C_p, T_r, T_s, h_{fr}, and h_{fs} represent the specific heat capacity of the fluid, the temperatures at the end face of the respective rotating seal and stationary seal, the convective heat transfer coefficient at the boundary of the rotating seal, and the convective heat transfer coefficient at the boundary of the stationary seal, respectively. The internal and external boundary conditions of Equation (14) in the fluid film are as described in Equation (15). The external boundary conditions include the internal temperature and external temperature, applied, respectively, at the inner radius and the outer radius. The periodic boundary condition implies that the temperature is continuous.

$$\begin{aligned} T(r = r_i, \theta) &= T_i \\ T(r = r_o, \theta) &= T_o \\ T(r, \theta) &= T(r, \theta + 2\pi) \end{aligned} \qquad (15)$$

where T_i and T_o represent the internal temperature and external temperature, respectively. To develop the finite element formulation of the energy equation, a weighting function is applied; partial integration is performed, and Green's theorem is applied to yield Equation (16).

$$\int_\Omega \rho C_p \left[\left(\frac{\omega r h}{2} \frac{\partial w}{r \partial \theta} T - \frac{G_\theta h^3}{\mu} \frac{\partial p}{r \partial \theta} \frac{\partial w}{r \partial \theta} T \right) - \left(\frac{G_r h^3}{\mu} \frac{\partial p}{\partial r} \frac{\partial w}{\partial r} T \right) \right] d\Omega$$
$$- \int_\Omega \left[\frac{G_\theta h^3}{\mu} w \left(\frac{\partial p}{r \partial \theta} \right)^2 + \frac{G_r h^3}{\mu} w \left(\frac{\partial p}{\partial r} \right)^2 \right] d\Omega - \int_\Omega w \frac{\mu(\omega r)^2}{h} d\Omega \qquad (16)$$
$$- \int_\Omega w \left[\{ h_{fr}(T_r - T) \} + \{ h_{fs}(T_s - T) \} \right] d\Omega = 0$$

The temperature of a four-node element can be defined through the nodal temperature T_e, arbitrary vector η_e, and shape function $\mathbf{N}$. With substitution and rearrangement, it is expressed as Equation (17).

$$\eta_e^T \int_\Omega \rho C_p \left[\begin{array}{c} \left(\frac{\omega r h}{2} \frac{\partial \mathbf{N}}{r \partial \theta} \mathbf{N}^T \mathbf{T}_e - \frac{G_\theta h^3}{\mu} \frac{\partial p}{r \partial \theta} \frac{\partial \mathbf{N}}{r \partial \theta} \mathbf{N}^T \mathbf{T}_e \right) \\ - \left(\frac{G_r h^3}{\mu} \frac{\partial p}{\partial r} \frac{\partial \mathbf{N}}{\partial r} \mathbf{N}^T \mathbf{T}_e \right) \end{array} \right] d\Omega$$
$$- \eta_e^T \int_\Omega \left[\frac{G_\theta h^3}{\mu} \mathbf{N} \left(\frac{\partial p}{r \partial \theta} \right)^2 + \frac{G_r h^3}{\mu} \mathbf{N} \left(\frac{\partial p}{\partial r} \right)^2 \right] d\Omega - \eta_e^T \int_\Omega \mathbf{N} \frac{\mu(\omega r)^2}{h} d\Omega \qquad (17)$$
$$- \eta_e^T \int_\Omega \mathbf{N} \left[\{ h_{fr}(T_r - \mathbf{N}^T \mathbf{T}_e) \} + \{ h_{fs}(T_s - \mathbf{N}^T \mathbf{T}_e) \} \right] d\Omega = 0$$

Here, the pressure gradient calculated from Equations (1)–(10) is used in Equation (17). The convective heat transfer coefficient at the lubrication surface can be calculated using the relationship expressed in Equations (18) and (19).

$$h_{fr} = h_{fs} = \frac{3 \mathrm{Pr}^{1/3} k}{h} \qquad (18)$$

$$\mathrm{Pr} = \frac{C_p \mu}{k} \qquad (19)$$

where Pr and k represent the Prandtl number and thermal conductivity, respectively.

The convective heat transfer coefficient at the outer diameter of the mechanical seal was derived through the empirical correlation of the Nusselt number in Equation (20), and the Nusselt number is calculated using Equation (21) [25].

$$Nu = 0.133 \mathrm{Re}^{2/3} \mathrm{Pr}^{1/3} \qquad (20)$$

$$Nu = \frac{h}{k/r_o} \quad (21)$$

The temperature of the fluid film is calculated by assembling the local finite element equations to construct the global finite element equation, applying the boundary condition, and solving the global finite element equation.

2.3. Heat Conduction and Elastic Deformation of the Seal Structure

The heat conduction equations for the stationary seal and the rotating seal are expressed as Equations (22) and (23), respectively; for the rotating seal, it includes the rotational component in the θ direction.

$$\frac{1}{r^2}\frac{\partial^2 T_s}{\partial \theta^2} + \frac{\partial}{\partial r}\frac{\partial T_s}{\partial r} + \frac{\partial^2 T_s}{\partial z^2} = 0 \quad (22)$$

$$\frac{k_r}{\rho_r C_{pr}}\left[\frac{1}{r^2}\frac{\partial^2 T_r}{\partial \theta^2} + \frac{\partial}{\partial r}\frac{\partial T_r}{\partial r} + \frac{\partial^2 T_r}{\partial z^2}\right] = \omega\frac{\partial T_r}{\partial \theta} \quad (23)$$

Here, T_s, T_r, k_r, ρ_r, and C_{pr} represent the temperature of the stationary seal, the temperature of the rotating seal, the thermal conductivity of the rotating seal, the density, and the specific heat capacity of the rotating seal, respectively.

Figure 2 illustrates the boundary conditions of the heat conduction equations for the lubricating fluid and the seal structure. An isothermal boundary was applied to the outer radius of the stationary seal, and a convective boundary with air was applied to the inner radius. For the rotating seal, a convective boundary with water coolant was applied to the outer radius, and an isothermal boundary was applied to the inner radius. On the lubrication surface, the heat flux, which is obtained by dividing the generated heat in Equation (13) by the element area, is applied to the lubrication surface. For the other parts, an insulating boundary was applied to calculate the heat transfer between the lubricating fluid and the solid [26,27].

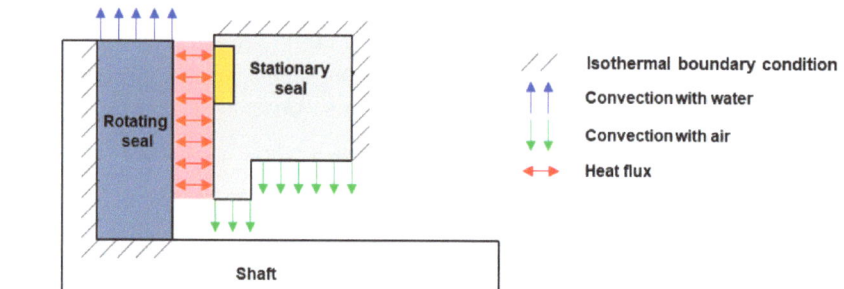

Figure 2. Heat transfer boundaries of a mechanical seal.

Figure 3 shows the pressure boundary conditions for the analysis of elastic deformation of the seal structure. External pressure was applied to the outer diameter of the stationary seal, and the internal pressure was applied to its inner diameter. Additionally, spring force was applied to the outer diameter of the stationary seal. External pressure was also applied to the outer diameter of the rotating seal, and the internal pressure was applied to its inner diameter. The pressure calculated from the Reynolds equation was applied to the lubricating surface. The other parts were subjected to a fixed boundary condition because they are fixed to the shaft and housing.

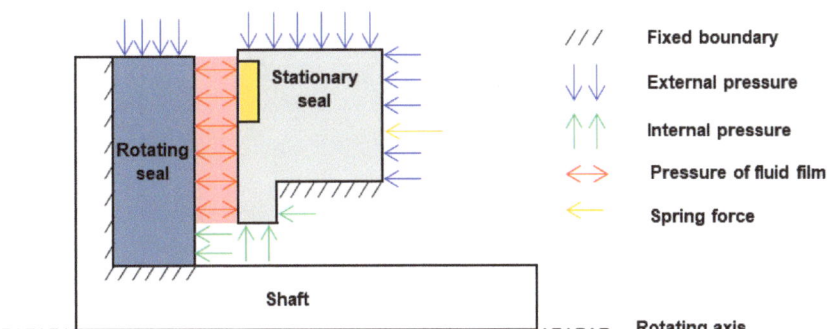

Figure 3. Pressure boundaries of a mechanical seal.

2.4. Numerical Algorithm

To calculate the lubrication characteristics at the high temperature and pressure encountered under ELAP operating conditions, it is necessary to consider the variation in the viscosity coefficient according to temperature. The viscosity coefficient depends on temperature and pressure. For liquid, it decreases with increasing temperature until a phase change occurs, after which the viscosity coefficient increases. However, under high pressure (where no phase change occurs), the viscosity coefficient increases with increasing pressure at the same temperature. In this study, Freeprop (available online: https://webbook.nist.gov/chemistry/) was used to determine the viscosity coefficient of each finite element depending on temperature and pressure.

Figure 4 shows the numerical procedure of this study. Lubrication analysis in the fluid film was conducted with a finite element program developed using C++. First, the Reynolds number was calculated on each finite element, and lubrication analysis was conducted based on laminar and turbulent flow. Subsequently, thermal analysis to determine the temperature of the fluid film was conducted using the finite element program developed using C++. In this state, we applied the heat generation obtained from the fluid film analysis to solve the energy equation. An iteration process was required to find the film thickness that satisfies the equilibrium state under the given operating conditions by calculating the opening force based on the initial film thickness (clearance) and comparing it to the closing force. The process was repeated until the force equilibrium state was achieved. The bisection method was used as the iteration method in this study. For this purpose, the minimum and maximum film thicknesses were defined, and the analysis was repeated until the difference between the opening force and the closing force for the calculated film thickness was less than 1×10^{-4}. Then, analyses of heat conduction and elastic deformation of the seal structure were conducted using ANSYS software 2022 R1. In this state, we applied the pressure from the fluid film analysis and the temperature distribution of the heat conduction analysis to solve the elastic deformation of the seal structure. These processes were repeated until the difference in total deformation of all nodes between the previous step and the current step was less than 1×10^{-4}.

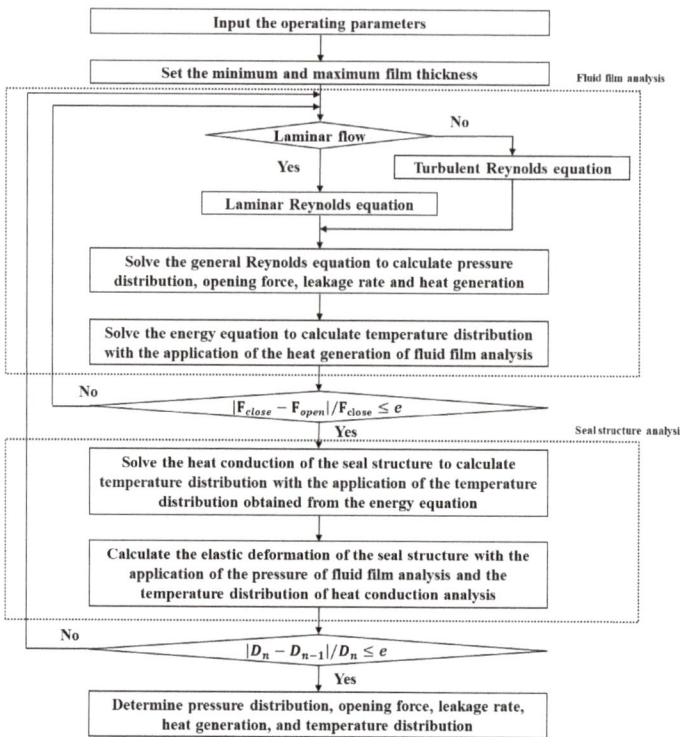

Figure 4. Numerical procedure of thermo-fluid–structural coupled analysis of a mechanical seal.

3. Numerical Verification

To validate the developed finite element programs for lubrication and thermal analyses and the proposed numerical procedure, we performed a thermo-fluid coupled analysis of the face seal, as analyzed by Tournerie et al. [28]. The external radius, internal radius, and tapering angle of this face seal were 45 mm, 40 mm, and 0.0001 rad, respectively. The rotating speed, pressure difference, fluid density, initial viscosity, and fluid thermal conductivity were 300 rad/s, 2 MPa, 850 kg/m^3, 0.001 Pas, and 0.14 W/mK, respectively. Our finite element model was constructed with 3000 quadrilateral elements. Figure 5 shows a comparison of our temperature distributions along the radial direction with those of Tournerie for film thicknesses of 2, 5, and 10 μm. It shows that our simulated temperature distributions are very close to those of Tournerie, within a discrepancy of 3%. Additionally, our simulated leakage rate for the film thickness of 10 μm is 1.182×10^{-6} m^3/s, very close to Tournerie's leakage rate of 1.175×10^{-6} m^3/s, with a discrepancy of 0.59%.

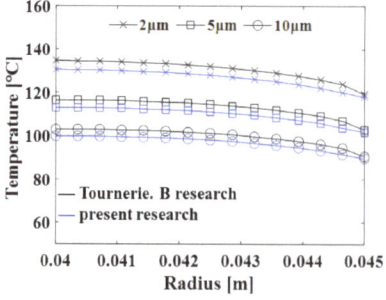

Figure 5. Temperature distribution according to film thickness.

4. Results and Discussion
4.1. Analysis Model

We applied the proposed method of the thermo-fluid–structural coupled analysis to investigate the lubrication characteristics of the mechanical seal under normal and ELAP operating conditions. Figure 6a shows the mechanical structure of a mechanical seal with a dodecagonal groove. The finite element model of the mechanical seal was developed with 144,000 quadrilateral elements. Figure 6b is the finite element model corresponding to the boxed area in Figure 6a, and it shows groove and ridge areas. The groove is engraved in the stationary seal, while the rotating seal has a plain shape. Table 1 shows the design variables and operating conditions used for the lubrication analysis of the mechanical seal. Table 2 shows the pressure and temperature boundary conditions of the mechanical seal under normal and ELAP conditions. The pressure differences between normal and ELAP operating conditions were 64 bar and 176 bar, respectively. We assumed that the fluid film in the clearance remained in a liquid state. Figure 7 shows the three-dimensional model to analyze the heat conduction and elastic deformation of the seal structure. In this research, we developed one-quarter of the three-dimensional model and applied a symmetric boundary condition. The gray area is made of silicon carbide, where lubrication occurs, while the rest is carbon graphite. We did not include the shaft and housing in the finite element model in the analysis of elastic deformation and heat conduction because they were much stiffer than that of the carbon graphite and because the thermal conductivity of the shaft and housing was almost equal to that of the carbon graphite, and they were located far from the heat source. Table 3 shows the material properties of the seal structure in the analysis of heat conduction and elastic deformation.

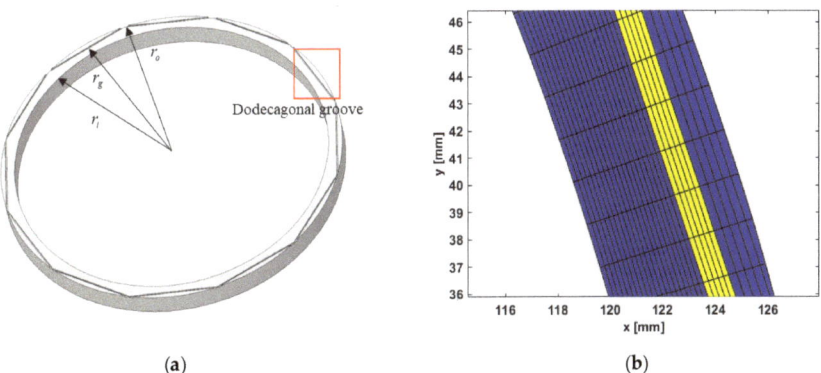

Figure 6. 3D Geometry (**a**) and FE model (**b**) of a mechanical seal for RCP for lubrication and thermal analyses.

Table 1. Design variables and operating conditions of the mechanical seal.

Parameter	Value
Outer radius, r_o [mm]	131
Groove radius, r_g [mm]	128.7
Inner radius, r_i [mm]	125
Balance radius, r_b [mm]	127
Radial groove length [mm]	1
Number of grooves [ea]	12
Groove depth [mm]	1
Fluid specific heat capacity, C_p [J/kgK]	4184
Fluid thermal conductivity, k [W/mK]	0.592
Rotating speed, ω [rpm]	1200

Table 2. Pressure and temperature boundary conditions of the mechanical seal under normal and ELAP operating conditions.

Parameter	Normal Condition	ELAP Condition
External pressure, p_o [bar]	64	176
Internal pressure, p_i [bar]	1	1
Seal fluid temperature, T_o [K]	313	583
Ambient temperature, T_i [K]	313	583
Fluid density, ρ [kg/m^3]	995.02	709.76
Fluid viscosity, μ [μPas]	655.38	85.44

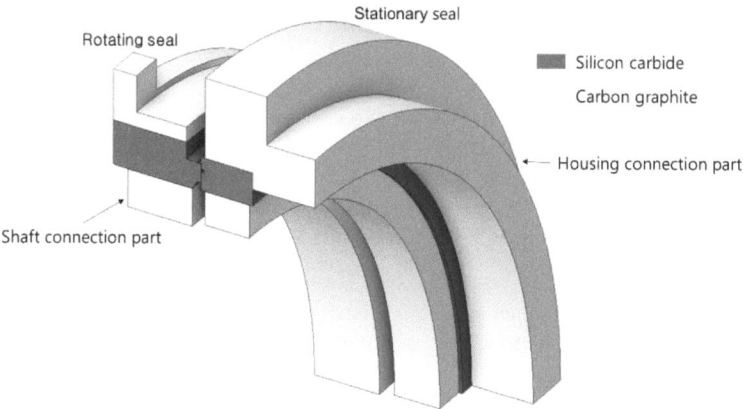

Figure 7. 3D Geometry of a mechanical seal for structural elastic deformation analysis.

Table 3. Material properties of the seal structure in the analysis of the heat conduction and elastic deformation of the seal structure.

Parameter	Carbon Graphite	Silicon Carbide
Density, ρ_r [kg/m^3]	1800	3100
Young's modulus [GPa]	25	400
Poisson's coefficient	0.2	0.17
Specific heat capacity, C_{pr} [J/kgK]	710	400
Thermal conductivity, k_r [W/mK]	15	150
Linear thermal expansion coefficient [/10^{-6} °C]	4	4.3

4.2. Characteristics of a Mechanical Seal under Normal Operating Conditions

A thermo-fluid–structural coupled analysis was conducted to analyze the characteristics of a mechanical seal under normal operating conditions. In this normal operating condition, the proposed numerical procedure in Figure 4 was converged after nine iterations, with an average clearance of 7.31 μm. Figure 8 shows the Reynolds number along the radial direction at 0° and 10°. Because there is no groove along the radius at 0°, the Reynolds number is lower than Re = 1000, and no turbulence occurs. However, there is a groove along the radius at 10°, and the Reynolds number is higher than Re = 1000 in the groove area, causing turbulence in the groove area. Figure 9a shows the pressure distribution on the lubrication surface. Under normal operating conditions, the opening force, maximum pressure, and leakage rates were 22,946 N, 64.03 bar, and 8.63 × 10^{-5} m^3/s, respectively. The maximum pressure is very close to the external pressure because the wedge effect caused by the groove was weak. Figure 9b shows the temperature distribution of the fluid film. The maximum temperature of 335 K occurred at the inner diameter of the seal, and the minimum temperature of 321 K occurred within the groove due to the lubricating fluid within the groove depth.

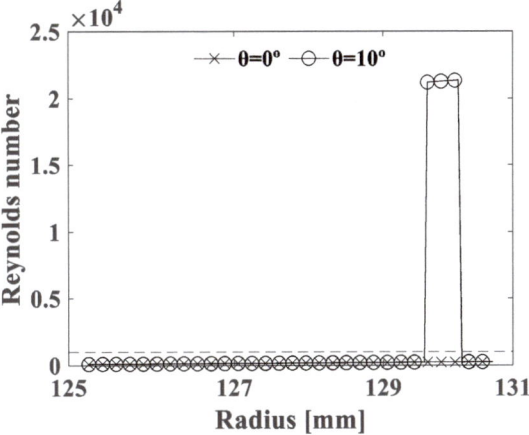

Figure 8. Reynolds number of the fluid film at θ = 0° and θ = 10° under normal operating conditions.

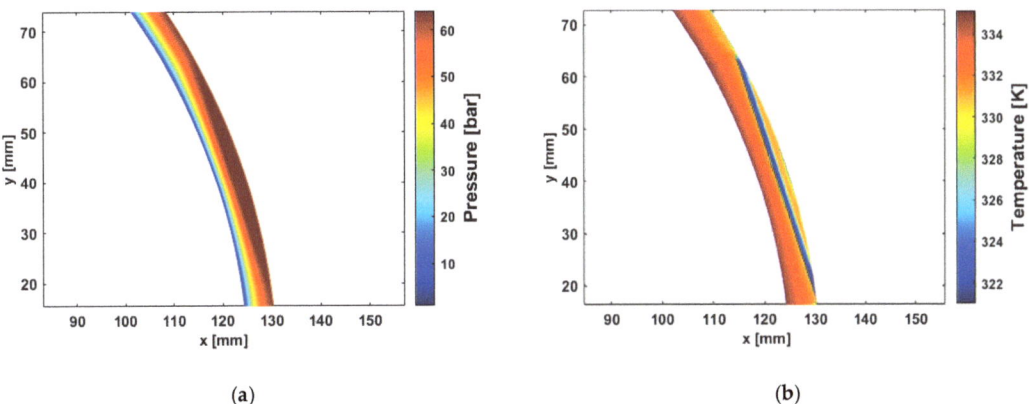

Figure 9. Pressure (**a**) and temperature distributions (**b**) in the fluid film of the mechanical seal under normal conditions.

Figure 10 shows the temperature distribution of the seal structure due to heat conduction. The maximum temperature of 335.4 K occurs at the middle lower part of the stationary seal, where a relatively high temperature is observed due to the low convective heat transfer coefficient of air. At the middle upper part of the rotating seal, a relatively low temperature is observed due to the high convective heat transfer coefficient of water. Figure 11 shows the elastic deformation of the mechanical seal under normal operating conditions. The maximum deformation of 4.5 μm occurs at the top of the stationary seal because the pressure of 64 bar at the outer diameter of both the stationary and rotating seals is much greater than the pressure of 1 bar at the inner diameter. Figure 12 shows the clearance along the radial direction due to temperature, pressure, and their interaction. When pressure is only applied, the clearance changes from a maximum of 4.92 μm at the outer radius to a minimum of 3.46 μm near the middle, and the clearance at the outer radius is slightly greater than that at the inner radius because of high external pressure. When the temperature is applied, the clearance changes from a maximum of 7.48 μm at the inner radius to a minimum of 4.7 μm near the middle, and the clearance at the inner radius is greater than that at the outer radius because of the high inner temperature. Temperature changes the clearance more significantly than pressure. When both pressure and temperature are applied, clearance changes from the maximum of 9.77 μm at the inner radius to the minimum of 6.47 μm near the middle, and the clearance at the inner radius is greater than that at the outer radius, mostly due to high inner temperature.

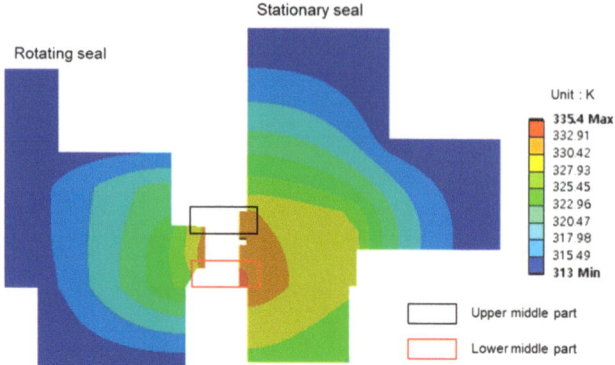

Figure 10. Temperature distribution of the mechanical seal under normal operating conditions.

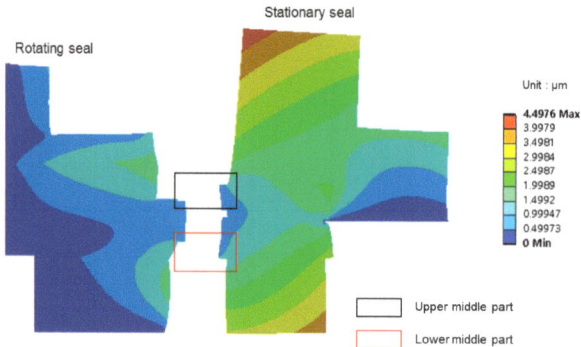

Figure 11. Elastic deformation of the mechanical seal under normal operating conditions.

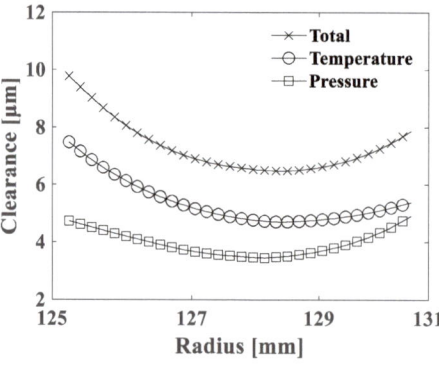

Figure 12. Clearance along the radial direction due to temperature, pressure, and their interaction under normal operating conditions.

4.3. Characteristics of a Mechanical Seal under ELAP Operating Conditions

The characteristics of a mechanical seal under ELAP operating conditions were analyzed by using the proposed thermo-fluid–structural coupled analysis. As shown in Figure 4, the proposed numerical procedure was converged to an average clearance of 4.91 μm after 23 iterations. Because the groove depth affects the Reynolds number more significantly than clearance change, turbulence occurs only at the groove area in the ELAP operating condition. Figure 13a shows the pressure distribution on the lubrication surface, and it shows that the maximum pressure is very close to the external pressure because the wedge effect caused by the groove is weak. Under ELAP operating conditions, the opening force, maximum pressure, and leakage rate are 61,564 N, 176 bar, and 9.7×10^{-5} m^3/s, respectively. Figure 13b shows the temperature distribution of the fluid film. The maximum temperature of 660 K occurs at the inner diameter of the seal, and the minimum temperature of 635 K occurs within the groove due to the lubricating fluid within the groove depth. According to the NIST, the coolant (water) remains in a liquid state up to 628.29 K at 176 bar and up to 372.76 K at 1 bar. Vaporization of the fluid film may occur where the temperature of the fluid film rises over the boiling temperature of water, as shown in Figure 13. Figure 14 shows the temperature distribution of the seal structure due to heat conduction. The maximum temperature of 669.3 K occurs at the middle lower part of the stationary seal with the distribution of a relatively high temperature due to a low convective heat transfer coefficient of air. A relatively low temperature occurred at the middle upper part of the rotating seal due to the high convective heat transfer coefficient of water.

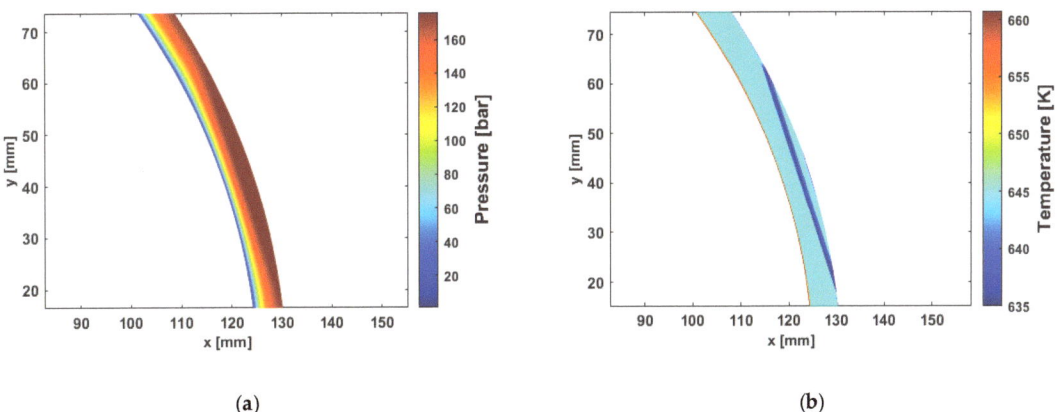

Figure 13. Pressure (**a**) and temperature distributions (**b**) in the fluid film of the mechanical seal under ELAP operating conditions.

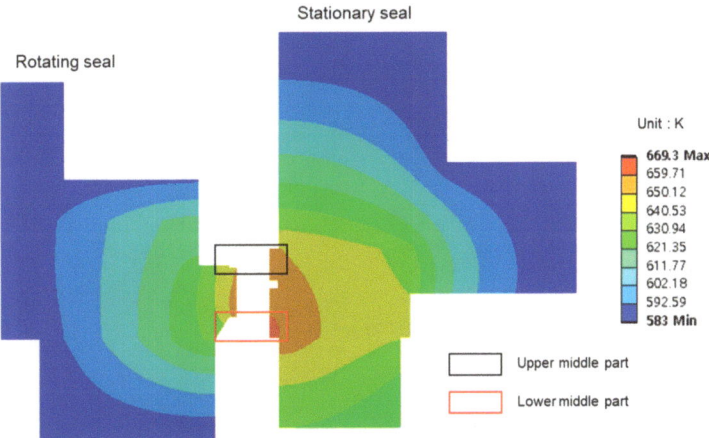

Figure 14. Temperature distribution of the mechanical seal under ELAP operating conditions.

Figure 15 shows the elastic deformation of the mechanical seal under ELAP operating conditions, and the deformed shape of the mechanical seal under ELAP operating conditions is very similar to that under normal operating conditions. However, the maximum deformation of 60.5 μm occurs at the top of the stationary seal because the high pressure of 176 bar at the outer diameter of both the stationary and rotating seals is much greater than the pressure of 1 bar at the inner diameter. Figure 16 shows the clearance along the radial direction due to temperature, pressure, and their interaction, respectively. The maximum clearances due to pressure, temperature, and their interaction are 4.59 μm at the outer radius, 5.99 μm at the inner radius, and 7.37 μm at the inner radius. The minimum clearances due to pressure, temperature, and their interaction are 3.09 μm, 3.8 μm, and 3.69 μm near the middle, and it shows that the pressure additionally decreases the minimum clearance near the middle when both temperature and pressure are considered. In total, elastic deformation of the seal structure happens near clearance, and the temperature has a more significant influence than pressure in ELAP operating conditions. Figure 17 shows the clearance of the fluid film along the radial direction under normal and ELAP operating conditions, and it shows that the clearance due to high temperature and pressure under ELAP operating conditions is much smaller than that under normal operating conditions. However, the leakage rate under ELAP operating conditions (9.7×10^{-5} m^3/s) slightly increased in comparison to that under normal operating conditions (8.63×10^{-5} m^3/s) due to the high internal pressure.

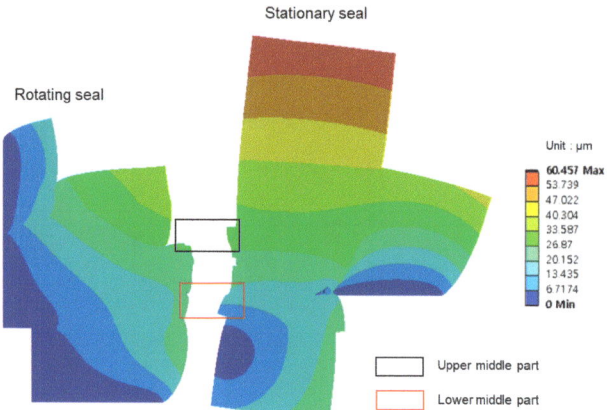

Figure 15. Elastic deformation of the mechanical seal under ELAP operating conditions.

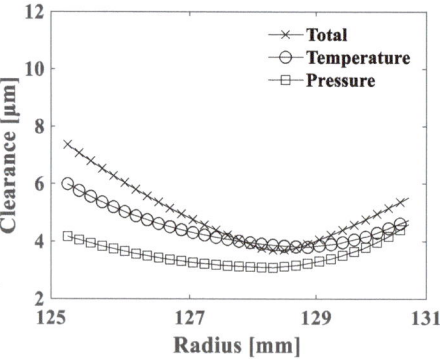

Figure 16. Clearance along the radial direction due to temperature, pressure, and their interaction under ELAP operating conditions.

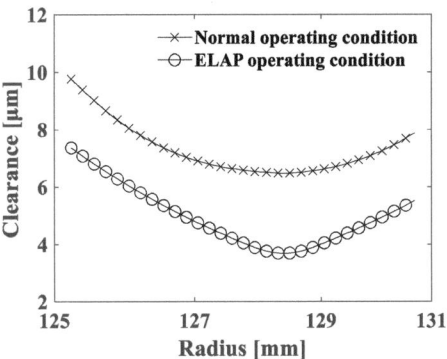

Figure 17. Clearance of the fluid film along radial direction under normal and ELAP operating conditions.

5. Conclusions

In this paper, we proposed a thermo-fluid–structural coupled analysis method to predict the pressure and temperature distributions, the change in the fluid film clearance due to elastic deformation of the mechanical seal, and the resulting leakage rate under normal and ELAP operating conditions of the RCP. We developed a finite element program for the general Reynolds equation, including the turbulence effect to calculate the pressure, opening force, and leakage rate of fluid lubricant and the two-dimensional energy equation to calculate the temperature distribution of the fluid lubricant. The accuracy of the developed program was verified by comparing the simulated temperature distribution and leakage rate with those of previous research. Heat conduction and elastic deformation due to pressure and temperature changes at the seal structure were analyzed using the ANSYS program. The results show that temperature more significantly affects the elastic deformation of the seal structure near clearance compared to pressure under both normal and ELAP operating conditions. High temperature and pressure of the coolant under ELAP operating conditions deformed the seal structure and resulted in a much smaller clearance of the fluid film than in normal operating conditions. Even in the small clearance of ELAP operating conditions, the leakage rate increased due to high internal pressure and low viscosity. A high temperature of the coolant in the seal structure under ELAP operating conditions rises over the boiling temperature of the coolant, which may cause the vaporization of the coolant. The effect of two-phase flow in the fluid film under ELAP operating conditions will be a future research topic to accurately analyze the fluid film and the thermo-fluid–structural coupled analysis of the seal structure under ELAP operating conditions. This research will contribute to the development of robust mechanical seals for RCP by accurately predicting the characteristics of mechanical seals, especially under ELAP in the RCP.

Author Contributions: Conceptualization, G.J.; methodology, Y.P. and G.H.; software, Y.P. and G.H.; validation, Y.P.; formal analysis, T.K. and G.H.; investigation, Y.P. and M.K.; resources, M.K. and T.K.; data curation, J.C. and S.J.; writing—original draft preparation, Y.P. and G.H.; writing—review and editing, G.J.; visualization, Y.P. and G.H.; supervision, G.J.; project administration, G.J.; funding acquisition, G.J. All authors have read and agreed to the published version of the manuscript.

Funding: This work was supported by Korea Energy Technology Evaluation and Planning (KETEP) grant funded by the Ministry of Trade, Industry, and Energy (MOTIE) (20222B10100010, Development of silicon carbide composite material technology with improved surface lubricity, 2023). This work was supported by Korea Research Institute for Defense Technology planning and advancement (KRIT) grant funded by Defense Acquisition Program Administration (DAPA) (No. KRIT-CT-21-008, Predicting precise leakage rates of carbon seals and technology of designing aerodynamic patterns for improving lifetime, 2023).

Data Availability Statement: Data are contained within the article.

Conflicts of Interest: Authors Sanghyun Jun, Jeongmook Choi, Taegyu Kim and Minsoo Kang were employed by the company Department of Engineering, Flowserve KSM. The remaining authors declare that the research was conducted in the absence of any commercial or financial relationships that could be construed as a potential conflict of interest.

Nomenclature

D_n	Total elastic deformation of all nodes at the n-th iteration
F_{close}	Closing force [N]
F_{open}	Opening force [N]
G_r	Fluid state coefficients for radial flow
G_θ	Fluid state coefficient for circumferential flow
h	Film thickness [mm]
h_c	Minimum film thickness [mm]
h_g	Groove depth [mm]
h_d	Deformation amount [mm]
p	Pressure of fluid film [bar]
T	Temperature of fluid film [K]
U	Fluid velocity [m/s]
w	Weighting function
μ	Viscosity [Pa·s]
ρ	Density [kg/m^3]
η	Arbitrary vector
$\mathbf{N}$	Shape function vector
$\mathbf{P}_e$	Element pressure vector [bar]
$\mathbf{T}_e$	Element temperature vector [K]
Ω	Interested domain
ω	Rotating velocity [rad/s]
Re	Reynolds number
Pr	Prandtl number
Nu	Nusselt number

References

1. Yang, J.H.; Wang, J.R.; Shih, C.; Huang, C.F.; Chen, S.W. The simulation and study of ELAP event with URG and FLEX mitigation strategies for PWR by using TRACE code. *Kerntechnik* **2019**, *84*, 72–83. [CrossRef]
2. Thomas, S.; Brunetiere, N.; Tournerie, B. Numerical modeling of high pressure gas face seals. *J. Tribol.* **2006**, *128*, 396–405. [CrossRef]
3. Thomas, S.; Brunetiere, N.; Tournerie, B. Thermoelastohydrodynamic behavior of mechanical gas face seals operating at high pressure. *J. Tribol.* **2007**, *129*, 841–850. [CrossRef]
4. Brunetiere, N.; Thomas, S.; Tournerie, B. The parameters influencing high-pressure mechanical gas face seal behavior in static operation. *Tribol. Trans.* **2009**, *52*, 643–654. [CrossRef]
5. Nyemeck, A.P.; Brunetiere, N.; Tournerie, B. A mixed thermoelastohydrodynamic lubrication analysis of mechanical face seals by a multiscale approach. *Tribol. Trans.* **2015**, *58*, 836–848. [CrossRef]
6. Mosavat, M.; Moradi, R.; Takami, M.R.; Gerdroodbary, M.B.; Ganji, D.D. Heat transfer study of mechanical face seal and fin by analytical method. *Eng. Sci. Technol. Int. J.* **2018**, *21*, 380–388. [CrossRef]
7. Su, H.; Rahmani, R.; Rahnejat, H. Thermohydrodynamics of bidirectional groove dry gas seals with slip flow. *Int. J. Therm. Sci.* **2016**, *110*, 270–284. [CrossRef]
8. Su, W.; Liu, W.; Yan, L.; Zhang, Y. Thermal performances of the wave-tilt-dam seal in a reactor coolant pump for sustainable development. *Sustain. Energy Technol. Assess.* **2022**, *52*, 102042. [CrossRef]
9. Brunetiere, N.; Tournerie, B.; Frene, J. TEHD lubrication of mechanical face seals in stable tracking mode: Part 1—Numerical model and experiments. *J. Tribol.* **2003**, *125*, 608–616. [CrossRef]
10. Brunetiere, N.; Tournerie, B.; Frene, J. TEHD lubrication of mechanical face seals in stable tracking mode: Part 2—Parametric study. *J. Tribol.* **2003**, *125*, 617–627. [CrossRef]
11. Brunetiere, N.; Rouillon, M. Fluid flow regime transition in water lubricated spiral grooved face seals. *Tribol. Int.* **2021**, *153*, 106605. [CrossRef]
12. Mo, H.; Hu, Y.; Quan, S. Thermo-hydrodynamic lubrication analysis of slipper pair considering wear profile. *Lubricants* **2023**, *11*, 190. [CrossRef]

13. Grun, H.; Feldmeth, S.; Bauer, F. Multiphase computational fluid dynamics of rotary shaft seals. *Lubricants* **2022**, *10*, 347. [CrossRef]
14. Liu, Y.; Liu, W.; Li, Y.; Liu, X.; Wang, Y. Mechanism of a wavy-tilt-dam mechanical seal under different working conditions. *Tribol. Int.* **2015**, *90*, 43–54. [CrossRef]
15. Galenne, E.; Pierre, D.I. Thermo-elasto-hydro-dynamic modeling of hydrostatic seals in reactor coolant pumps. *Tribol. Trans.* **2007**, *50*, 466–476. [CrossRef]
16. Su, W.T.; Wang, Y.H.; Feng, X.D.; Li, X.B. Thermal-liquid–solid coupling characteristics of wavy-end-face mechanical seal for reactor coolant pump. *Nucl. Eng. Des.* **2023**, *414*, 112545. [CrossRef]
17. Wang, J.L.; Chen, X.Y.; Binama, M.; Su, W.T.; Wu, J. A numerical study on mechanical seal dynamic characteristics within a reactor coolant pump. *Front. Energy Res.* **2022**, *10*, 879198. [CrossRef]
18. Srivastava, G.; Chiappa, P.; Shelton, J.; Higgs, C.F. A thermo-elasto-hydrodynamic lubrication modeling approach to the operation of reactor coolant pump seals. *Tribol. Int.* **2019**, *138*, 487–498. [CrossRef]
19. Meng, X.; Qiu, Y.; Ma, Y.; Peng, X. An investigation into the thermo-elasto-hydrodynamic effect of notched mechanical seals. *Nucl. Eng. Technol.* **2022**, *54*, 2173–2187. [CrossRef]
20. Huang, W.; Liao, C.; Liu, X.; Suo, S.; Liu, Y.; Wang, Y. Thermal fluid-solid interaction model and experimental validation for hydrostatic mechanical face seals. *Chin. J. Mech. Eng.* **2014**, *27*, 949–957. [CrossRef]
21. Hirs, G.G. A bulk-flow theory for turbulence in lubricant films. *J. Lubr. Technol.* **1973**, *95*, 137–145. [CrossRef]
22. Hirs, G.G. A systematic study of turbulent film flow. *J. Tribol.* **1974**, *96*, 118–126. [CrossRef]
23. Ma, C.; Bai, S.; Peng, X. Thermo-hydrodynamic characteristics of spiral groove gas face seals operating at low pressure. *Tribol. Int.* **2016**, *95*, 44–54. [CrossRef]
24. Yu, B.; Hao, M.; Xinhui, S.; Wang, Z.; Fuyu, L.; Yongan, L. Analysis of dynamic characteristics of spiral groove liquid film seal under thermal–fluid–solid coupling. *Ind. Lubr. Tribol.* **2021**, *73*, 882–890. [CrossRef]
25. Becker, K.M. Measurements of convective heat transfer from a horizontal cylinder rotating in a tank of water. *Int. J. Heat Mass Transf.* **1963**, *6*, 1053–1062. [CrossRef]
26. Wu, D.; Jiang, X.; Yang, S.; Wang, L. Three-dimensional coupling analysis of flow and thermal performance of a mechanical seal. *J. Therm. Sci. Eng. Appl.* **2014**, *6*, 196–204. [CrossRef]
27. Parviz, M.; Nori, A.O.; Robert, L.P.; Larry, E.J. Experimental and computational investigation of flow and thermal behavior of a mechanical seal. *Tribol. Trans.* **1999**, *42*, 731–738. [CrossRef]
28. Tournerie, B.; Danos, J.C.; Frene, J. Three-dimensional modeling of THD lubrication in face seals. *J. Tribol.* **2000**, *123*, 196–204. [CrossRef]

Disclaimer/Publisher's Note: The statements, opinions and data contained in all publications are solely those of the individual author(s) and contributor(s) and not of MDPI and/or the editor(s). MDPI and/or the editor(s) disclaim responsibility for any injury to people or property resulting from any ideas, methods, instructions or products referred to in the content.

Article

Criteria for Evaluating the Tribological Effectiveness of 3D Roughness on Friction Surfaces

Oleksandr Stelmakh [1], Hongyu Fu [1], Serhii Kolienov [2], Vasyl Kanevskii [3], Hao Zhang [1,*], Chenxing Hu [1] and Valerii Grygoruk [2]

1 School of Mechanical Engineering, Beijing Institute of Technology, Beijing 100081, China
2 Educational Scientific Institute of High Technologies, Quantum Radiophysics Department, Taras Shevchenko National University of Kyiv, 01601 Kyiv, Ukraine
3 Chuiko Institute of Surface Chemistry, National Academy of Sciences of Ukraine, 03164 Kyiv, Ukraine
* Correspondence: hao_zhang@bit.edu.cn

Abstract: A new technique for finishing the surfaces of friction pairs has been proposed, which, in combination with the original test method, has shown a significant influence of the initial roughness configuration (surface texture) on friction and wear. Two types of finishing processing of the shaft friction surfaces were compared, and it was found that the friction and wear coefficients differ by more than 2–5 and 2–4 times, respectively. Based on a new methodology for analyzing standard roughness parameters, the tribological efficiency criteria (in the sense of reducing friction and wear) are proposed for the initial state of the friction surface of a radial plane sliding bearing shaft relative to the friction direction, which is consistent with its frictional characteristics. Comparison of the laboratory test results with the surface tribological efficiency criteria showed that these criteria are very promising for controlling existing technologies and optimizing new technologies for friction surface finishing in various friction systems.

Keywords: roughness; friction surface; wear; angular spectrum; bearing area curve

1. Introduction

From a tribology perspective, mechanical devices consist of various friction units with distinct contact geometry and friction kinematics. The overall efficiency and operability of the product depend on the efficiency, reliability, and durability of each of these friction units, especially in case of critical wear [1]. A special place among them is occupied by aviation fuel control equipment, on which flight safety directly depends. These units use fuel as a lubricating medium, making it impossible to lubricate tribological contacts with modern oils that have highly effective additive packages. Therefore, traditional approaches are used to increase the reliability and service life of such units. These include selecting a material combination for the parts of the friction pairs, using various chemical and thermal technologies for processing friction surfaces, and applying various effective coatings through various physico-chemical methods [2,3].

The technology of finishing friction surfaces is the most common method to improve tribological properties in the manufacture of friction parts, especially fuel control equipment on surfaces, which can retain boundary-lubricating layers. This texturing technique is widely used in various sliding bearings, as evidenced by numerous publications [4–9] and their references. However, there are significantly fewer publications dedicated to shaft texturing [10–13].

The microgeometry of the rough surfaces of contacting parts changes rapidly at the start of the tribological interaction process [14]. At the same time, the intensive processes that occur during the initial stage of run-in, such as microplastic deformation, local thermal flares (triboplasma), and desorption of boundary layers, significantly affect the behavior of the tribological contact, including wear, friction force, and durability [15]. Therefore, the reliability and service life of a product are not only determined by the accuracy of its size or shape but also by the state of the micro- or nano-geometry of the surface layer.

The adhesion–deformation or molecular–mechanical theory of friction and wear has been widely used to describe the mechanisms and causes of wear of rubbing surfaces [16]. According to this theory, the boundary layers of lubricant are almost solid bodies that are destroyed by tangential mechanical stresses and heat flows. However, this contradicts the fundamental theories of boundary layer adsorption. Based on the identified fundamental phenomena of extrusion and rarefaction in boundary layers of lubricant, we propose an adhesion–deformation–hydrodynamic (ADH) model of friction and wear [17]. The ADH model explains that the primary cause of adhesion interaction and destruction of friction surfaces is the desorption of boundary layers in diffuser elastically deformed discrete micro- and macro-curved contacts. Rarefaction occurs there under the influence of negative gradients of contact stresses. Therefore, the diffuser elastically deformed regions are the most dangerous, as rarefaction always occurs there during friction. Intense desorption of the boundary layers causes quasi-dry friction conditions, leading to primary adhesion and subsequent tearing of the material from the bearing surface.

To prevent the destructive effects of friction in diffuser regions, the ADH model of friction and wear proposes creating a structure of protrusions and indentations on the surface of the shaft. This structure could increase the pressure in the diffuser zones of microcontacts. This approach was tested in [18]. The authors suggest that this is due to the lateral flow of the boundary lubricant from areas of high pressure, such as the confusor zones, to areas of lower pressure, such as the diffuser microcontact zones, through bypass channels on the bearing.

This paper presents an approach based on the ADH model to create such closed micro-crater zones on the friction surface of the shaft, in the depressions of which, with natural pre-compression in the confusor and transition zones of the elastically deformed contact, it is possible to increase the pressure of the boundary lubricant. It is assumed that in the rarefied zones of tribological contacts, pressure increases due to the flow of previously compressed lubricant fragments from closed cavities through the contact contours. This results in a decrease in the intensity of their desorption in accordance with the Langmuir and BET (Brunauer–Emmett–Teller) adsorption isotherms.

To implement the proposed approach, two objectives must be met: to develop a method for creating micro-craters on the surface of the shaft and to determine methods for evaluating their 3D configuration relative to the direction of friction.

The technological process for creating the friction surfaces of sliding bearing shafts involves sequentially grinding, fine-tuning, and polishing at high rotational speeds using special lapping and abrasive media of various grain sizes. The resulting mirrored surfaces are believed to have better tribological properties when friction occurs in the bearing. However, it has been observed that such tribological sliding systems experience significant adhesive–deformation wear during operation. To improve wear resistance and decrease friction force, we propose creating micro-crater formations on the mirrored surfaces, as suggested by the ADH model of friction and wear. The primary objective in creating surfaces with uniformly distributed micro-crater formations on the shafts is to utilize a specific abrasive medium and finishing modes.

The task also involves determining the tribological properties of surfaces after finishing and establishing their relationship with the initial roughness parameters. The tribological properties of the surface are determined through laboratory tests using specialized equipment, for example, as in [17]. These tests ensure that identical samples experience the same type of tribological contact under the same initial conditions throughout the friction process of the model shaft along the model bearing, including during a single revolution.

To determine the surface roughness parameters, a well-known control and analysis methods according to ISO 4287 [19], ISO 13565 [20], and ISO 25178 [21] are utilized. Modern measuring equipment enables the acquisition of information regarding the volumetric configuration of nano-roughness and its corresponding parameters in 3D space. As demonstrated in [22], this information provides a more comprehensive understanding of the statistical properties of roughness than existing standard roughness parameters.

Preliminary research suggests that these systems can be used to measure local surface defects, including assessing volumetric wear of machine parts. This is particularly important in laboratory tribological tests [18]. The 3D parameters or characteristics obtained from this information should allow for both qualitative and quantitative evaluation of the variation in roughness properties in different directions for shaft surfaces, whether obtained by traditional or proposed new finishing techniques.

The purpose of this paper is therefore (1) to experimentally confirm the possibility of reducing the frictional force and increasing the wear resistance of frictional surfaces by applying a surface-finishing technology that provides the necessary configuration of the nano-geometry of the roughness of the shaft, and (2) to search for parameters of the volumetric roughness configuration that allow for the evaluation of the tribological efficiency of the surface in terms of reducing friction and wear, taking into account the direction of friction.

2. Objects, Materials, and Research Method

The Timken linear contact scheme was chosen for laboratory modeling of friction pairs because it corresponds to the operation of the sliding friction unit, which is widely used in mechanical engineering, including plunger pumps, multiple spool pairs, and other friction units of fuel control equipment of various types of engines. This scheme is also the most problematic in terms of wear resistance. To ensure good repeatability of the results and avoid misinterpretation of data after testing, it is extremely important to ensure several important requirements regarding the preparation of test samples and the test methodology itself. We used a technique for processing the surfaces of shafts assembled in one cassette that allows us to create a certain texture on their rubbing surfaces with a high degree of uniformity. Also, when physically modeling tribocontacts, our test methodology assumes the need to ensure constant contact and constant contact stresses at any time of friction.

The BIT-TRIBO-01 laboratory system, shown in Figure 1, was chosen from the available Timken friction-testing machines [23]. The BIT-TRIBO-01 friction laboratory setup is designed to determine the wear and friction resistance of lubricants and their additives, structural materials, and coatings. It also evaluates the effect of friction surface texture on their friction performance by testing a model friction system with a single linear contact (see Figure 2). Its notable feature is that this single contact is complete and constant within a circle of the axis, which closely corresponds to its theoretical ideas. On this device, the counter-sample 1 (model shaft) slides over the surface of the planar sample 2 while maintaining a constant contact length (indicated by the white contact line) (see Figure 2). At the same time, displacement of the contact line relative to the sensitive element of the friction force measurement sensor is almost completely eliminated. The design of the device is such that all axes along which the contact may have changed (OX, OY, and OZ) intersect at the centroid of the counter-sample 1—at point O (Figure 2). Figure 2 shows the axis of action of load N at the center of linear contact (OY axis). Figure 2 also shows the OX_1 and OZ_1 axes associated with the orientation of the linear contact. At the same time, all axes are coordinated to the centroid O of the model shaft. Therefore, at the slightest displacement of the contact, all axes simultaneously deviate relative to the centroid of the counter-sample O, maintaining perpendicularity to each other and providing constant and complete single linear contact.

Figure 1. Laboratory device BIT-TRIBO-01 for tribological tests.

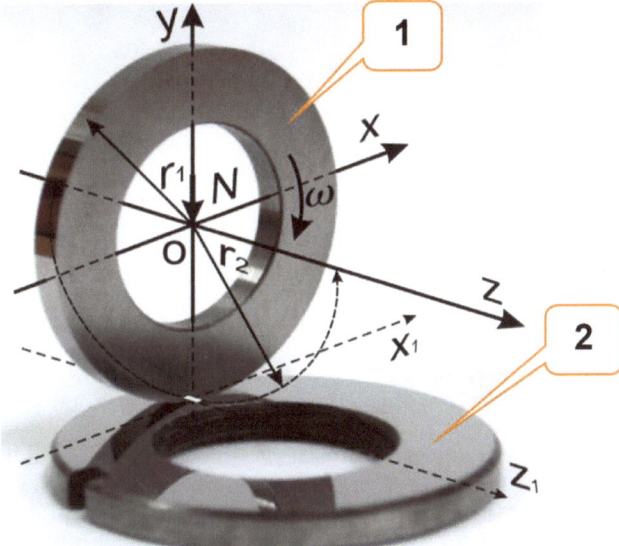

Figure 2. Friction pair: 1—counter-sample (model shaft); 2—fixed sample (model bearing).

The BIT-TRIBO-01 device achieved high reproducibility of friction and wear laboratory test results due to the high stability of the initial contact conditions. Additionally, this device allows for observation of a uniform secondary structure on the friction surfaces of counter-samples that corresponds to the properties of the friction pair material, lubricant, and surface texture under specified friction conditions. This is a difficult observation to make for other devices with Timken schemes.

Flat stationary specimens (model flat bearings) were made of steel G52986 with a hardness of HRC 59–62 and the nano-sized roughness of a model flat bearing. The surface roughness, determined by the Ra parameter, was less than 0.01 μm and achieved through sequential multistage polishing on cast iron lapping plates with different diamond pastes ranging from 40 μm to 0.5 μm in grain size.

Two types of friction pairs were used for tribological tests within the framework of this paper, differing only in the friction surfaces of the model shafts. The first type had surfaces obtained through the traditional method of grinding and polishing with diamond paste AFM. The second type had surfaces obtained through additional finishing of previously polished shafts using silicon carbide and the same laps.

Counter-samples that simulate the shaft were prepared from steel G52986 with HRC 59-62 hardness. The surfaces of the samples were prepared using glass laps with increased flatness (no more than 0.5 μm on an area of 120×40 mm^2). The laps were made using the well-known three-plate method and manually in a suspension of silicon carbide (5 μm) in mineral oil with a ratio of 1:3. More than a dozen control samples were collected in cassettes, fixed, and ground in processing centers (see Figure 3a). The cassettes were manually polished using various diamond paste AFMs and rotated at a frequency of 1000 rpm. This followed the traditional step-by-step diamond-polishing technology with a grain size ranging from 40 μm to 0.5 μm. As a result, we obtained several highly polished model shafts.

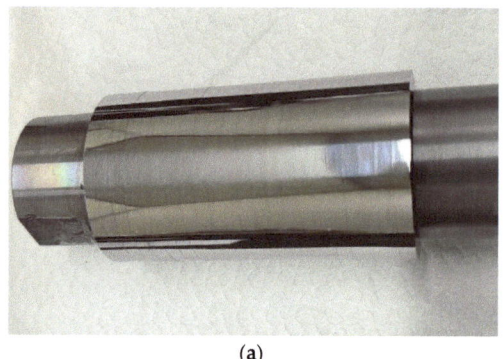

(a) (b)

Figure 3. Appearance of a cassette tape with 22 counter-samples. After the usual diamond slurry finishing is completed, the counter-samples in the cassette tape have a mirrorlike surface state (**a**). One counter-sample extracted from the cassette tape after additional Si-finishing has a matte diffusing surface (**b**).

To create micro-crater formations on the surface of the model shafts, we utilized an additional finishing technology called Si-technology. This involved finishing with an oil suspension of silicon carbide.

It has been experimentally established that when silicon carbide with a maximum grain size of 5 μm is in a free state in an oil emulsion in a ratio of 1:3 and is located between a flat glass lap and the processed steel surface of the shaft along the line, a micro-crater texture forms on the steel surface. Given the plate-like structure of silicon carbide crystallites and their relatively high hardness (ranked fourth after diamond), it is reasonable to assume that large silicon carbide crystals may crumble into smaller fractions during finishing, resulting in craters on the surface of the shaft.

After polishing the diamonds, the cassettes were washed thoroughly with petroleum ether, a mixture of low-molecular-weight hydrocarbons (C_5H_{12} and C_6H_{14}), and were then wiped with moistened filter paper before being dried. The same procedure was repeated for the cassettes after additional finishing with Si-technology.

Five samples were taken from the central part of the cassettes, which were processed differently and tested to avoid the influence of edge effects after finishing. The model shafts were prepared and installed on the friction machine according to the described methods, with radial deviations from the axis of rotation not exceeding 3 μm. Upon contact with the flat sample, it was fully immersed in aviation kerosene.

The hardness of the samples was determined using a standardized DECHUAN hardness tester (Laizhou, China). Radial runout of the model shafts after installation on the spindle was eliminated using a special device, which measured micro-displacements when the spindle shaft was slowly rotated using a MITUTOYO (Kanagawa, Japan) dial indicator with a division value of 1 μm. Measurements of deviations in the shape of the surfaces of the samples were carried out on a "KEYENCE" (Osaka, Japan) coordinate-measuring machine.

Test conditions: The linear sliding speed was 0.3 m/s and the axial load was 110 N. Based on G. Hertz's formulas [24] for this linear contact of surfaces made of G52986 steel, the maximum initial design contact stresses were 70 MPa, which is realistic for fuel equipment. However, it should be noted that the theory of G. Hertz assumed perfectly smooth surfaces without boundary layers. The calculated initial contact width was approximately 0.16 mm, which is over 50 times greater than the radial runout of the shaft, which was 0.003 mm.

Friction and wear test method: Tribological comparison tests were performed using Timken test methods based on ASTM D2509 [25] for greases and ASTM D2782 [26] for oils. However, in our case, we compared not lubricating media but the anti-wear and anti-friction properties of surfaces having different three-dimensional states in the same environment—aviation kerosene at the same initial contact loads, at the same sliding speed, for the same four-stage tests, with the same time, and in the same external environmental conditions.

After applying a load of 110 N and filling the chamber with aviation kerosene TS-1, the shaft rotated at a tangential speed of 0.3 m/s. The first stage of friction lasted for 15 min. Subsequently, two more such tests (2nd and 3rd stages) were carried out using the same model shaft, but on new sections of the surface of the flat model bearing, each lasting 15 min. Finally, the fourth stage of friction was carried out for 3 h. It was possible to construct accurate wear rate dependencies by assuming complete running-in occurred in the first three stages. The similar wear values after friction in the second and third stages indicate that the additional work in kerosene was completed. The tests monitored the volumetric temperature of the kerosene and friction force, and wear tracks were measured and investigated afterward.

Surface roughness study: We used the laser-scanning profilograph–profilometer for the surface measurement and determination of 2D roughness parameters as defined by the ISO 4287 [19], ISO 25178 [21], and ISO 13565 [20]. The profilograph–profilometer had the following parameters: field of view 300×300 μm^2, lateral resolution 0.5 μm, and vertical resolution 0.1 nm.

If the surface roughness is described in accordance with ISO 4287 [19], the data obtained on the average value and distribution of the values of the roughness parameters along a plane do not provide information on the angular characteristics of these parameters. However, raster surface scanning and additional data processing made it possible to obtain the angular distribution of the surface roughness parameters, taking into account the direction of friction. This paper introduces some angular characteristics that allow us to describe the standard roughness parameters.

For example, Figure 4 shows the results of the calculation for the parameter Ra of the standard surface roughness. The calculation is presented in the form of a standard regular profile with $Rz = 0.1$ μm. The relief of the profile is shown in Figure 4a, taking into account the angular distribution of the average value of this parameter, as shown in Figure 4c. Additionally, a histogram of the distribution of the Ra parameter along raster lines is presented in Figure 4b, and its dependence on the direction is shown in Figure 4d. Figure 4b shows a histogram of the distribution of the Ra parameter value for angle $\varphi = 0°$. The profile is measured in the direction perpendicular to the grooves on the surface of the test sample, as shown in Figure 4a. The blue curves in Figure 4c show the standard deviation of the Ra parameter from its average value. Figure 4d shows the quantitative distribution of surface profiles based on the Ra parameter value and the direction of the profile study. The red color Indicates the maximum number of profiles (N). The contours of the color image characterizing the distribution of the Ra parameter in Figure 4d correlate with the curves shown in Figure 4c. Following this example, one can analyze the angular dependences of the different parameters of the test sample surfaces such as Ra, Rz, and Rk.

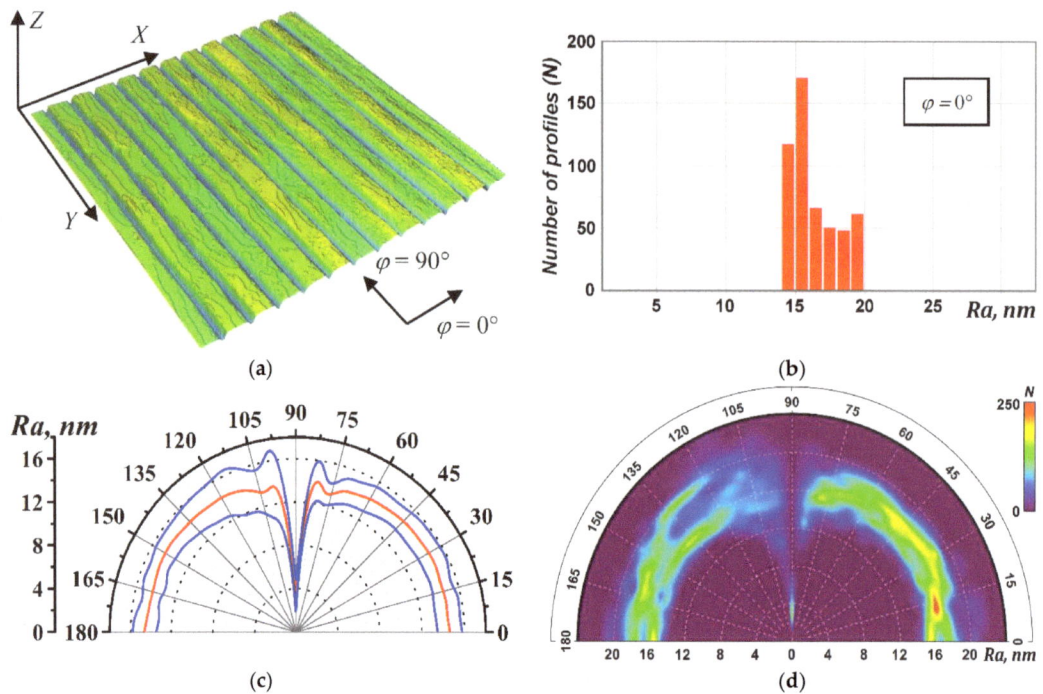

Figure 4. Angular dependence (by angle φ) distribution of the average value (red line) and the boundaries of the range of change (blue lines) of the parameter Ra (**c**) for the standard surface roughness with a regular profile (**a**), including considering histograms of the distribution of this parameter along raster lines (**b**) and along the angle directions of profile research (**d**).

3. Analysis of the Results

During multiple tribological laboratory tests of friction pairs with various finishing work surfaces, it was discovered that these pairs have fundamentally different tribological properties, all other factors being equal (in the same environment, at the same initial contact loads, at the same sliding speed, during the same duration of the friction process, and under the same external normal environmental conditions). Friction pairs in which the working surfaces of the model shafts were adjusted using Si-technology exhibited abnormally low values for the friction coefficient and wear rate. At the same time, these friction pairs exhibited a slight variation in tribological characteristics during the burnishing stage and maximum resistance to setting, with the critical specific load being at least 200 MPa.

Figure 5 show some results of wear tests on friction pair samples (5 control samples from different batches) in four stages (I, II, III, and IV). The model shaft surfaces were finished using various technologies. The temporal dependencies of wear of a flat bearing are presented in Figure 5a, while the histograms of wear of a flat bearing depending on the batch number are shown in Figure 5b. The wear is presented as the depth I of the trace formed on the surface of the flat bearing as result of the material removing during the friction process.

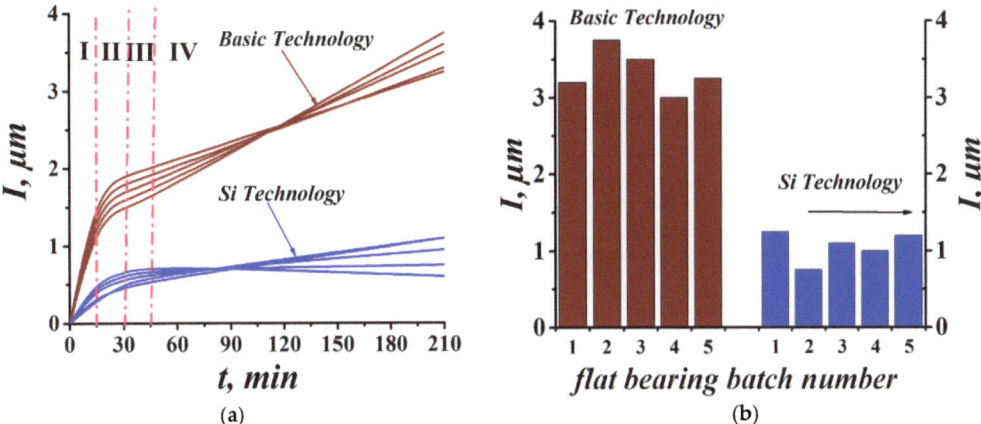

Figure 5. The time dependences of wear of a flat bearing (**a**); wear histograms of a flat bearing depending on the batch number (**b**) for two finishing technologies.

Significant differences were observed in the behavior of friction coefficient oscillograms during the friction of shafts prepared using traditional processing methods and during the friction of the same shafts processed using Si-technology in the most informative fourth stage of the tests (see Figure 6). The results show that there were both quantitative and qualitative differences between the two technologies. The friction coefficient C_f decreased by 2–5 times in the case of the Si-technology compared to the traditional technology both in terms of amplitude and frequency of oscillations. Additionally, there were differences in the amplitude and frequency of oscillations, indicating a decrease in the intensity of the adhesive forces in the interaction of the friction surfaces in contact (submicron adhesions) in the case of Si-technology. Figure 6 shows dotted lines that limit the range of trends in oscillograms of friction coefficients when testing five batches of model shafts under other equal friction conditions with different processing technologies on a stationary sample. The minor discrepancies between the friction force values during the five identical tests for each case of the surface condition of the model shafts (no more than 10%) indicate a high level of reproducibility of the experiments performed.

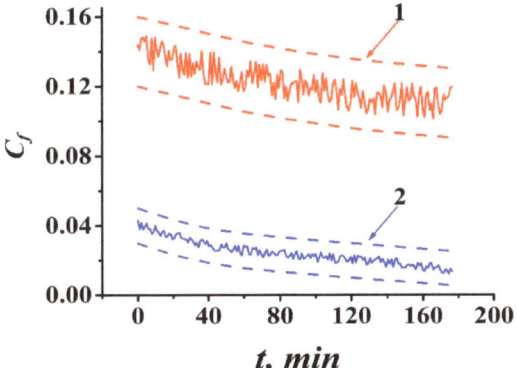

Figure 6. Change in the coefficient of friction C_f over time during sliding friction of model shafts at the fourth stage of testing using various technologies for finishing shaft surfaces: traditional processing method (1); processing using Si-technology (2). The coefficients of friction trends during the testing of five batches of model shafts are shown with dashes.

The graphs in Figure 6 show that samples treated with Si-technology have a friction coefficient reduced by up to five times as well wear resistance increased by two to three times (Figure 5b). The linear roughness parameters (ISO 4287) obtained from the surfaces under study using one profilogram for samples with Si-technology surface treatment differed from those with traditional surface treatment, showing a decrease. The surfaces of model shafts manufactured using Si-technology had, on average, 30–50% lower Ra and Rz parameter values compared to those manufactured using traditional technology. However, this alone does not account for the significant differences in wear rates and friction coefficients. Therefore, a search was carried out for more informative parameters using a profilograph–profilometer.

The study investigated cylindrical surface sections of a series of model shafts with surface finishing using both traditional technology and Si-technology. Figure 7 shows typical surface reliefs of the indicated areas with a size of 300 × 300 μm², which were obtained using a profilograph–profilometer. The obtained images (Figure 7b,c) clearly show that the three-dimensional configuration of roughness on the cylindrical surfaces of the investigated shafts has different characteristics. On the surface of the shaft processed using traditional technology, grooves in the form of a structure are clearly visible. These grooves run perpendicular to the axis of the model shaft and coincide with the direction of the X axis in Figure 7. Similarly, grooves are visible on the surface of the shaft processed using Si-technology, but they are less noticeable and lack a clearly defined direction. In addition, it is worth mentioning that the surface irregularities on the shaft, which were processed using Si-technology, are distributed more uniformly.

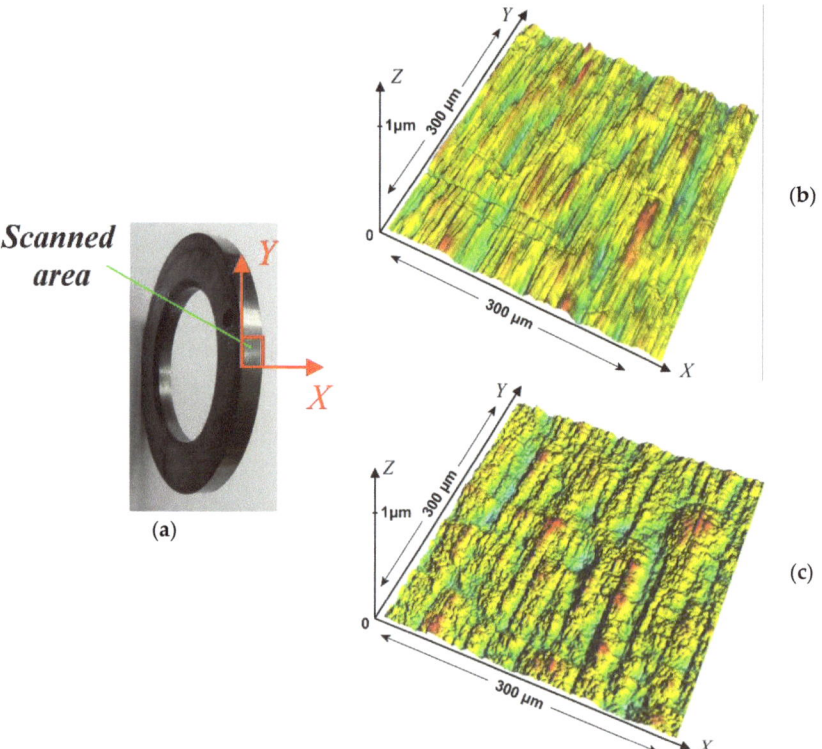

Figure 7. The reliefs of cylindrical surface section (**a**) of model shaft samples processed using traditional technology (**b**) and Si-technology (**c**) were scanned by a profilograph–profilometer.

The angular spatial spectrum of the test surfaces analyzed due to their pronounced dependence on the direction was considered first. It was found that the angular spatial spectrum has almost identical characteristics in different areas of the cylindrical surface of samples processed using the same technology. However, the angular spectra of the surfaces of the samples processed by different technologies differed significantly. Figure 8 shows typical graphs of angular spectra for the test surfaces of model shafts. The direction with an angle of 0° in the graphs of the angular spectra corresponds to the X-axis in Figure 7.

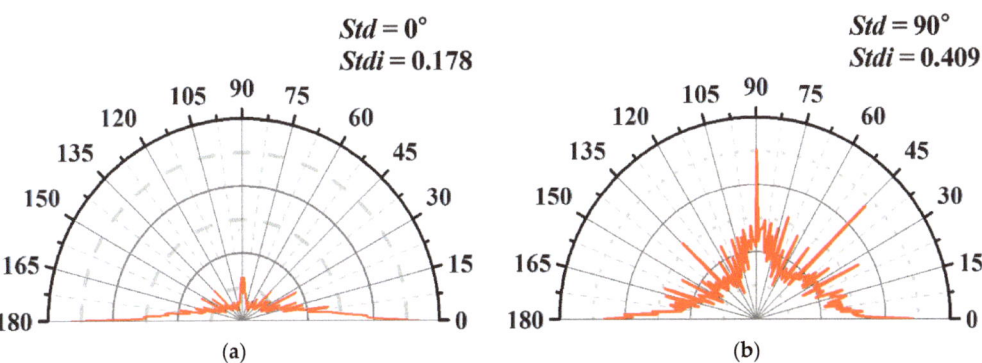

Figure 8. Angular spectra of the surfaces of model shaft cylinders processed using traditional technology (**a**) and Si-technology (**b**).

The ISO 25178 [21] provides a quantitative characterization of the angular spatial spectrum of the test surface using the parameters *Std* and *Stdi*. The texture direction, *Std*, is defined as the angle of the dominating texture of the rough surface. The texture direction index, *Stdi*, is a measure of how dominant the dominating direction is. The *Stdi* value is always between 0 and 1. Surfaces with very dominant directions will have *Stdi* values close to zero, and if all directions are similar, *Stdi* will be close to 1. Figure 8 shows the values of *Std* and *Stdi* parameters for different technologies used to finish the surfaces of model shafts. The *Std* parameter confirms the main direction of the surface texture, which is consistent with the conclusions drawn from Figure 7. With traditional surface-processing technology (Figure 8a), directional roughness is clearly expressed. The grooves on the surface are parallel to the Y axis in Figure 7a, the maximum values of which are achieved perpendicular to the grooves on the surface in the direction of 0° (along the X axis in Figure 7a). Similar reasoning can be repeated regarding Si-technology. Figure 8b shows that the angular spectrum has the largest values in several selected directions, namely, 0°, 45°, and 90°. In this case, the *Std* parameter shows the greatest dominance in the direction of 90°. Additionally, the roughness direction with Si-technology is significantly less dominant compared to with traditional technology, as evidenced by the values of the *Stdi* parameter. With traditional technology, the *Stdi* is much closer to zero (*Stdi* = 0.178) than with Si-technology, where it is closer to unity (*Stdi* = 0.409). This is supported by the shape of the angular spectrum graph. With traditional technology, the angular spectrum is narrow, with a clear maximum in the direction of roughness dominance. However, with Si-technology, the angular spectrum is significantly wider.

Figure 9 shows typical angular dependences of the histogram of the distribution of parameters *Ra*, *Rz*, and *Rk* over the surface for model shaft samples processed using traditional technology (Figure 9a,c,e) and Si-technology (Figure 9b,d,f). The orientation of the shaft samples with different surface-polishing technologies and the test areas of their surfaces coincide with the designations shown in Figure 7a. This case does not consider the fact that the average values of the parameters are slightly different for both technologies, as previously noted. The obtained images show that for both surface pre-treatment technologies the parameters *Ra*, *Rz*, and *Rk* have similar angular dependencies within the same technology. However, these dependencies differ between different surface treatment technologies. This applies to both the average values of the parameters and to the distributions of parameters across surface profiles. With traditional technology, there is a wide and non-uniform scatter of all parameters (40–60% of the average value), which depends on the angle and on individual surface profiles in a given direction. For Si-technology, the dispersion of parameter in terms of angle and individual profiles in a given direction is significantly smaller (25–35% of the average value) and more consistent (the peaks of the predominant parameter values are less fragmented in terms of angle compared to the case of standard technology). It is important to note that the level of roughness with Si-technology is approximately the same over the entire surface of the sample and is not significantly affected by the direction, as the direction of roughness is weakly expressed.

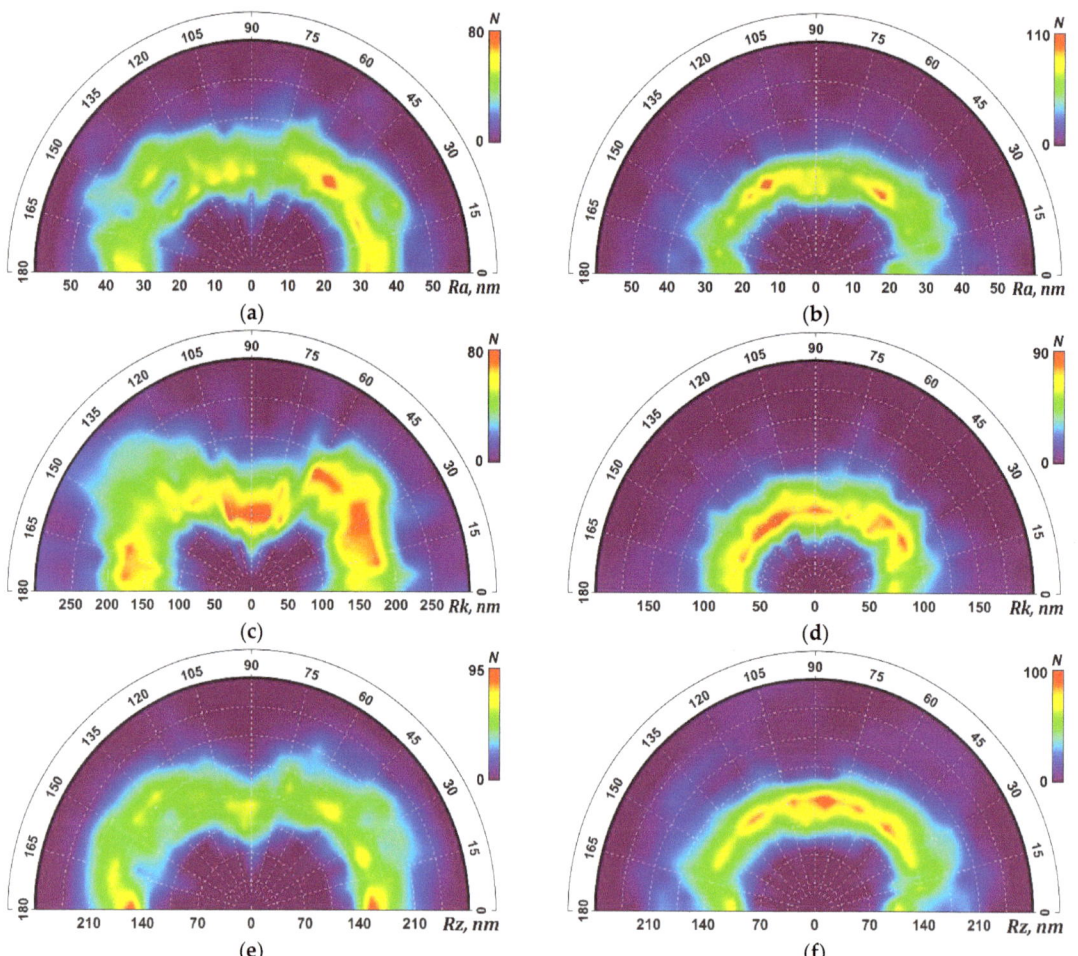

Figure 9. Angular dependences of the histogram of the distribution of the parameters Ra (**a**,**b**), Rz (**c**,**d**), and Rk (**e**,**f**) over the surface for samples of model shafts, the surfaces of which are processed using traditional technology (**a**,**c**,**e**) and Si-technology (**b**,**d**,**f**).

The ISO 13565 [20] identifies the bearing area curve (BAC) as a significant characteristic of a rough surface. The BAC, also known as the Abbott–Firestone curve, was proposed as a general measure of surface quality for machined parts. It is a simple and practical tool for assessing surface quality. The BAC is an integral characteristic of contacting rough surfaces and provides a visual representation of the primary operational characteristics of the contact, including rigidity, wear resistance, and tightness. The BAC determines several crucial parameters, such as *Rk*, *Rpk*, and *Rvk*, which establish the limits of the profile core, the average height of the protrusions, and the average depth of the depressions, respectively. Additionally, the parameters *Mr1* and *Mr2* determine the proportion of surface protrusions and depressions, respectively.

Figure 10 shows typical images illustrating the change in BAC based on the direction. The color-coded amplitude values of the points on the BAC are shown for both surface-processing technologies of model shafts. The orientation of shaft samples with different surface-polishing technologies and the areas of surface examination coincides with the designations shown in the Figure 7a. The images obtained also illustrate the presence of a dominant roughness direction on surfaces manufactured using traditional technology (Figure 10a). The average depth of the depressions and the height of the protrusions increase in these surfaces. This is indicated by the increase in BAC values in Figure 10a, as well as in the *Rk* values in Figure 10c, in the direction of the dominant roughness. In contrast, surfaces manufactured using Si-technology (Figure 10b) do not exhibit a pronounced dominant direction, which is consistent with the previously obtained characteristics. Figure 10c,d present typical graphs of the parameters *Mr1* and *Mr2* depending on the direction for two finishing technologies. The graphs show that the proportion of protrusions and depressions on the surface of shafts processed using traditional technology can vary significantly depending on the direction. In most cases, depressions predominate in the direction of maximum roughness. On shaft surfaces processed using Si-technology, depressions are more prevalent in the direction of greatest roughness, while the proportion of protrusions and depressions is roughly equal in other directions. It is worth noting that the parameters *Rpk* and *Rvk* exhibit similar behavior (Figure 10c,d). For shaft surfaces processed using traditional technology, in the direction of greatest roughness, the ratio *Rvk/Rpk* is usually greater than one in the direction of greatest roughness. This ratio varies significantly across different samples and areas of the surface. For shaft surfaces processed using Si-technology, the ratio *Rvk/Rpk* remains relatively constant. Specifically, this ratio is greater than 1 and close to 1.3 in angle ranges from 0 to 50° and from 150° up to 180°. In the range of angles from 50° up to 150°, the ratio is close to 1.

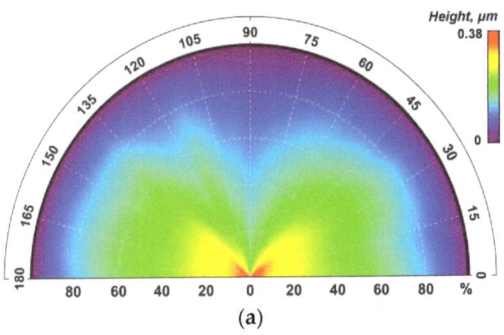

(a)

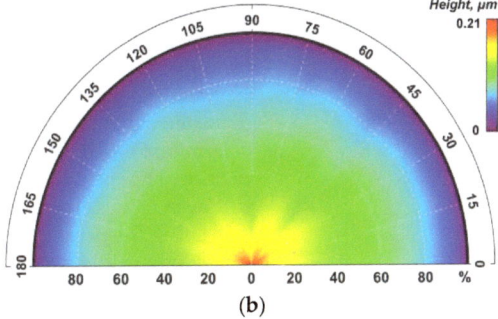

(b)

Figure 10. *Cont.*

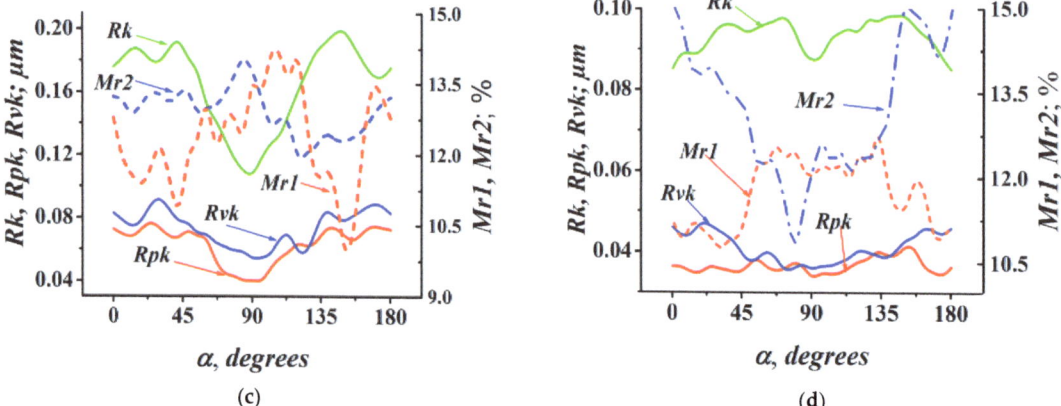

Figure 10. The 3D images of the BAC dependencies were averaged over the surface and determined based on the direction of study of the surface profile for samples of model shafts manufactured using traditional technology (**a**) and using Si-technology (**b**). The parameters *Rk*, *Rpk*, *Rvk*, *Mr1*, and *Mr2* were determined based on the direction of study of the surface profile for samples of model shafts manufactured using traditional technology (**c**) and using Si-technology (**d**).

The angular dependencies of standard surface roughness parameters can identify certain directions on the surface where these parameters exhibit features that make it possible to describe the fundamental differences between the surface-finishing technologies of the model shafts considered here.

4. Criteria for Evaluating the Tribological Efficiency of Roughness

The analysis of the above angular dependencies of standard roughness parameters enables the formulation of criteria for evaluating the tribological efficiency of roughness in the sense of reducing friction and wear.

When comparing the images in Figure 9, it can be seen that the distributions of average values of standard roughness parameters over an angle (as shown in Figure 4c) and over individual surface profiles (similarly, as shown in Figure 4b) for Si-technology surface finishing appear more symmetrical relative to angles 0° and 90° than for traditional technology. It is assumed that this feature contributes to Si-technology's superior tribological friction efficiency. Therefore, when evaluating the tribological efficiency of a surface, the symmetry of the surface texture relative to the direction of friction can be considered a characteristic of the surface. This characteristic depends on the direction of friction and is determined based on a standard roughness parameter (such as *Ra* or *Rz*). To obtain this characteristic objectively, we will consider the coefficient of symmetry of the surface texture in the direction of friction. The calculation is proposed to be performed as follows.

Using the parameter *Ra* as an example, it is evident from Figure 9a that traditional technology exhibits a significant scatter of values for individual profiles taken in one direction relative to its average value. Additionally, the maximum of the distribution of this parameter also changes significantly depending on the direction in which it is determined. For Si-technology (as shown in Figure 9b), the *Ra* parameter values are less spread out over the surface in different directions of its determination. Furthermore, the maximum distribution of this parameter is characterized by a lower dependence on the observation angle. Thus, it can be concluded that for Si-technology, there exists a dominant value of the parameter *Ra* over the entire test surface, as determined along various profiles and directions. This dominant value can be used to calculate the dominance coefficient, which characterizes the degree of dominance of this parameter over the surface in a given direction. We propose calculating the dominance coefficient using the following formula:

$$\Omega_{Ra}(\alpha) = C_p(\alpha) \cdot e^{-\left(\frac{\Delta Ra(\alpha)}{Ra(\alpha)}\right)^2}, \qquad (1)$$

where for the selected direction of friction the following α values are determined: $C_p(\alpha) = N_p(\alpha)/N$—ratio of the maximum number of profiles $N_p(\alpha)$ in the direction α with the same value Ra (determined from the parameter distribution histogram in Figure 9) to the total number of profiles N on a surface, $\Delta Ra(\alpha)$—the standard deviation of the Ra parameter over all surface profiles in the direction α, $Ra(\alpha)$—the average value of the parameter over all surface profiles in the direction α. The coefficient $C_p(\alpha)$ in Formula (1) describes the maximum amplitude distribution of parameter values Ra over the surface in a given direction α. The second factor in Formula (1) describes the influence of the width of the distribution function of the parameter Ra over the surface in the direction α.

By utilizing the data shown above to characterize the dominance of the roughness parameter Ra on the surface, it is possible to calculate the symmetry coefficient of the distribution of this roughness parameter in relation to the chosen direction of friction. To do this, you need to determine the degree of symmetry (similarity) of the functional dependence $\Omega_{Ra}(\alpha)$ on both sides (within the range of $\pm 90°$) relative to the selected direction α using a normalized autocorrelation function. We propose determining the symmetry coefficient with respect to the selected direction of friction as follows:

$$S_{Ra}(\alpha) = \frac{\int_0^{\pi/2} \Omega_{Ra}(\alpha - \theta) \cdot \Omega_{Ra}(\alpha + \theta) \, d\theta}{\int_0^{\pi} \Omega_{Ra}^2(\theta) \, d\theta}. \qquad (2)$$

As a result, we obtain the following graphs (Figure 11) for the coefficient of symmetry of the surface texture depending on the direction of friction within the range of angles $\alpha = (-90° \ldots +90°)$.

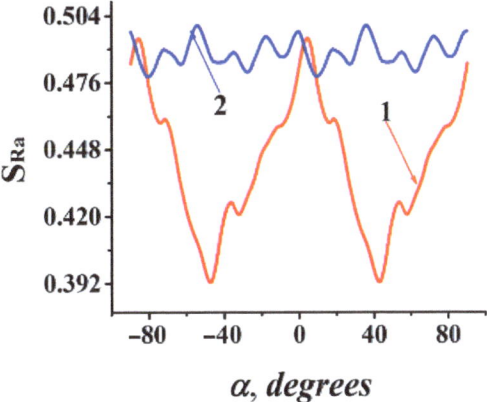

Figure 11. Angular dependencies of the symmetry coefficient SR_a of surface texture on the direction of friction α, obtained for traditional technology (curve 1) and Si-technology (curve 2) of model shaft surface finishing.

The graphs indicate that traditional technology exhibits maximum symmetry in two directions, i.e., 0° and 90°. This is due to the dominant direction of the roughness on the surface, which takes the form of regular grooves that run perpendicular to the axis of the model shaft. Therefore, the roughness in traditional technology is most symmetrical in the aforementioned directions. For Si-technology, symmetry is uniform and maximal in all directions. This characteristic can serve as a criterion for determining the optimal surface texture for friction in a specific direction.

The second criterion is related to the analysis of the BAC and its directional dependence. As shown in Figure 10, Rk roughness parameters determined from the BAC exhibit several features that are characteristic of the surface-finishing technologies used for model shafts. This analysis utilized all Rk parameters and their angular dependencies. To evaluate the tribological efficiency of the friction surface in a specific direction in terms of reducing friction and wear using Rk parameters, the dependencies shown in Figure 10 were used as a basis.

The dependencies presented in Figure 10 clearly show that for traditional technology, the parameter Rk has a minimum value only in the direction perpendicular to the shaft axis, which coincides with the direction of friction. The parameter itself varies within a wide range, from 0.1 μm to 0.2 μm. In addition to the observations made along the shaft axis, there is also a noticeable divergence in the values of Rpk and Rvk when measured perpendicular to the shaft axis (90°). The fractional components of the protrusions and depressions corresponding to the parameters $Mr1$ and $Mr2$ also show a similar divergence. It is worth noting that for Si-technology, the parameter Rk exhibits two insignificant minima at 0° and 90°, and its value is less than that of traditional technology, varying within a significantly smaller range (0.085 to 0.1). In the direction perpendicular to the shaft axis, similar values of the parameters Rpk and Rvk, as well as $Mr1$ and $Mr2$, respectively, are observed.

It is known that the samples prepared using the Si-technology have significantly higher friction efficiency (in the sense of reducing friction and wear) in the direction perpendicular to the axis of the model shaft compared to traditional technology. To account for the behavior of the parameters Rk, Rpk, Rvk, $Mr\,1$, and $Mr\,2$ described above, a friction roughness efficiency coefficient is introduced. This coefficient varies in the range of 0 to 1. In this case, a value of 0 corresponds to the maximum ineffective roughness (in the sense of increasing friction), while a value of 1 corresponds to the maximum effective roughness (in the sense of reducing friction). When selecting a suitable functional dependence for calculating this coefficient, it is important to consider that in order to achieve maximum efficiency of the tribological interaction between the surfaces of the friction pair, the coefficient should quickly approach unity. The coefficient of roughness efficiency with respect to friction should tend to zero, except when the parameters Rpk and Rvk, as well as the parameters $Mr1$ and $Mr2$, respectively, have close values. In the direction of maximum friction efficiency, the parameters Rk, Rpk, and Rvk should take a minimum value. Based on the given conditions, the coefficient of roughness efficiency with respect to friction should tend to zero. Therefore, we propose representing the function describing the angular dependence of the roughness efficiency coefficient with respect to friction as follows:

$$\chi(\alpha) = \exp\left\{1 - \frac{Rk(\alpha) \cdot Rpk(\alpha) \cdot Rvk(\alpha)}{[Rk \cdot Rpk \cdot Rvk]_{\min}\left(1 - \sqrt{\frac{|Rpk(\alpha)-Rvk(\alpha)|}{Rpk(\alpha)+Rvk(\alpha)} \cdot \frac{|Mr1(\alpha)-Mr2(\alpha)|}{Mr1(\alpha)+Mr2(\alpha)}}\right)^2}\right\}, \quad (3)$$

where the parameters Rk, Rpk, Rvk, $Mr1$, and $Mr2$ are determined for a given direction at the angle α, and the parameter $[Rk \cdot Rpk \cdot Rvk]_{\min}$ represents the minimum value of the product of the parameters Rk, Rpk, and Rvk in the angle α range from 0 to 180°. If we take α as an angle relative to the axis of the model shaft, the dependencies $\chi(\alpha)$ for both the traditional technology and the additional Si-technology for finishing the surfaces of model shafts will have the form shown in Figure 12.

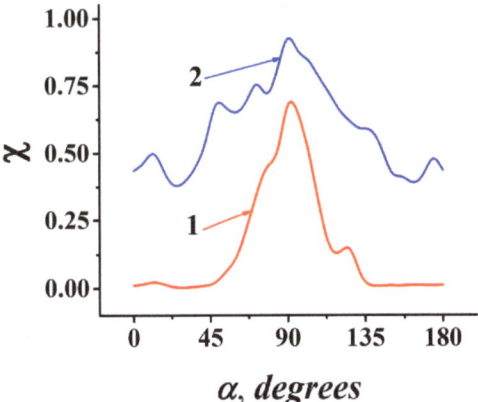

Figure 12. Angular dependencies of the roughness efficiency coefficient with respect to friction χ on the direction of friction α obtained for traditional technology (curve 1) and Si-technology (curve 2) for finishing the surfaces of model shafts.

Using the obtained dependencies, a criterion for the efficiency of roughness in terms of reducing friction and wear can be established. It is observed that in the direction perpendicular to the axis of the model shaft ($\alpha = 90°$), both surface-finishing technologies exhibit a pronounced maximum. This indicates that surface wear and friction force will be minimized when rubbing in this direction. This statement is consistent with tribological test data and the general understanding that the surface's tribological efficiency will be better in the direction where the surface roughness parameters have the lowest value. However, it is important to note that the maximum roughness efficiency coefficient with respect to friction $\chi(\alpha)$ for Si-technology (Figure 12, curve 2) is much closer to 1 than the maximum of the similar coefficient for traditional technology (Figure 12, curve 1). Based on the tribological test data, it can be assumed that the surfaces of friction pairs exhibit the best tribological efficiency when rubbing in the direction with a higher roughness efficiency coefficient $\chi(\alpha)$. A coefficient closer to 1 indicates greater tribological efficiency of the friction pair.

5. Conclusions

The original technique for performing tribological tests of friction pairs in combination with a new method of additional finishing to the surfaces of friction pairs has shown a significant influence of the configuration of the initial relief of rough surfaces on the parameters of their tribological interaction (friction and wear). The surface of the shaft was prepared using both traditional diamond paste polishing techniques and additional finishing with silicon carbide. Tribological testing was performed on a single permanent linear contact with other factors being equal. In a comparative experiment on sliding friction of linear contact friction pairs composed of model shafts with different states of friction surfaces on a fixed plane, significant differences in friction performance were found between shafts with different surface textures. Additional finishing of the shaft surface with silicon carbide resulted in a 2–5 times reduction in friction the coefficient and a 2–4 times reduction in wear. In particular, the friction coefficient decreased from a value of 0.1 to a value of less than 0.02. Along with this, the standard parameters of the initial roughness of the surfaces of friction pairs processed using different technologies did not show significant differences.

The abnormally low values of friction coefficient and wear strength observed during the additional finishing process on the shaft surface are consistent with the adhesive deformation fluid dynamics model of friction and wear. This is due to the random micro cracks on the polished nano surface. The new technology applied forms a uniform surface relief. The reduction in the degree of rarefaction and micro-adhesive interaction of local areas of surfaces in the elastic deformation contact area of the diffuser can explain this phenomenon.

A new technique for analyzing rough friction surfaces in polar coordinates has been developed. This technique enables the acquisition of roughness angular characteristics and the evaluation of their impact on the frictional characteristics of the surface relative to the friction direction using standard roughness parameters. The technique was developed considering the frictional test results of the model friction pair and the study of the influence of various machining techniques on the three-dimensional-state characteristics of roughness generated on the shaft surface. Criteria for evaluating the tribological efficiency of rough surfaces (in the sense of reducing friction and wear) in a given friction direction are proposed based on roughness angular characteristics. These criteria are consistent with the results of frictional tests and enable the control of existing technologies for machining friction pair working surfaces. They also facilitate the development of new technologies to increase the wear resistance of high load friction sliding systems operating under boundary fuel lubrication conditions. This method shows promise and could be instrumental in optimizing various friction system friction surface-finishing technologies.

Author Contributions: Conceptualization, O.S.; Methodology, S.K.; Software, H.F.; Formal analysis, V.K. and V.G.; Data curation, H.F.; Writing—original draft, O.S.; Writing—review & editing, C.H.; Project administration, H.Z.; Funding acquisition, H.Z. All authors have read and agreed to the published version of the manuscript.

Funding: The work presented in this paper was supported by a Ukrainian–Chinese bilateral project (contract #M/9-2023), the Beijing Natural Science Foundation (grant No. IS23048), and the National Natural Science Foundation of China (grant No. 52306038).

Data Availability Statement: Data underlying the results presented in this paper are not publicly available at this time but may be obtained from the authors upon reasonable request.

Conflicts of Interest: The authors declare no conflicts of interest.

References

1. Aymard, A.; Delplanque, E.; Dalmas, D.; Scheibert, J. Designing metainterfaces with specified friction laws. *Science* **2024**, *383*, 200–204. [CrossRef] [PubMed]
2. Zhu, Y.; Qu, H.; Luo, M.; He, C.; Qu, J. Dry friction and wear properties of several hard coating combinations. *Wear* **2020**, *456*, 203352. [CrossRef]
3. Wang, X.; Zhang, R.; Zhou, T.; Wei, X.; Liaw, P.; Feng, R.; Wang, W.; Li, R. Microstructural Evolution in Chroming Coatings Friction Pairs under Dry Sliding Test Conditions. *Adv. Tribol.* **2018**, *2018*, 5962153. [CrossRef]
4. Chang, T.; Guo, Z.; Yuan, C. Study on influence of Koch snowflake surface texture on tribological performance for marine water-lubricated bearings. *Tribol. Int.* **2019**, *129*, 29–37. [CrossRef]
5. Gropper, D.; Harvey, T.J.; Wang, L. Numerical analysis and optimization of surface textures for a tilting pad thrust bearing. *Tribol. Int.* **2018**, *124*, 134–144. [CrossRef]
6. Shinde, A.B.; Pawar, P.M. Multi-objective optimization of surface textured journal bearing by Taguchi based Grey relational analysis. *Tribol. Int.* **2017**, *114*, 349–357. [CrossRef]
7. Song, F.; Yang, X.; Dong, W.; Zhu, Y.; Wang, Z.; Wu, M. Research and prospect of textured sliding bearing. *Int. J. Adv. Manuf. Technol.* **2022**, *121*, 1–25. [CrossRef]
8. Yang, L.; Ding, Y.; Cheng, B.; He, J.; Wang, G.; Wang, Y. Investigations on femtosecond laser modified micro-textured surface with anti-friction property on bearing steel GCr15. *Appl. Surf. Sci.* **2017**, *434*, 831–842. [CrossRef]
9. Zhang, Y.; Chen, G.; Wang, L. Effects of thermal and elastic deformations on lubricating properties of the textured journal bearing. *Adv. Mech. Eng.* **2019**, *11*, 1687814019883790. [CrossRef]
10. Filho, I.F.; Bottene, A.; Silva, E.; Nicoletti, R. Static behavior of plain journal bearings with textured journal—Experimental analysis. *Tribol. Int.* **2021**, *159*, 106970. [CrossRef]
11. Galda, L.; Sep, J.; Olszewski, A.; Zochowski, T. Experimental investigation into surface texture effect on journal bearings performance. *Tribol. Int.* **2019**, *136*, 372–384. [CrossRef]
12. Sinanoğlu, C.; Nair, F.; Karamış, M.B. Effects of shaft surface texture on journal bearing pressure distribution. *J. Am. Acad. Dermatol.* **2005**, *168*, 344–353. [CrossRef]
13. dos Anjos, L.F.; Jaramillo, A.P.; Buscaglia, G.C.; Nicoletti, R. Improving the load capacity of journal bearings with chevron textures on the shaft surface. *Tribol. Int.* **2023**, *185*, 108561. [CrossRef]

14. Wang, P.; Liang, H.; Jiang, L.; Qian, L. Effect of nanoscale surface roughness on sliding friction and wear in mixed lubrication. *Wear* **2023**, *530–531*, 204995. [CrossRef]
15. Wong, V.W.; Tung, S.C. Overview of automotive engine friction and reduction trends–Effects of surface, material, and lubricant-additive technologies. *Friction* **2016**, *4*, 1–28. [CrossRef]
16. Svenningsson, I.; Tatar, K. Exploring the mechanics of adhesion in metal cutting. *Int. J. Adv. Manuf. Technol.* **2023**, *127*, 3337–3356. [CrossRef]
17. Stel'makh, A.U. Adhesion-hydrodynamic (AHD) model of friction. *J. Frict. Wear* **2014**, *35*, 316–326. [CrossRef]
18. Stelmakh, A.U.; Pilgun, Y.V.; O Kolenov, S.; Kushchev, A.V. Reduction of friction and wear by grooves applied on the nanoscale polished surface in boundary lubrication conditions. *Nanoscale Res. Lett.* **2014**, *9*, 226. [CrossRef] [PubMed]
19. ISO 4287; Surface Texture: Profile Method. International Organization for Standardization: Geneva, Switzerland, 1997.
20. ISO 13565; Geometrical Product Specifications (GPS)—Surface Texture: Profile method; Surfaces Having Stratified Functional Properties. International Organization for Standardization: Geneva, Switzerland, 1998.
21. ISO 25178; Surface Texture: Areal. International Organization for Standardization: Geneva, Switzerland, 2021.
22. Kundrak, J.; Gyani, K.; Bana, V. Roughness of ground and hard-turned surfaces on the basis of 3D parameters. *Int. J. Adv. Manuf. Technol.* **2007**, *38*, 110–119. [CrossRef]
23. Zhang, H.; Zhu, W.; Stelmakh, O.; Du, W.; Fu, H.; Hu, C.; Sun, B.; Wang, X. A Constant Line Contact Friction and Wear Testing Machine. CN202210685448.6. 20 September 2022. Available online: https://www.zhangqiaokeyan.com/patent-detail/06120114998949.html (accessed on 20 April 2024).
24. Wang, B.; Lyu, Q.; Jiang, L.; Chen, Y.; Guo, Z. An Extended Hertz Model for Incompressible Mooney–Rivlin Half-Space Under Finite Spherical Indentation. *Int. J. Appl. Mech.* **2022**, *14*, 2250103. [CrossRef]
25. D2509; Standard Test Method for Measurement of Load-Carrying Capacity of Lubricating Grease (Timken Method). ASTM International: West Conshohocken, PA, USA, 2021.
26. D2782; Standard Test Method for Measurement of Extreme-Pressure Properties of Lubricating Fluids (Timken Method). ASTM International: West Conshohocken, PA, USA, 2020.

Disclaimer/Publisher's Note: The statements, opinions and data contained in all publications are solely those of the individual author(s) and contributor(s) and not of MDPI and/or the editor(s). MDPI and/or the editor(s) disclaim responsibility for any injury to people or property resulting from any ideas, methods, instructions or products referred to in the content.

Article

Influence of Polymer Flow on Polypropylene Morphology, Micro-Mechanical, and Tribological Properties of Injected Part

Martin Ovsik *, Klara Fucikova, Lukas Manas and Michal Stanek

Faculty of Technology, Tomas Bata University in Zlin, Vavreckova 5669, 760 01 Zlín, Czech Republic; k_fucikova@utb.cz (K.F.); lmanas@utb.cz (L.M.); stanek@utb.cz (M.S.)
* Correspondence: ovsik@utb.cz

Abstract: This research investigates the micro-mechanical and tribological properties of injection-molded parts made from polypropylene. The tribological properties of polymers are a very interesting area of research. Understanding tribological processes is very crucial. Considering that the mechanical and tribological properties of injected parts are not uniform at various points of the part, this research was conducted to explain the non-homogeneity of properties along the flow path. Non-homogeneity can be influenced by numerous factors, including distance from the gate, mold and melt temperature, injection pressure, crystalline structure, cooling rate, the surface of the mold, and others. The key factor from the micro-mechanical and tribological properties point of view is the polymer morphology (degree of crystallinity and size of the skin and core layers). The morphology is influenced by polymer flow and the injection molding process conditions. Gained results indicate that the indentation method was sufficiently sensitive to capture the changes in polypropylene morphology, which is a key parameter for the resulting micro-mechanical and tribological properties of the part. It was proven that the mechanical and tribological properties are not equal in varying regions of the part. Due to cooling and process parameters, the difference in the indentation modulus in individual measurement points was up to 55%, and the tribological properties, in particular the friction coefficient, showed a difference of up to 20%. The aforementioned results indicate the impact this finding signifies for injection molding technology in technical practice. Tribological properties are a key property of the part surface and, together with micro-mechanical properties, characterize the resistance of the surface to mechanical failure of the plastic part when used in engineering applications. A suitable choice of gate location, finishing method of the cavity surface, and process parameters can ensure the improvement of mechanical and tribological properties in stressed regions of the part. This will increase the stiffness and wear resistance of the surface.

Keywords: polypropylene; gate distance; micro-mechanical properties; tribological properties; structure; crystallinity; surface quality

Citation: Ovsik, M.; Fucikova, K.; Manas, L.; Stanek, M. Influence of Polymer Flow on Polypropylene Morphology, Micro-Mechanical, and Tribological Properties of Injected Part. *Lubricants* **2024**, *12*, 202. https://doi.org/10.3390/lubricants12060202

Received: 15 April 2024
Revised: 31 May 2024
Accepted: 2 June 2024
Published: 4 June 2024

Copyright: © 2024 by the authors. Licensee MDPI, Basel, Switzerland. This article is an open access article distributed under the terms and conditions of the Creative Commons Attribution (CC BY) license (https://creativecommons.org/licenses/by/4.0/).

1. Introduction

Injection molding is one of the most commonly used manufacturing methods to produce polymer parts. It is characterized by a high degree of automation, high productivity, and good volumetric stability of injected parts. The polymer is exposed to thermal and mechanical effects during the injection molding cycle. The polymer experiences a transition from a molten state to a rubber, glass, or crystalline state. The final physical, optical, and mechanical properties of the injected part closely correlate with the created micro-structure.

During the melting phase of the injection molding process, the polymer experiences high shear stress, normal pressure, and a thermal gradient. The high pressure at the wall leads to the creation of a highly oriented lamellar micro-structure, commonly called the skin layer. On the other hand, the low pressure in the core leads to the creation of a spherulitic micro-structure. This difference in morphology between the surface and core has already been investigated in numerous works [1–6]. This phenomenon is commonly called skin–core morphology, which can be observed by a polarized optical microscope (Figure 1). Injection-molded semi-crystalline parts generally contain 2–5 layers [7–9].

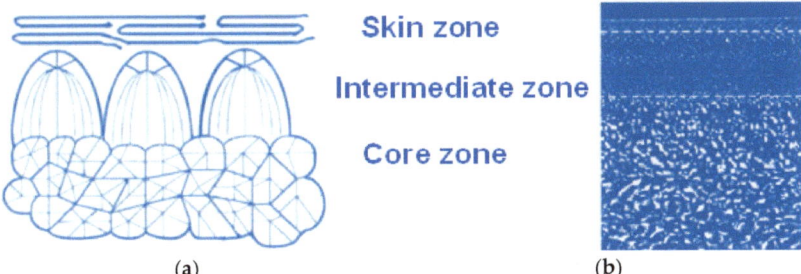

Figure 1. Morphological structure: (**a**) scheme of skin–core structure; (**b**) optical microscope—skin–core structure.

Crystal growth occurs at temperatures below the melting point (T_m) and above the glass transition point (T_g). Higher temperatures interfere with the molecular arrangement, while lower temperatures lead to the freezing of molecular chains' movement. Secondary crystallization can occur under T_g, although in the time scale of months and years. This process influences the mechanical properties of polymers and decreases their volume due to the more compact arrangement of polymer chains [10,11].

The growth of crystalline regions occurs in the direction of the biggest temperature gradient. In the case of a strong gradient, this growth is unidirectional and has a dendritic character. If the temperature distribution is isotropic and static, then the lamellae grow radially and create bigger quasi-spherical aggregates called spherulites. Spherulites' size ranges from 1 to 100 μm [10]. During observation by a polarized optical microscope, spherulites create a great number of colored figures, including the typical Maltese cross [10–12].

The crystallization mechanism of the polymer melt is quite important for the injection molding of plastic parts. Different types of crystallization occur, for example, in extrusion during the production of fibers and films. The crystallization theory indicates that the thickness of lamellae is always less than the expected length of molecules. This means that molecules in crystals can be folded numerous times. During crystallization, only a part of the melt is deposited in the crystalline phase, while the remainder freezes in the amorphous phase, which envelops the crystalline regions. A common polyolefin is a semi-crystalline material with a relatively complex interior structure [10].

The problem of crystallization during injection molding was investigated by Le et al. [13], who focused on the influence of pressure on crystallization kinetics. The detected change in crystallization temperature enabled the identification of the pressure dependence of crystallization kinetic parameters Tm and T_g, which are used in the Hoffman–Lauritzen equation. Crystallization is the ability of a material to create rigid structures, which give the material its specific properties.

Liu et al. [14] compared the morphology of iPP samples prepared by conventional injection molding and micro-injection molding. The samples were studied using PLM, SEM, DSC, and WAXD. The results showed that micro-injected samples contained a much higher percentage of oriented shear layers. Varying crystallization in different points of the injected part was investigated by Sun et al. [15] and Pantani et al. [16], who focused on the creation of skin–core structure in the injected parts. The change in polypropylene morphology is described by the arrangement and size of spherulites in dependence on distance from the wall. The problem of crystallization was researched by more authors [17,18].

The correct setting of injection molding process parameters is a key factor not only for the stable manufacturing process but also for the expected final properties of injected parts. The most important parameters that significantly influence the entire injection molding process are injection speed, injection pressure, holding pressure, duration of holding pressure, melt temperature, and mold temperature. The individual process parameters act simultaneously and influence each other. Therefore, a change to one parameter affects other parameters [1].

Furthermore, the influence of process parameters on the mechanical properties of the injected part has been studied in numerous other publications. Wang et al. [19] focused on the effect of process parameters (especially injection speed) on the mechanical properties of micro-injected PP samples. It was found that increasing injection speed led to higher hardness, which increased more in the perpendicular direction of flow than in the flow direction. A similar study was conducted by Glogowska et al. [20], who injected samples with differing process conditions. The samples were then grinded, injected again, and their mechanical properties measured. Sykutera et al. [21] investigated the influence of process conditions on the polymer viscosity, which was measured directly in the cavity. In general practice, viscosity is measured in rheometers, but the goal of this study was to provide real values coming straight from the manufacturing process. Studies [22–25] dealt with the effect of multiple process conditions (pressure and temperature) on final flow length.

In conclusion, the submitted study focuses on the influence of flow length (distance from the gate) and process conditions on the creation of crystalline morphology (skin–core structure). The final structure affects the micro-mechanical properties of the surface layer of injection-molded polypropylene. The literary research on the problem of the influence of distance from the gate on the mechanical properties of injection-molded parts was conducted, and no existing publication concerning this topic was found. In most cases, the mechanical properties and sometimes hardness were measured only locally and subsequently taken as results for the overall part. However, there was no study that investigated the problem of varying properties along the polymer flow path of injection-molded products. The aforementioned studies were concerned with partial research, mainly with the influence of process parameters in injection molding on mechanical properties and the effect of tool quality on flow length and surface replication. Furthermore, other studies focused on changes in crystallinity, but once again, mostly locally. Some investigated properties were researched only for micro-injection molding, which is quite different from regular molding. In the field of micro-injection molding, several studies were concerned with the influence of surface roughness on the flow length of the polymer. The results of some studies were similar to macro-injection molding, although the dimensions of the final part were in the range of micrometers; thus, the effect of surface roughness on polymer flow was much greater. These results cannot be directly applied to those of macro-injection molding due to the significant difference in size. As the mechanical properties of injection-molded products are not uniform along the flow length, this research was designed to specifically target the non-homogeneity of the properties of injection-molded parts. This non-homogeneity can be influenced by numerous factors, for example, the distance from the gate, the temperature of the mold and melt, injection pressure, crystalline structure, degree of cooling, the surface of the mold, and others. This problem has a significant effect on injection molding in general practice.

2. Materials and Methods

The preparation of the experiment was inspired by the practical requirements of manufacturing injection-molded technical parts. Individual designs were checked by injection molding simulation. Information gained from the simulation was used to choose the most suitable technological parameters.

2.1. Injection-Molded Material (Polypropylene)

This research, as well as the selection of material (polypropylene), is inspired by general practice requirements, as polypropylene is commonly used for injection molding of technical parts. The testing was conducted on polypropylene, which is a semi-crystalline thermoplastic with the trade name Borealis BJ380MO provided by Borealis (Linz, Austria). The selected material is commonly used in the automotive industry, from which the request for testing properties along the flow length originated. Polypropylene is nowadays being pushed out by more expensive construction materials. On the other hand, polypropylene is still quite useful, especially due to its wide range of applications and processing parameters. The material properties were taken from the provided material sheet, as can be seen in Table 1.

Table 1. Basic properties of injection-molded material.

Properties	Unit	Value
MFI	g/10 min	80
Density	kg/m^3	905
Elastic modulus	GPa	1.3
Melt temperature	°C	210–260
Mold temperature	°C	20–60
Holding pressure	MPa	Min. 20

2.2. Injection Molding

The test samples were prepared using the injection molding machine Allrounder 470 E 1000-290 Golden Edition, manufactured by Arburg (Losburg, Germany). The mold tempering was conducted by the oil tempering unit Regloplas 150 Smart, manufactured by Regloplas (St. Gallen, Switzerland). The process conditions were set based on values gained from the injection molding simulation and material sheet (Table 2). The test samples were manufactured as rectangular blocks with dimensions of 6 × 1 × 240 mm. The selection of injection molding parameters for this research is once again based on the requirements of industrial practice. This research is only a part of extensive work regarding the problem of the polymer flow path and its final properties. The chosen injection molding parameters for this research were taken closer to the lower boundary of recommended parameters given by the manufacturer with regards to the economic perspective of the length of the injection molding cycle (melt temperature 215 °C, mold temperature 30 °C). The recommended melt temperature range is 215–255 °C, and the mold temperature is 30–50 °C.

Table 2. Technological parameters of the injection molding process.

Technological Parameters	Unit	Value
Injection pressure	bar	800
Holding pressure	bar	640
Holding pressure duration	s	1
Cooling time	s	20
Clamping force	kN	1000
Mold temperature	°C	30
Melt temperature	°C	215
Screw zone 1	°C	215
Screw zone 2	°C	210
Screw zone 3	°C	205
Screw zone 4	°C	200
Screw zone 5	°C	200

The cavity of the testing injection mold was in the shape of four grooves with varying lengths, which were determined according to previously conducted injection molding analysis that focused on the estimation of process parameters that could lead to the complete filling of the mold. The length of individual cavities was 208 mm, 158 mm, 68 mm, and 38 mm (Figure 2a). The cavity had a rectangular cross-section with a dimension of 6 × 1 mm and a length of 208 mm (Figure 2b). The connectivity of the cavity with the runner was ensured by a cold slug, which was rotated by 90° for each length. The shape plates were changeable and fit into a universal frame. These plates were tempered through drilled channels.

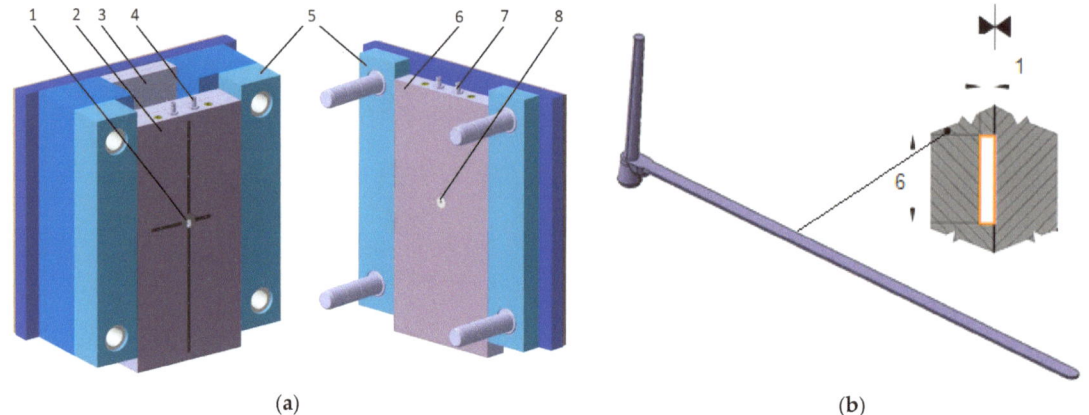

Figure 2. Injection mold cavity: (**a**) Test mold, 1—sprue holder, 2—mold plate, 3—ejection plates, 4—tempering system right, 5—mold frame, 6—test mold plate, 7—tempering system left, 8—sprue insert; (**b**) cross-section of mold cavity.

The cavity of the injection mold was manufactured by fine milling, as the required quality of the surface was in the range of Ra 1.6 µm to Ra 2.2 µm. The conditions of the manufacturing were set according to the requirements.

2.3. Injection Molding Simulation

The injection molding simulation was performed in order to provide data that could be compared with real results and to help with the setting of the process parameters. The simulation was performed in MoldFlow Synergy, which was made by Autodesk (San Rafael, CA, USA). The conditions were set according to the material sheet in a way that closely resembled conditions during injection molding.

The material for simulation was selected from the software database, which offers the same type of polypropylene (Borealis BJ380MO) that was used to produce the injection-molded specimens. The parameters of the selected material can be seen in Table 3.

Table 3. Basic properties of polypropylene (Borealis BJ380MO).

Properties	Unit	Value
MFI	g/10 min	80
Density	kg/m^3	900
Elastic modulus	GPa	1.3
Melt temperature	°C	210–260
Mold temperature	°C	20–60
Ejection temperature	°C	117

The imported model was hatched by a 3D network made of 4-sided bodies (Figure 3), which provided an adequate representation of the part in its entire volume and subsequent display of results throughout the entire thickness of the part. The gate was made in the form of a conical sprue. The generated 3D network was subsequently checked for all requirements, and then the simulation was conducted (Table 4).

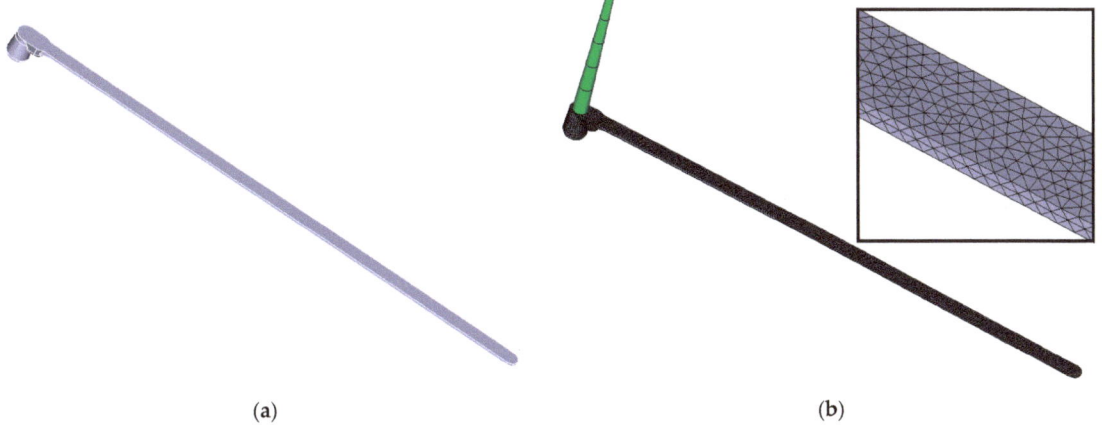

Figure 3. (**a**) Imported 3D model; (**b**) meshed model includes gate system.

Table 4. Properties of mesh.

Properties	Unit	Value
Mesh type	-	Tetrahedral
Global edge length on surface	mm	1
Entity counts	-	111,686
Aspect ratio	-	4.61

Figure 4 displays the tempering system and the mold block. The trajectory of tempering channels, including the inlet and outlet of the tempering medium, is derived from the real mold concept. The core, cavity, and other mold plates were simplified as blocks for the sake of simulation. Geometry defined in this way was subsequently also hatched by a 3D network and then checked. After this step, it was necessary to set all other conditions of the simulation, which can be seen in Table 5.

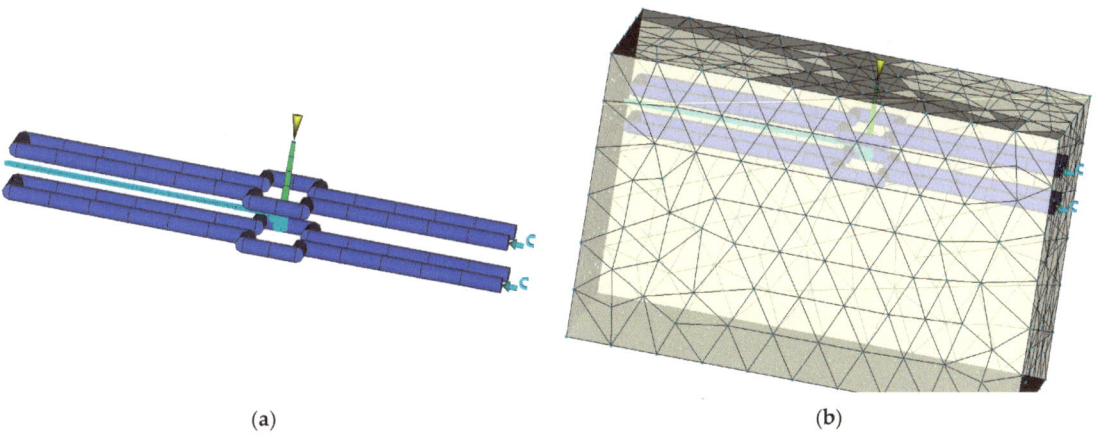

Figure 4. (**a**) Cooling system; (**b**) mold block.

Table 5. Process settings for simulation.

Properties	Unit	Value
Molding material	-	Borealis BJ380MO
Mold material	-	Tool steel P-20 (1.2311)
Injection molding machine	-	Allrounder 470e 143 tons 16.4 oz
Melt temperature	°C	215
Ejection temperature	°C	117
Cycle time	s	30
Mold temperature	°C	30
Coolant	-	oil
Flow rate	lit/min	50

After the simulation was started, all necessary calculations ran according to preset conditions. The complex Moldflow analyses provided numerous important results that were significant for the visualization of events taking place in individual phases of the manufacturing cycle and allowed the evaluation of qualitative parameters of the product. The goal of the simulations was to analyze the time of filling, injection pressure, and orientation in the skin–core layer. The results of these analyses provide important information for the stabilization of the injection molding process and the prediction of polymer behavior in the cavity.

2.4. Tribological Properties

Tribological properties were measured using the MicroCombi tester MCT^3 from Anton Paar (Graz, Austria). The measurements were carried out using the micro-indentation test, which allows the coefficient of friction and abrasion resistance of the tested surfaces to be determined. The principle of this measurement is based on the straightforward movement of the indentor (Rockwell cone) with a tip angle of 120° and a tip radius of 100 μm along the surface of the test specimen. The process parameters can be seen in Table 6. Tribological properties were measured on a MicroCombi tester, which was also used in the works of other authors [26,27].

Table 6. Measurement parameters for tribological properties.

Measurement Parameters	Unit	Value
Applied load	N	1
Speed	mm/min	10
Length	mm	5
Acquisition Rate	Hz	30

Ahead of the process, the indentor moves along the surface with a defined force to initiate the measurement device. Following the initiation, the indentor penetrates the test sample with normal force F_n, which leads to material deformation and the creation of an imprint. The sensors record the friction force F_t, which is proportional to the normal force, and a so-called pre-scan, which is concerned with the surface profile of the test sample before the penetration depth P_d (penetration depth) and post-scan of the imprint R_d (residual depth), which are important for polymer relaxation evaluation. Valuable information about the tribological properties of the material can be obtained from the difference in profile depths ($P_d - R_d$). Finally, the critical loads can be accurately measured using the acoustic emission AE and friction coefficient μ. Figure 5 shows a schematic diagram of the micro-tribometer.

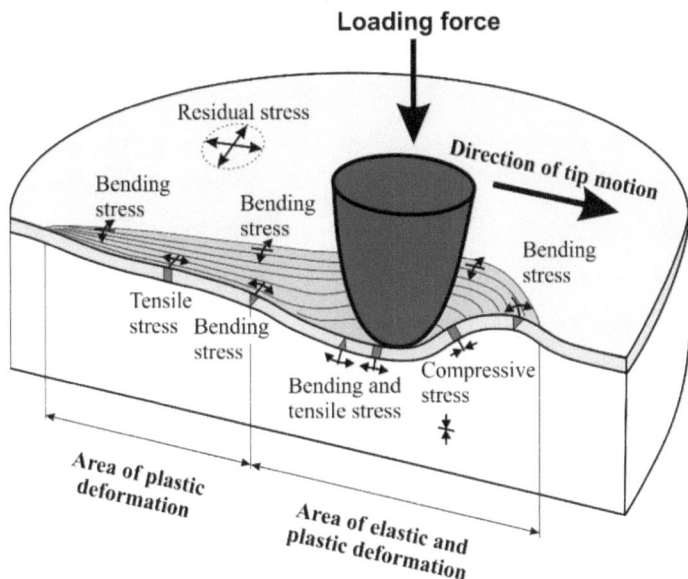

Figure 5. Schematic diagram of the micro-tribometer.

2.5. Micro-Mechanical Properties

The micro-mechanical properties measurement was conducted by depth sensing indentation (DSI) on a Micro-Combi tester (MCT^3) manufactured by Anton Paar (Graz, Austria).

A four-sided diamond pyramid with a top angle of 136° (Vickers indentor) was used as the indenting body. The measurement was conducted by the DSI method (depth sensing indentation), and the data gained were evaluated by the Oliver and Pharr method. The measurement was conducted according to the ČSN EN ISO 14577 standard [28]. Two loading forces were used to observe the changes in the structure (skin–core layer). The use of loading force 1 N led to a 20 µm depth of indentation, while the application of 5 N led to a 100 µm depth of penetration. The measurement parameters can be seen in Table 7. Since it was a micro-hardness measurement, which works with low indenting force with depth in the range of µm, the only evaluated parameters were indentation hardness, indentation modulus, and indentation creep.

Table 7. Parameters of mechanical property measurement.

Measurement Parameters	Unit	Measurement of Skin Layer	Measurement of Skin + Core Layer
Applied load	N	1	5
Maximum load duration	s	90	90
Loading and de-loading speed	N/min	2	10
Poisson's ratio	-	0.4	0.4

In order to determine mechanical properties and structure changes at varying distances from the surface, the test samples were cut, fixed in resin, and polished. The test samples prepared in this way can be measured in a direction perpendicular to the surface. Individual cuts were conducted in five sections evenly distributed along the flow direction (gate location 0 mm, and then 79 mm, 158 mm, 198 mm, and 220 mm from the gate).

As described in the ISO 14577 standard [28], the following parameters were evaluated: indentation hardness, modulus, and creep. The calculation of the individual values has been carried out using the Oliver and Pharr method (Figure 6).

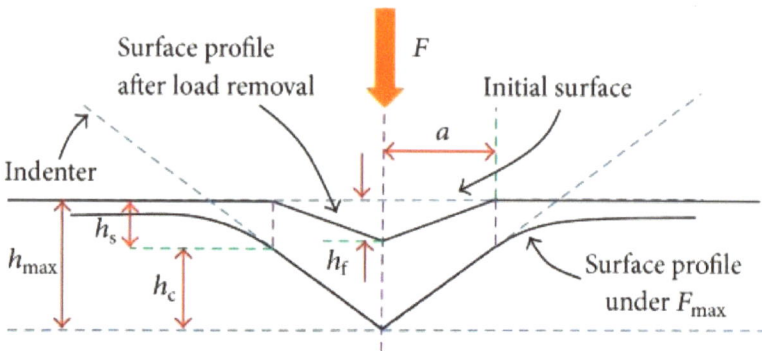

Figure 6. Schematic representation of the indentation processes shows the decrease in indentation depth during loading (according to Oliver and Pharr).

Indentation hardness (H_{IT}) can be defined as the ability of a material to resist plastic deformation. Indentation hardness H_{IT} (Equation (1)) is expressed by the ratio between the applied load F_{max} and the contact area (A_p) between the indenter and the specimen at the maximum depth and load. The contact area A_p (Equation (2)) is specified by the shape constant of the indentor (24.50 for the Vickers indenter) and the depth of contact of the indentor with the test specimen h_c [29,30].

$$H_{IT} = \frac{F_{max}}{A_p} \quad (1)$$

$$A_p = 24.50 \cdot h_c^2 \quad (2)$$

Another material property that can be obtained from indentation testing using the DSI method is the indentation modulus E_{IT} (Equation (3)). The indentation modulus E_{IT} is comparable with Young's modulus of the material. In general, the indentation modulus can be determined from the slope of the tangent line used to calculate the indentation hardness H_{IT}. As described in the ISO 14577 standard, the reduced modulus, E_r, is used to account for the fact that the elastic displacements occur in both the indenter and the sample. The instrumented elastic modulus in the test material, E_{IT}, can be calculated from E_r (Equation (5)). The calculations involve Poisson's ratio (v_s), which is usually between 0.2 and 0.4 for metallic materials and 0.3 and 0.4 for polymeric materials [29,30].

The plane strain modulus E^* is calculated from the following Equation (4), where E_i is the elastic modulus of the indenter (diamond 1141 GPa), E_r is the reduced modulus of the indentation contact, and v_i is Poisson's ratio of the indenter (0.07) [29,30].

Reduced modulus E_r is calculated from the following Equation (5), where C is contact pliability and A_p is contact area, $\sqrt{A_p} = 4.950 * h_c$ for the Vickers indentor [29,30].

$$E_{IT} = E^* \cdot (1 - v_s^2) \quad (3)$$

$$E^* = \frac{1}{\frac{1}{E_r} - \frac{1-v_i^2}{E_i}} \quad (4)$$

$$E_r = \frac{\sqrt{\pi}}{2 \cdot C \sqrt{A_p}} \quad (5)$$

Indentation creep is defined as the relative change in the indentation depth while the applied load is kept constant. It is defined in the ISO 14577 instrumented indentation standard as C_{IT} (Equation (6)), where h_1 is the depth at the beginning of the creep test, and h_2 is the depth at the end of the creep test [29,30].

$$C_{IT} = \frac{h_2 - h_1}{h_1} 100 \qquad (6)$$

2.6. Differential Scanning Calorimetry (DSC)

The behavior of the tested samples during melting and solidification was observed by the differential scanning calorimeter DSC Q20 (TA Instruments, New Castle, DE, USA). The weight of the sample was 6 mg, which was prepared by a microtome device. The speed of heating and cooling was set to 10 °C/min. The DSC measurement was divided into two parts. The first part contained primary heating from T_0 to T_1, followed by staying at constant temperature (isotherm 1 min), cooling to T_0, staying at constant temperature (isotherm 1 min), and heating to T_1. The observed properties were evaluated during the first heating phase. The first phase of heating was observed in order to determine the changes induced by the previous processing of the material (injection molding). The first phase of heating allowed the observation of the history of polypropylene's processing and its influence on the structure and macromolecules, while the second phase no longer showed this history.

Polypropylene Borealis BJ380MO is not 100% crystalline; its crystallinity is around 44%, as reported by the authors in the [31] article. The crystallinity must be recalculated according to Formula (7). Crystallinity was calculated from the heat flow by the following equation:

$$X_c = \frac{\Delta H_m}{\Delta H_m^{100}} \times 100 \qquad (7)$$

where X_c is crystallinity (%), ΔH_m is heat flow (J/g), and ΔH_m^{100} is heat flow for 100% crystalline polypropylene (207 J/g), which was found in the literature [32].

The process parameters of the DSC test for the determination of the crystallinity of polypropylene were set according to previously conducted studies [32–36]. The authors of these publications used DSC to determine the crystallinity of polypropylene and demonstrated the use of this method to find the crystallinity with sufficient sensitivity.

2.7. Polarized Optical Microscope

Tested samples were cut into slices with a 20 µm thickness in the transverse direction on a rotational microtome Leica RM 2255 (Deer Park, IL, USA). These cuts were examined by a polarized optical microscope, which helped in determining the changes in structure along the cross-section of the tested material.

3. Results

The goal of the experimental work was to find the influence of gate distance on observed parameters such as micro-mechanical properties (indentation hardness, indentation modulus, and indentation creep) and morphology. These properties were measured on the surface layer.

3.1. Injection Molding Simulation

The process conditions were set according to the MoldFlow analysis. Figure 7 shows the results of filling time (Figure 7a), injection pressure (Figure 7b), and skin–core structure (Figure 7c,d) for plate-shaped samples. The results were used not only to set the process parameters of injection molding on the machine itself but also to show the behavior of the polymer in the cavity and predict the orientation created in the skin and core layers. As can be seen in the results of layer orientation, a highly oriented layer was created in the surface layer, while an un-oriented structure was created in the core. This orientation of the skin–core layer has a significant effect on final properties, which can be seen in the results.

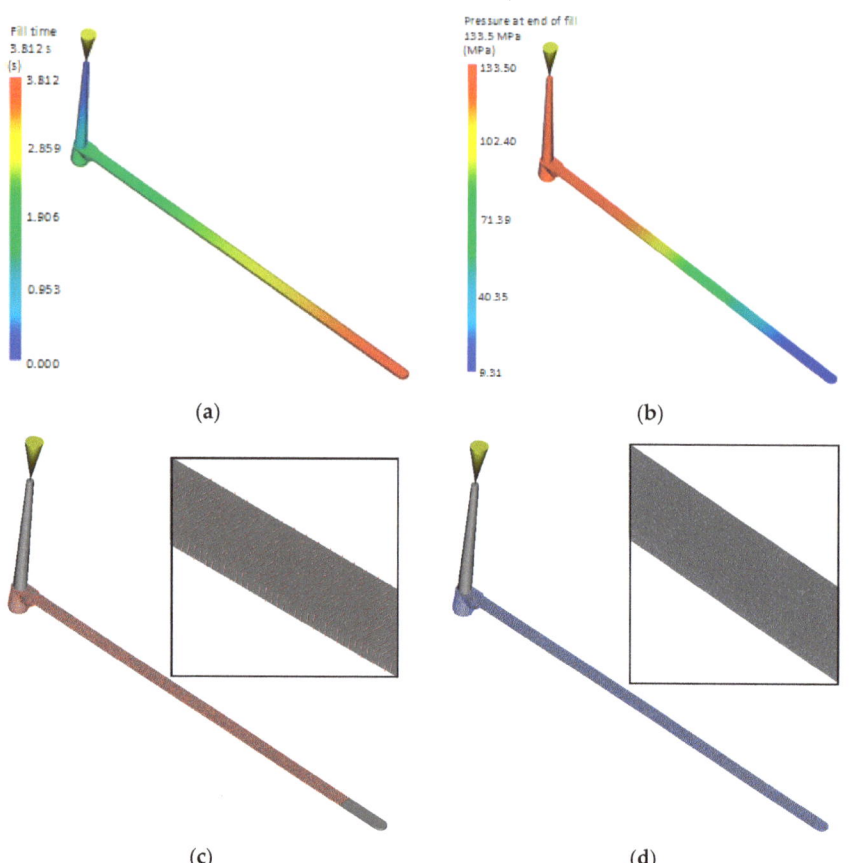

Figure 7. Results of simulation analysis: (**a**) filling time; (**b**) injection pressure; (**c**) core layers; (**d**) skin layers.

The orientation of the skin result (Figure 7d) provides a good indication of how molecules will be oriented on the outside of the part, showing the average principal alignment direction for the whole local area at the end of the filling. The magnitudes of these vectors are normalized to one and are displayed multiplied by the given scale factor. Skin orientation is determined by the velocity direction when the melt front first reaches a given location. The orientation of the skin result is useful for estimating the mechanical properties of a part. For example, the impact strength is typically much higher in the direction of molecular orientation at the skin. The skin orientation generally represents the direction of strength. For plastic parts that must withstand high impact or force, the gate location can be designed to give a skin orientation in the direction of the impact or force.

The orientation at the core result (Figure 7c) provides a good indication of how molecules will be oriented at the part core, showing the average principal alignment direction for the whole element. The core orientation for each triangular element is perpendicular to the velocity vector before the center layer reaches the transition temperature. This is the most probable orientation in the core region of a part. The other possible orientation is in the direction of the velocity vector. Core fiber orientation may be different from skin fiber orientation. Because the melt freezes very quickly when it contacts the mold for the first time, the velocity vector provides the most probable molecular orientation at the skin.

3.2. Tribological Properties

Using the scratch test, it is possible to examine the tribological properties of the surface layer of the tested injection-molded polypropylene, such as frictional force (Figure 8a), acoustic emission (Figure 8b), friction coefficient (Figure 8c), and surface profile after indentation (Figure 8d). The tribological properties were examined at five locations, which were specified at different distances from the point of the injection gate to the end of the part. The measured distance X was determined to be 5 mm for all samples. These tests can help with the prediction of surface resistance to abrasion and thus complement measurements of micro-mechanical properties such as hardness, modulus, etc.

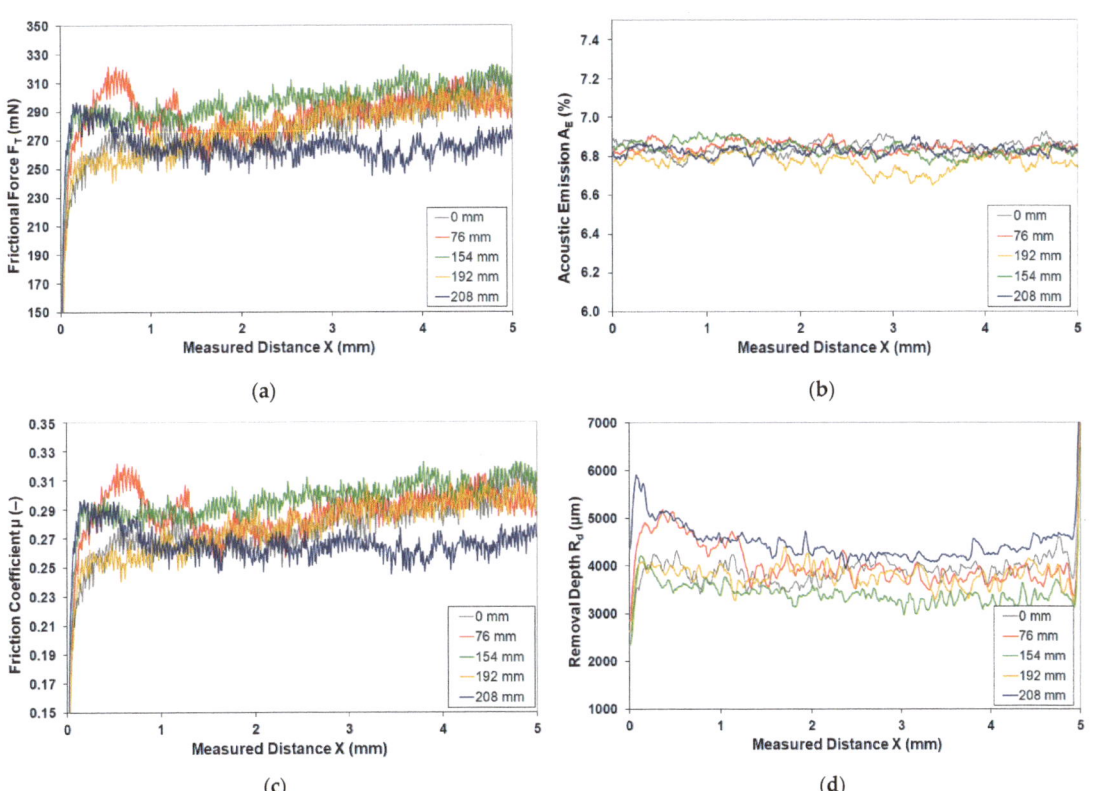

Figure 8. Tribological properties: (**a**) frictional force; (**b**) acoustic emission; (**c**) friction coefficient; (**d**) removal depth.

The results of tribological tests show that the highest values of observed properties, e.g., friction force and friction coefficient, were found at a distance of 154 mm. Acoustic emission showed similar values for all measured distances. On the contrary, observed tribological properties improved at the gate (0 mm) and at 208 mm from the gate, where the lowest values of tribological properties were measured. For example, the friction coefficient showed a value of 0.27 at a distance of 0 mm (start of the part), 0.30 at a distance of 154 mm (center of the part), and 0.25 at a distance of 208 mm (end of the part). The difference in the coefficient of friction between the individual points of the part was up to 20%.

As the results show, the tribological properties of injection-molded PP are not homogeneous along the length of the part, but due to different flow conditions, the tribological properties change, which is reflected in an increase in the wear resistance of the surface layer. This enhancement was caused by structural changes that directly impacted micro-mechanical properties.

3.3. Micro-Mechanical Properties of Surface Layer

This part was focused on the observation of micro-mechanical properties (indentation hardness, indentation modulus, and indentation creep) in the surface layer at various distances from the gate. The depth of mechanical property measurement reached approximately 20 μm (Figure 9a,c,e). The structures and micro-mechanical properties deeper in the test samples were measured with a loading force of 5 N, which penetrated up to 100 μm (Figure 9b,d,f). This measurement was supplemented by morphology (crystallinity) observations and the thickness of the skin layer.

The test samples prepared by injection molding demonstrated that distance from the gate significantly affects mechanical properties, such as indentation hardness and modulus. The hardness decreased with distance from the gate, up until the measurement point located at 154 mm. The total decrease was 15%. Further measurement points demonstrated the opposite trend, and the values measured at the end of the sample were 17% higher than at the gate and 34% higher than at the middle (154 mm) (Figure 9a).

The indentation modulus (Figure 9c) characterizes the rigidity of the part, and it can be determined from the de-loading phase of the indentation cycle. The reported results (20 μm) of the indentation modulus display similar tendencies to indentation hardness. When compared, the modulus values measured at the end of the sample were 24% higher than at the gate. On the other hand, the indentation modulus measured in the middle of the sample was 55% smaller than at the gate.

Indentation creep is another important property of the polymer part, as it characterizes the resistance of the material against long-term strain. Most polymer parts are used in applications that are exposed to continuous and long-term strain. The results of indentation creep show that flow path and flow length have a significant effect on the values of indentation creep (Figure 9e). Once again, the indentation creep decreased by 20% until the measurement point was located at 154 mm and then improved by 36%. The indentation creep value measured at the end of the sample was 14% higher in comparison with the gate.

The measurement of the mechanical properties of the surface layer at individual distances from the gate shows that the length of polymer flow and subsequent cooling of the polymer within the cavity have a significant influence on the mechanical and morphological properties of injection-molded products. Within the surface layer, the mechanical properties (indentation hardness, indentation modulus, and indentation creep) were not homogenous along the flow direction. The maximum values were measured at the gate (0 mm) and at the end of the injection-molded product (208 mm), while the minimum values were measured at the middle of the sample (154 mm), which resulted in the decline of mechanical properties. The measurements at a penetration depth of 100 μm displayed an opposite trend in values of indentation hardness. An increasing indentation hardness was observed, up to 154 mm from the gate. The increase in indentation hardness from the gate (0 mm) to the distance of 154 mm was approximately 43%. On the other hand, a significant decrease in mechanical properties, up to 15%, was observed at the end of the sample. When comparing the gate and end of the sample, it can be seen that there was a 24% increase in indentation hardness between these points. (Figure 9b).

The values of indentation modulus in 100 μm depth once again demonstrated an opposite trend to those found in 20 μm. Towards the center of the part, an increasing trend in values of the indentation modulus was observed. The difference between the gate (0 mm) and middle (154 mm) was 120%. Towards the end of the part, the indentation modulus decreases by 47% (Figure 9d).

Values of indentation creep in 100 μm displayed improvement towards the center of the part by up to 18%. Towards the end of the sample, the values of indentation creep were the same as those measured in the gate location (Figure 9f).

An opposite trend in mechanical properties was observed in deeper layers of the material in comparison with the surface layer. This trend was also dependent on the distance from the gate. An increasing trend in mechanical properties was observed in test samples along the flow direction, i.e., the mechanical properties rose from the gate up to

154 mm from the gate and, on the other hand, declined from 154 mm from the gate up to the end of the specimen.

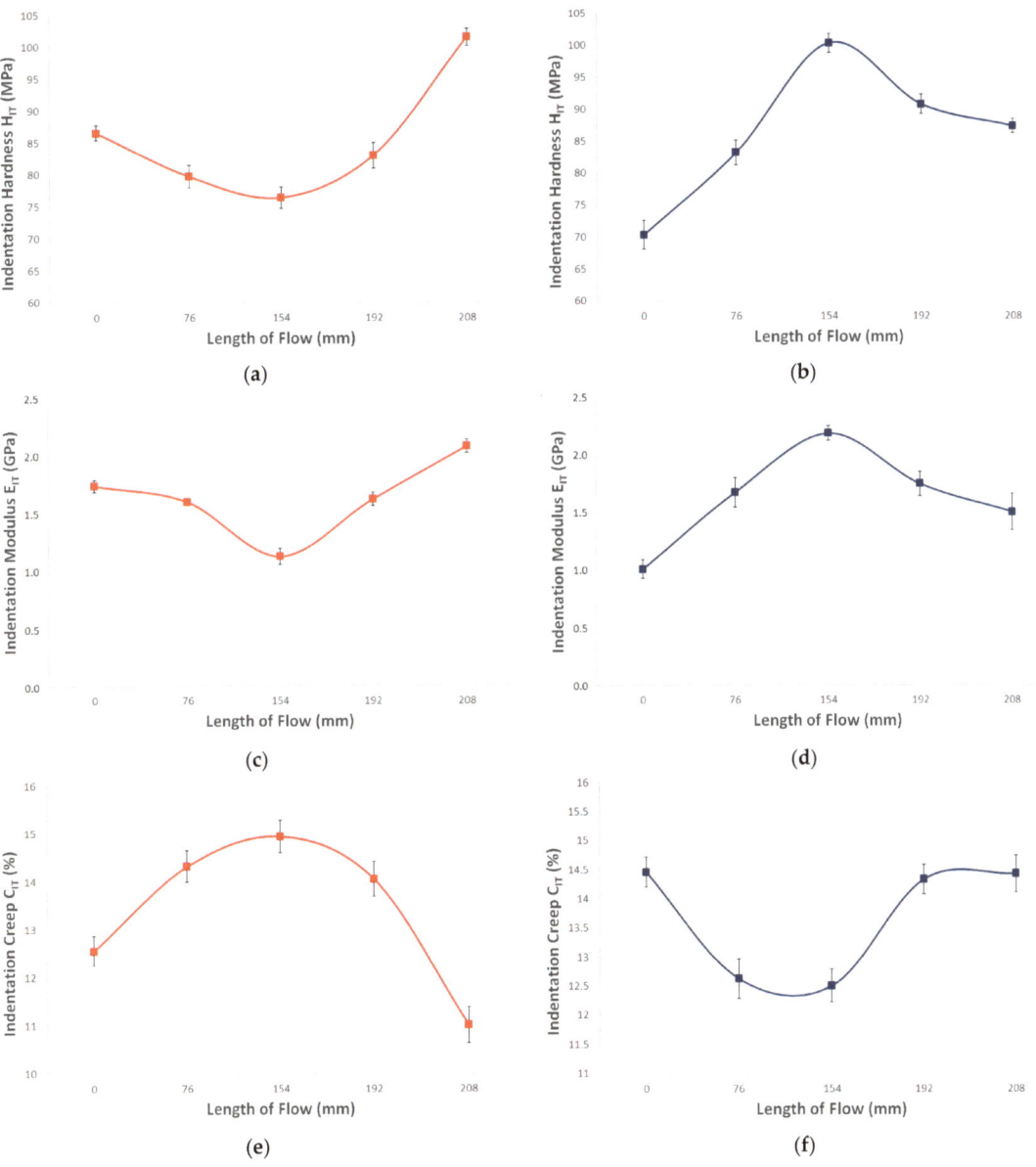

Figure 9. Influence of measurement depth and distance from the gate on mechanical properties: (**a**) indentation hardness—1 N (penetration depth 20 µm), (**b**) indentation hardness—5 N (measurement depth 100 µm), (**c**) indentation modulus—1 N (depth of measurement 20 µm), (**d**) indentation modulus—5 N (depth of measurement 100 µm), (**e**) indentation creep—1 N (depth of measurement 20 µm), (**f**) indentation creep—5 N (depth of measurement 100 µm).

The observed changes in mechanical properties are given by the distribution of the crystalline phase and the change in the arrangement of the structure (size and shape of spherulites) along the flow direction. This is significantly influenced by the magnitude of holding pressure and mold temperature. The aforementioned findings correspond with the behavior of polymers within the cavity, in which the polymer experiences fountain flow, i.e., the polymer moves from the middle of the cavity to the cold surface of the mold, where the polymer melts quickly, cools down, and creates a frozen layer. As is described in the following sub-paragraph, property changes at varying distances from the gate were caused by different skin layer thicknesses along the part. Molecules in this layer are more oriented, and the total content of the amorphous phase is higher than in the core. The thickness of the skin layer is lowest at the gate and at the end of the sample, while a thicker skin layer can be found in the middle of the sample.

3.4. Micro-Mechanical Properties across Sample Depth

In this sub-paragraph, the influence of gate distance in varying depths of the test sample on the mechanical properties was observed. The measurements were conducted on cut slices of test samples picked out at differing distances from the gate. The goal of this measurement was to prove varying mechanical properties in the cross-section of the test samples. The presentation of results was conducted using 3D columns in order to clearly display the trends of mechanical properties at two levels (along the specimen and in individual depths). The average values and standard deviation presented in these graphs can be seen in Tables 8–10.

Table 8. Statistical parameters of indentation hardness (MPa).

Length of Flow mm	Statistical Parameters MPa	Distance from Surface (mm)				
		0	0.25	0.5	0.75	1
0	$\bar{x}$	71.85	75.21	84.22	76.15	73.74
	s	0.72	0.81	0.81	0.17	0.74
76	$\bar{x}$	71.23	75.74	87.78	79.71	69.84
	s	0.55	0.75	0.71	0.58	0.48
154	$\bar{x}$	68.30	77.69	90.53	79.97	67.56
	s	0.65	0.70	0.82	0.58	0.93
192	$\bar{x}$	67.46	81.07	86.41	79.96	66.70
	s	0.93	0.65	0.19	0.11	0.51
208	$\bar{x}$	70.07	77.95	84.38	78.79	70.08
	s	0.66	0.45	0.65	0.87	0.96

Table 9. Statistical parameters of indentation modulus (GPa).

Length of Flow mm	Statistical Parameters GPa	Distance from Surface (mm)				
		0	0.25	0.5	0.75	1
0	$\bar{x}$	1.36	1.54	1.68	1.56	1.38
	s	0.04	0.03	0.05	0.08	0.05
76	$\bar{x}$	1.28	1.65	1.81	1.73	1.29
	s	0.05	0.04	0.01	0.03	0.06
154	$\bar{x}$	1.26	1.72	1.92	1.79	1.21
	s	0.09	0.08	0.06	0.07	0.03
192	$\bar{x}$	1.30	1.68	1.84	1.69	1.28
	s	0.03	0.01	0.01	0.02	0.03
208	$\bar{x}$	1.33	1.38	1.62	1.41	1.33
	s	0.06	0.08	0.04	0.01	0.09

The measured results indicate that the distance from the surface has a significant influence on mechanical properties. The results from all gate distances show that the indentation hardness is significantly higher at the center of the part than at the surface. These differences were, for example, 17% at the gate and 33% at 152 mm from the gate (Figure 10a). A similar trend was measured for the indentation modulus, which rose towards the center (Figure 10b). The difference between the surface at the gate and in the middle of the sample was 24%, while it was 53% at 0.5 mm depth. For indentation creep, a similar improvement between surface and center was measured, up to a 39% increase (Figure 10c).

Table 10. Statistical parameters of indentation creep (%).

Length of Flow mm	Statistical Parameters %	\multicolumn{5}{c	}{Distance from Surface (mm)}
		0	0.25	0.5	0.75	1
0	$\bar{x}$	12.56	11.92	11.54	12.05	12.56
	s	0.31	0.32	0.34	0.36	0.31
76	$\bar{x}$	14.32	11.37	11.21	12.13	14.32
	s	0.32	0.13	0.09	0.24	0.32
154	$\bar{x}$	14.94	10.87	10.69	11.70	14.94
	s	0.34	0.19	0.11	0.08	0.34
192	$\bar{x}$	14.05	11.64	11.46	12.05	14.05
	s	0.36	0.07	0.15	0.31	0.36
208	$\bar{x}$	11.00	12.32	11.94	12.05	11.00
	s	0.37	0.32	0.34	0.36	0.37

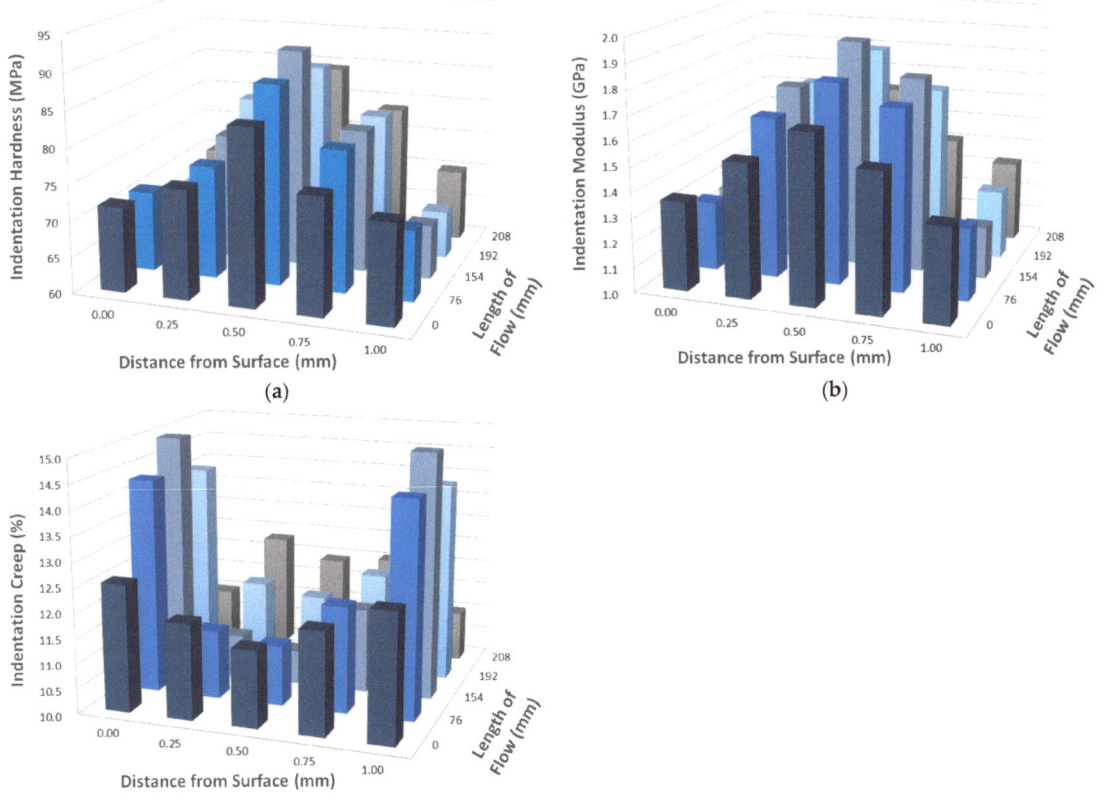

Figure 10. Influence of distance from the surface of the test sample and gate distance on mechanical properties: (a) indentation hardness; (b) indentation modulus; (c) indentation creep.

These results confirm the findings discussed in Sections 3.2 and 3.3 while also showing that the polymer structure is not the same across the entire cross-section of the tested sample.

3.5. Surface Quality

Results of surface replication (Figures 11 and 12) indicate that the surface quality of injection mold replicates on test samples to a limited degree and differently at individual points of the part. Milled surfaces displayed deviances in surface qualities that could be caused by the direction of milling. According to the results, the surface quality changes over the course of the flow length. The surface quality of the test sample near the gate was Ra 1.1 μm, while the mold had 2.1 μm. The surface quality at the end of the test sample (225 mm from the gate) was similar to the gate. Surface qualities at gate distances of 77 mm and 154 mm increased to 1.4 μm in comparison with mold, which displayed Ra 1.8 μm. After this measurement point, surface quality decreased all the way towards the end. The surface quality at the gate was Rz 7.5 μm for the test sample and Rz 11 μm for the injection mold, while at the end, the surface quality was Rz 4.2 μm at the gate and Rz 10.5 μm for the mold.

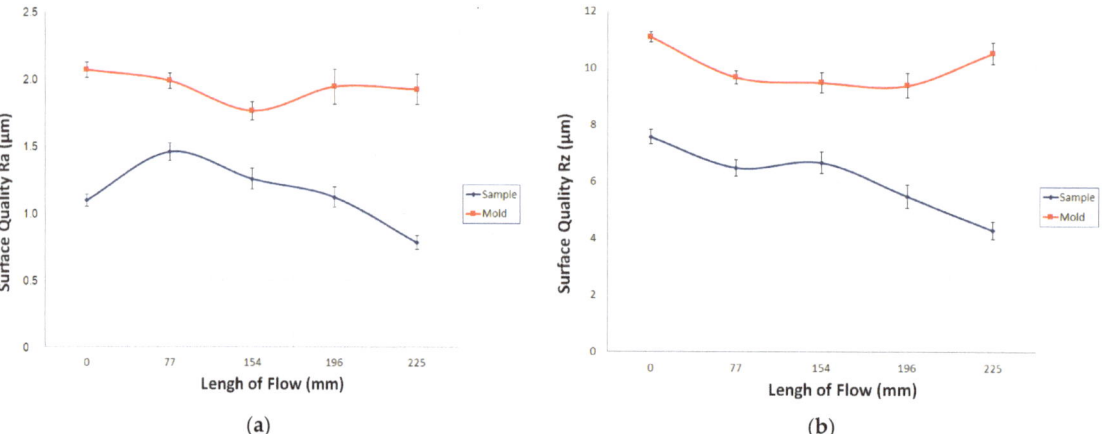

Figure 11. Influence of surface replication at varying distances from the gate: (**a**) surface quality Ra, (**b**) surface quality Rz.

The surface quality of the mold replicated with better surface quality on the test sample. The replication can be influenced by numerous factors, such as injection pressure, mold temperature, enclosed air, etc. The main parameter that influences surface replication is pressure drop, which manifests in a decrease in surface quality from 154 mm from the gate all the way to the end of the test sample.

The surface profile, as shown by the 3D surface image (Figure 12), shows that the difference in replication is significant. During flow, the polymer failed to fill the biggest irregularities in the mold due to the temperature profile. For all measured points, a positive trend in surface replication was measured. The results of the tool's and test sample's 2D surface quality profiles indicate that the highest irregularities of the tool surface were not replicated on top of the test sample, which manifested in differing surface quality at individual measurement points. These tendencies were most likely influenced by the pressure drop in the cavity, enclosed air, and melt temperature.

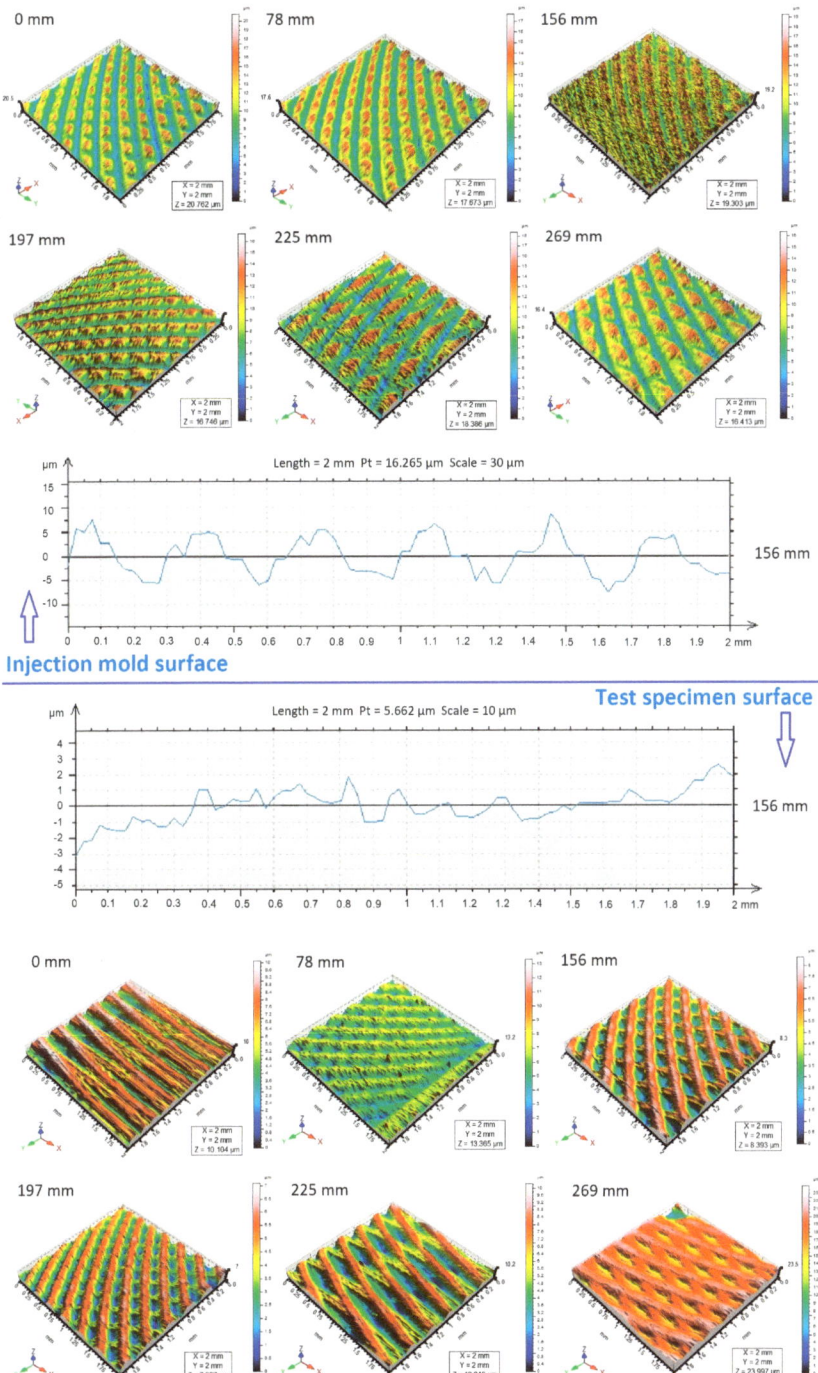

Figure 12. Replication of the tool's surface on the test sample—Ra 1.6 μm.

3.6. Influence of Gate Distance on Polypropylene Structure

This part served for the observation of morphology changes that occurred during the filling and cooling of test samples in the cavity. The individual structural changes were measured at the same distances from the gate as the mechanical properties.

3.6.1. Polarized Optical Microscope

The changes in surface layer (skin) thickness were observed along the length of the part by a polarized optical microscope. Microtome cuts with 20 μm thickness were made at individual distances from the gate. During injection molding, the polymer is forced to flow towards the cold surface of walls, where it cools and solidifies; this is called the fountain flow. This type of layer displays a high degree of orientation (Figure 13), which directly translates to specific polymer properties.

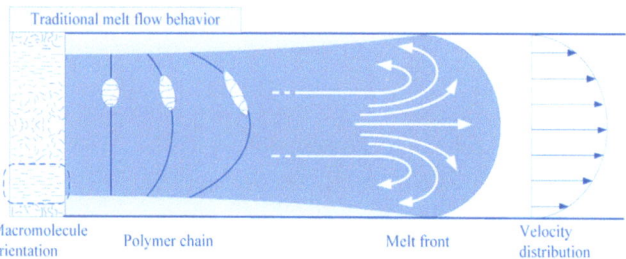

Figure 13. Mechanism of structure creation during injection molding.

Figure 14 illustrates cross-sectional views of different morphological structures on the surface and in the middle of an injection-molded tensile sample. As can be seen, the morphological structure changes with the thickness of the injected samples. The structure with high orientation appears in the skin layer, and the spherulitic structure with essentially no preferred orientation appears in the core layer. It has been reported that in the skin layer, because of the high shear stress and shear strain, the extended polymer chains lead to extended chain crystals. In the core layer, because of the absence of shear, the random polymer chains lead to lamellar, chain-folded crystals, and, finally, spherulites. Hence, the structure is related to flow-induced crystallization, and the spherulitic structure is related to quiescent crystallization.

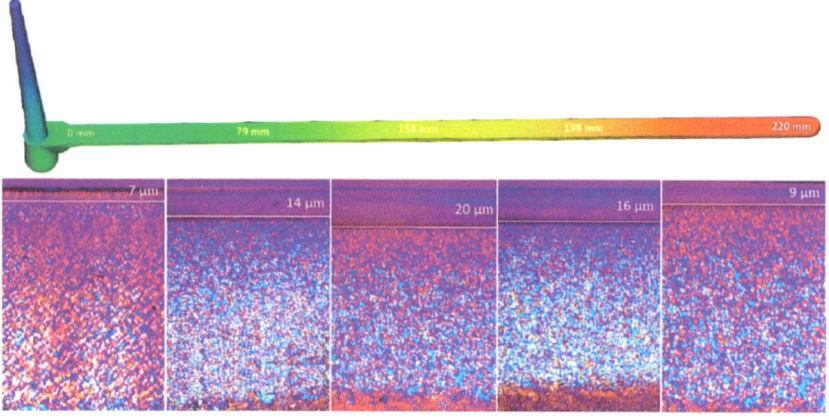

Figure 14. Changes in surface (skin) layer thickness at various distances from the gate.

As can be seen in Figure 14, the skin layer is not the same along the injection-molded part. The thickest skin layer can be found at gate distances of 158 mm in the sample (up to 20 µm), while the thinnest skin layer can be found at the end of the part. These differences in the thickness of the highly oriented layer are due to differences in the intensity and velocity of cooling of the polymer during mold filling. This is reflected in the resulting skin layer. The high orientation of macromolecules in the surface layer prevents the PLM light from passing, as seen in Figure 14. This surface layer has a significant influence on the varying mechanical properties of the part.

3.6.2. Differential Scanning Calorimetry

This sub-paragraph deals with observed changes in crystallinity (heat flow, Table 11) due to differing distances from the gate (Figure 15). Results of surface layer crystallinity indicate that the highest content of the crystalline phase is near the gate and at the end of the part. Towards the middle of the sample, the content of crystallinity decreases (Figure 15b). Crystallinity in the core layer (Figure 15d) points towards the opposite trend as observed in the skin layer. The highest crystallinity was measured at the center of the sample (158 mm). These results agree with the mechanical property measurements, which followed a similar trend.

Table 11. Heat flow ΔH_m (J/g).

Length of Flow mm	Heat Flow ΔH_m (J/g)		
	100% Crystalline Polypropylene	Skin	Core
0		88.78	84.45
76		87.48	88.69
154	207	84.19	92.01
192		88.28	86.42
208		91.45	86.51

It is obvious from the DSC measurements that the crystalline phase content, and thus micro-mechanical properties, change along the flow length (Figure 15). These changes correspond with changes in micro-mechanical properties. The crystallization rate is not uniform during polymer cooling, and so different structures are created, including shear-oriented lamellae and spherulites. These structures provide different mechanical properties.

The injection molding process is sensitive to polymer temperature, especially during cooling, during which the molecular chains orient in the direction of flow. In the core layer, the longer chains can remain in a stretched-out state, while the shorter chains are oriented randomly during filling. In the final structure, the prevalence of spherulites is significant.

The aforementioned results correspond with polymer behavior in the cavity, where the polymer flows by fountain flow from the middle towards the cold surface of the walls. The polymer melts, cools rapidly at the wall, and creates a solid layer. This significant cooling imposes a high degree of elongation orientation in the skin layer, while in other layers, the molecules have more time to relax. The combined effect of solidification and relaxation creates several regions with varying degrees of orientation (surface layer, shear layer, and core). The surface layer solidifies quickly with next to no relaxation and contains highly oriented molecules. This is caused by elongation deformation brought on by fountain flow. The degree of orientation corresponds with the flow length at the moment the mold is filled. The final orientation along the flow length is strongly affected by the holding pressure phase, as mentioned above. The degree of orientation and especially differences in crystalline morphology in individual layers of a part have a significant effect on research properties.

In technical practice, a part is generally understood to have uniform properties along its entire length. Although this does not correspond with reality, the properties can vary at different points. The findings of this work show that it is not possible to view one part as homogenous (from the mechanical and morphological point of view), but it is necessary to focus on specific points of the injected part. A suitable choice of gate location and process conditions, such as holding pressure or mold temperature, can result in improved local properties. This can be especially beneficial in parts with local straining, as they can be modified without requiring more expensive material.

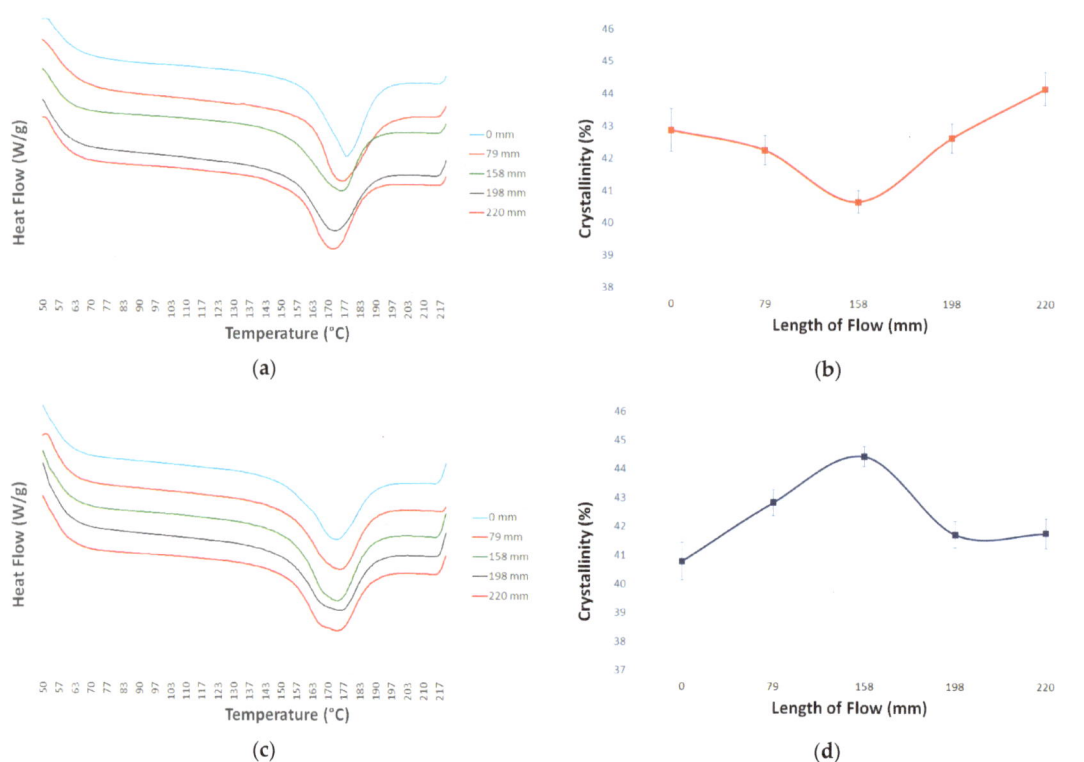

Figure 15. Measurement of crystallinity: (a) DSC characteristic—skin layer; (b) crystallinity of surface (skin) layer; (c) DSC characteristic—core layer; (d) crystallinity of core layer.

4. Discussion

This research aims to explain the tribological and mechanical behavior of polypropylene to support the beneficial introduction of those materials in actual applications. Generally, the designers have to take into consideration a set of tribological and mechanical parameters, not only one, including friction coefficient, wear, contact durability related to application, and the hardness of the surface. Polymers are very promising materials to be used for rubbing components in machines and devices. However, the selection of materials with appropriate tribological and mechanical properties is critical. Understanding the frictional and wear mechanisms controlled, in particular by the intensive and decisive transfer of material during the operation of polymeric tribosystems, is a very important task for tribologists. Low cost, corrosion resistance, damping of vibrations, ability to adapt to work in the presence of contamination, and many other advantages of the use of polymers in sliding (as well as rolling) systems open up a very interesting research area for tribology.

This research is part of large-scale research based on sub-parts from practice, where it has been found that the mechanical and tribological properties are not the same along the length but vary at different points in the injection-molded products. Based on this finding, test molds with different lengths, shapes, and cross-sections (straight cavity (Figure 2), spiral cavity, and cavity in the form of a tensile test body) were designed, in which the surface of the cavity was manufactured by different technologies with manual surface roughness (milled cavity Ra 1.6 μm, grinded cavity Ra 0.8 μm and Ra 0.45 μm, polished cavity Ra 0.1 μm, and coated cavity TiB$_2$). Subsequently, the tested materials (PP, PA6, etc.) and conditions of injection molding (melt temperature in the range of 215 to 255 °C, mold temperature 30–50 °C, and injection pressure 20–80 MPa) were changed. Based on these variations, it is possible to declare that the results demonstrated in this work can be used even in the case of different mold cavities and surfaces, materials, and processing parameters. The main condition is that the product has a shape with a longer flow path. Then, it can be said that the product's properties are not the same along the length of the part and are influenced by numerous factors that can be affected during injection molding. These changes have a significant effect on the distribution of the skin–core layer and, thus, the mechanical and tribological properties.

These findings are crucial for technical practice and can be used in the injection molding of polymer materials in the industry. Currently, in technical practice, the view of an injection-molded product's properties is quite simplified, with no regard for the non-homogeneity along the flow length. Based on the findings of this study, mold or process parameters can be modified to improve the properties of injection-molded products. The areas of the product that are more mechanically strained can be locally reinforced by these modifications. The biggest influence on mechanical and tribological properties along the flow is exerted by gate placement, which is closely followed by polymer cooling in the cavity. A local change in mold temperature in concrete areas could lead to significant changes in mechanical properties. Due to the aforementioned effects, there could be a change in the morphological structure of the polymer in the specified area, which could then manifest as changed mechanical and tribological properties. This publication opens new opportunities for modification of the injection molding process that could be used for complex applications with specific requirements.

This knowledge was applied to a practical part, which was a headlamp bezel, where the clamping points were subject to cracking. By changing the location of the injection gate, changing the melt and mold temperature, and modifying the tempering circuit, the mechanical properties were improved (up to 38%). This was achieved by more intensive cooling in the problem area and also by changing the injection gate location so that the polymer path was not too long, thus moving the stronger spot into the clamping point area. It is also possible to improve these properties by applying a TiB$_2$ coating to the mold cavity, which will improve the flowability of the polymer and increase the mechanical properties. As the following research shows, the application of the coating increased the mechanical properties by 33%.

The tribological, mechanical, surface, and morphological properties of the injection-molded polypropylene samples were investigated. It was found that the injection molding process and the gate location can be used to increase the abrasion resistance of polypropylene. We have also found that the relationship between tribological characteristics and morphological characteristics (skin–core layer, crystallinity, etc.) has a major influence on the final product. This can be attributed to the injection molding process, the location of the injection gate, and the behavior of the polymer during cooling, resulting in a higher surface resistance of the polypropylene. Thus, it can be concluded that in order to achieve optimum tribological and mechanical properties of injection-molded polypropylene, it is necessary to monitor the morphological properties of the material with a focus on crystallinity, which has a major influence on the aforementioned properties.

Gained results of varying properties along the flow path of injection-molded properties change the view of polymer behavior during injection molding and can have a significant effect on technical practice. For a complex description of this behavior, further investigation is necessary, especially with a focus on other polymer materials or process parameters.

5. Conclusions

This work deals with the influence of gate distance on the properties of injection-molded polypropylene parts. The importance and current research in the area of tribological and mechanical properties of injection-molded product surfaces have been extensively reviewed to provide an understanding of their importance and benefits relating to their use in industry. The effect of the technology itself, manufacturing processes, and related process parameters on the tribological properties of injection-molded polypropylene has been discussed in terms of the complex behavior of the polymer surface. Understanding the mechanisms of friction and wear on an injection-molded part is very important for designers and, when correlated with mechanical properties and especially morphological properties, opens up a very interesting area of research in the field of tribology. Prepared test samples were measured for their micro-mechanical properties (indentation hardness, indentation modulus, indentation creep), tribological properties (friction force, acoustic emission, friction coefficient), surface quality, and structural changes.

The test samples showed heterogenic behavior along the flow length as well as within individual depths of the part. In the surface layer (depth of measurement 20 μm), the mechanical properties decreased from the gate to the middle of the sample (158 mm). This decrease was in total 15% for indentation hardness and 55% for indentation modulus. The tribological properties also showed similar behavior to the micro-mechanical properties. The coefficient of friction has its maximum value at the beginning and end of filling and decreases towards the middle distance of filling. The coefficient of friction increases significantly with crystallinity. The difference in the tribological properties (coefficient of friction) between the individual points of the part was up to 20%. Towards the end of the sample, an opposite trend was observed. Deeper in the surface layer (100 μm), an opposite trend was found for mechanical properties, with its maximum at the center of the sample. The increase towards the center of the sample was 43% for indentation hardness and 120% for indentation modulus. Also, the replication of the tool surface on the test sample surface showed significant changes along the polymer flow. The measured results were probably influenced by the filling process as well as the process parameters of injection molding. These parameters affected the creation of the final skin–core structure and crystallinity, which varied along the flow length but also through the cross-section. The results indicate that the proper indentation method can catch changes in tribological and mechanical properties that were influenced by polypropylene morphology. Thus, morphology changes can be correlated with tribological and mechanical changes.

In conclusion, this work demonstrates that the properties of injection-molded parts are not uniform along the entire sample but change locally according to conditions within the mold. This significantly alters how tribological and micro-mechanical properties are looked upon in injection-molded parts. An important part is played by the way a mold is filled and how this, together with flow behavior, influences the final properties at specific points of a part.

Author Contributions: Conceptualization, M.O.; methodology, M.O. and M.S.; formal analysis, M.S. and K.F.; data curation, M.O. and M.S.; writing—original draft preparation, M.O.; visualization, M.O. and L.M.; project administration, M.O.; funding acquisition, M.S. All authors have read and agreed to the published version of the manuscript.

Funding: This article was written with the support of the project TBU at Zlin Internal Grant Agency (No. IGA/FT/2024/003).

Institutional Review Board Statement: Not applicable.

Informed Consent Statement: Not applicable.

Data Availability Statement: The data presented in this study are available on request from the corresponding author.

Conflicts of Interest: The authors declare no conflicts of interest.

References

1. Chu, J.; Kamal, M.R.; Derdouri, S.; Hrymak, A. Characterization of the microinjection molding process. *Polym. Eng. Sci.* **2010**, *50*, 1214–1225. [CrossRef]
2. Liu, Z.; Chen, Y.; Ding, W.; Zhang, C. Filling behavior, morphology evolution and crystallization behavior of microinjection molded poly(lactic acid)/hydroxyapatite nanocomposites. *Compos. Part A Appl. Sci. Manuf.* **2015**, *72*, 85–95. [CrossRef]
3. Wang, J.Y.; Bai, J.; Zhang, Y.Q.; Fang, H.G.; Wang, Z.G. Shearinduced enhancements of crystallization kinetics and morphological transformation for long chain branched polylactides with different branching degrees. *Sci. Rep.* **2016**, *6*, 26560. [CrossRef] [PubMed]
4. Schrauwen, B.A.G.; Von Breemen, L.C.A.; Spoelstra, A.B.; Govert, L.E.; Peters, G.W.M.; Meijer, H.E.H. Structure, deformation, and failure of flow-oriented semicrystalline polymers. *Macromolecules* **2004**, *37*, 8618–8633. [CrossRef]
5. Giboz, J.; Copponnex, T.; Mele, P. Microinjection molding of thermoplastic polymers: Morphological comparison with conventional injection molding. *J. Micromech. Microeng.* **2009**, *19*, 025023. [CrossRef]
6. Isayev, A.I.; Chan, T.W.; Shimojo, K.; Gmerek, M.J. Injection molding of semicrystalline polymers. I. Material characterization. *Appl. Polym. Sci.* **1995**, *55*, 807. [CrossRef]
7. Persson, J.; Zhou, J.; Ståhl, J. Characterizing the mechanical properties of skin-core structure in polymer molding process by nanoindentation. *Mater. Sci.* **2014**, *6*, 9.
8. Baerwinkel, S.; Seidel, A.; Hobeika, S.; Hufen, R.; Moerl, M.; Altstaedt, V. Morphology Formation in PC/ABS Blends during Thermal Processing and the Effect of the Viscosity Ratio of Blend Partners. *Materials* **2016**, *9*, 659. [CrossRef] [PubMed]
9. Bociaga, E.; Kula, M.; Kwiatkowski, K. Analysis of structural changes in injection-molded parts due to cyclic loading. *Adv. Polym. Technol.* **2018**, *37*, 2134–2141. [CrossRef]
10. Menges, G.; Haberstroh, E.; Michaeli, W.; Schmachtenberg, E. *Plastics Materials Science*; Hanser Verlag: Munich, Germany, 2002.
11. Carraher, C.E.; Seymour, R.B. *Polymerní Chemie Seymour/Carraher*; CRC Press: Boca Raton, FL, USA, 2003; pp. 43–45.
12. Ehrenstein, G.W.; Richard, P. Theriault. *Polymerní Materiály: Struktura, Vlastnosti, Aplikace*; Hanser Verlag: Munich, Germany, 2001; pp. 67–78.
13. Le, M.C.; Belhabib, S.; Nicolazo, C.; Vachot, P.; Mousseau, P.; Sarda, A.; Deterre, R. Pressure influence on crystallization kinetics during injection molding. *J. Mater. Process. Technol.* **2011**, *211*, 1757–1763. [CrossRef]
14. Liu, F.; Guo, C.; Wu, X.; Qian, X.; Liu, H.; Zhang, J. Morphological comparison of isotactic polypropylene parts prepared by micro-injection molding and conventional injection molding. *Polym. Adv. Technol.* **2011**, *23*, 686–694. [CrossRef]
15. Sun, H.; Zhao, Z.; Yang, Q.; Yang, L.; Wu, P. The morphological evolution and β-crystal distribution of isotactic polypropylene with the assistance of a long chain branched structure at micro-injection molding condition. *J. Polym. Res.* **2017**, *24*, 75. [CrossRef]
16. Pantani, R.; Coccorullo, I.; Speranza, V.; Titomanlio, G. Modeling of morphology evolution in the injection molding process of thermoplastic polymers. *Prog. Polym. Sci.* **2005**, *30*, 1185–1222. [CrossRef]
17. Lei, X.; Grueneberg, T.; Steuernagel, L.; Ziegmann, G.; Militz, H. Influence of particle concentration and type on flow, thermal, and mechanical properties of wood-polypropylene composites. *J. Reinf. Plast. Compos.* **2009**, *29*, 1940–1951.
18. Kocic, N.; Kretschmer, K.; Bastian, M.; Heidemeyer, P. The influence of talc as a nucleation agent on the nonisothermal crystallization and morphology of isotactic polypropylene: The application of the lauritzen-hoffmann, Avrami, and Ozawa theories. *J. Appl. Polym. Sci.* **2012**, *126*, 1207–1217. [CrossRef]
19. Wang, L.; Zhang, Y.; Jiang, L.; Yang, X.; Zhou, Y.; Wang, X.; Li, Q.; Shen, C.; Turng, L.S. Effect of injection speed on the mechanical properties of isotactic polypropylene micro injection molded parts based on a nanoindentation test. *J. Appl. Polym. Sci.* **2018**, *136*, 47329. [CrossRef]
20. Glogowska, K.; Sikora, J.; Dulebova, L. The effect of multiple processing of polypropylene on selected properties of injection moulded parts. *Adv. Sci. Technol. Res. J.* **2016**, *10*, 65–72. [CrossRef] [PubMed]
21. Sykutera, D.; Wajer, Ł.; Kosciuszko, A.; Szewczykowski, P.P.; Czyzewski, P. The influence of processing conditions on the polypropylene apparent viscosity measured directly in the mold cavity. *Macromol. Symp.* **2018**, *378*, 1700056. [CrossRef]
22. Lafranche, E.; Krawczak, P.; Ciolczyk, J.P.; Maugey, J. Injection moulding of long glass fiber reinforced polyamide 66: Processing conditions/microstructure/flexural properties relationship. *Adv. Polym. Technol.* **2005**, *24*, 114–131. [CrossRef]
23. Moritzer, E.; Heiderich, G.; Hirsch, A. Fiber length reduction during injection molding. *AIP Conf. Proc.* **2019**, *2055*, 070001.
24. Ramzy, A.Y.; El-sabbagh, A.M.; Steuernagel, L.; Ziegmann, G.; Meiners, D. Rheology of natural fibers thermoplastic compounds: Flow length and fiber distribution. *J. Appl. Polym. Sci.* **2013**, *131*, 39861. [CrossRef]
25. Liparoti, S.; Speranza, V.; De Meo, A.; De Santis, F.; Pantani, R. Prediction of the maximum flow length of a thin injection molded part. *J. Polym. Eng.* **2020**, *40*, 783–795. [CrossRef]
26. Xiong, J.; Guo, Y.; Kaschta, J.; Nie, F.; Mao, G.; Yang, J.; Zhou, Q.; Zhu, H.; Li, X. Temporal evolution of microstructure and its influence on micromechanical and tribological properties of Stellite-6 cladding under aging treatment. *J. Mater. Sci.* **2023**, *58*, 10802–10820. [CrossRef]
27. Bhushan, B.; Li, X. Micromechanical and tribological characterization of doped single-crystal silicon and polysilicon films for microelectromechanical systems devices. *J. Mater. Res.* **2011**, *12*, 54–63. [CrossRef]
28. ČSN EN ISO 14577; Metallic Materials—Instrumented Indentation Test for Hardness and Materials Parameters. ČSN ISO: Praha, Czech Republic, 2003.

29. Oliver, W.C.; Pharr, G.M. Measurement of hardness and elastic modulus by instrumented indentation: Advances in understanding and refinements to methodology. *J. Mater. Res.* **2004**, *19*, 3–20. [CrossRef]
30. Ovsik, M.; Stanek, M.; Stanek, M.; Dockal, A.; Vanek, J.; Hylova, L. Influence of Cross-Linking Agent Concentration/Beta Radiation Surface Modification on the Micro-Mechanical Properties of Polyamide 6. *Materials* **2021**, *14*, 6407. [CrossRef] [PubMed]
31. Panaitescu, D.M.; Vuluga, Z.; Sanporean, C.G.; Nicolae, C.A.; Gabor, A.R.; Trusca, R. High flow polypropylene/SEBS composites reinforced with differently treated hemp fibers for injection molded parts. *Compos. Part B Eng.* **2019**, *174*, 107062. [CrossRef]
32. Lanyi, F.J.; Wenzke, N.; Kaschta, J.; Schubert, D.W. On the Determination of the Enthalpy of Fusion of α-Crystalline Isotactic Polypropylene Using Differential Scanning Calorimetry, X-Ray Diffraction, and Fourier-Transform Infrared Spectroscopy: An Old Story Revisited. *Adv. Eng. Mater.* **2020**, *22*, 1900796. [CrossRef]
33. Schawe, J.E.K. Analysis of non-isothermal crystallization during cooling and reorganization during heating of isotactic polypropylene by fast scanning DSC. *Thermochim. Acta* **2015**, *603*, 85–93. [CrossRef]
34. Alariqi, S.A.S.; Kumar, A.P.; Rao, B.S.M.; Singh, R.P. Effect of γ-dose rate on crystallinity and morphological changes of γ-sterilized biomedical polypropylene. *Polym. Degrad. Stab.* **2009**, *94*, 272–277. [CrossRef]
35. Lima, M.F.S.; Vasconcellos, M.A.Z.; Samios, D. Crystallinity changes in plastically deformed isotactic polypropylene evaluated by x-ray diffraction and differential scanning calorimetry methods. *Polym. Phys.* **2002**, *40*, 896–903. [CrossRef]
36. Hernandez-Sanchez, F.; Castillo, L.F.; Vera-Graziano, R. Isothermal crystallization kinetics of polypropylene by differential scanning calorimetry. I. Experimental conditions. *J. Appl. Polym. Sci.* **2004**, *92*, 970–978. [CrossRef]

Disclaimer/Publisher's Note: The statements, opinions and data contained in all publications are solely those of the individual author(s) and contributor(s) and not of MDPI and/or the editor(s). MDPI and/or the editor(s) disclaim responsibility for any injury to people or property resulting from any ideas, methods, instructions or products referred to in the content.

Article

Comparison of Friction Properties of GI Steel Plates with Various Surface Treatments

Miroslav Tomáš [1,*], Stanislav Németh [2], Emil Evin [1], František Hollý [2], Vladimír Kundracik [2], Juliy Martyn Kulya [1] and Marek Buber [1]

[1] Department of Automotive Production, Faculty of Mechanical Engineering, Technical University of Košice, Mäsiarska 74, 040 01 Košice, Slovakia; emil.evin@tuke.sk (E.E.)
[2] USSE Research and Development, U.S. Steel Košice s.r.o., Vstupný Areál U.S. Steel, 044 54 Košice, Slovakia; snemeth@sk.uss.com (S.N.); fholly@sk.uss.com (F.H.); vkundracik@sk.uss.com (V.K.)
* Correspondence: miroslav.tomas@tuke.sk; Tel.: +421-55-602-3524

Abstract: This article presents the improved properties of GI (hot-dip galvanized) steel plates in combination with a special permanent surface treatment. The substrate used was hot-dip galvanized deep-drawn steel sheets of grade DX56D + Z. Subsequently, various surface treatments were applied to their surface. The coefficient of friction of the metal sheets without surface treatment, with a temporary surface treatment called passivation, and a thin organic coating (TOC) based on hydroxyl resins dissolved in water, Ti and Cr^{3+} were determined by a cup test. The surface quality and corrosion resistance of all tested samples were also determined by exposing them for up to 288 h in an atmosphere of neutral salt spray. The surface microgeometry parameters Ra, RPc and Rz(I), which have a significant influence on the pressing process itself, were also determined. The TOC deposited on the Zn substrate was the only one to exhibit excellent lubrication and anticorrosion properties, resulting in the lowest surface microgeometry values owing to the uniform and continuous layer of the thin organic coating compared to the GI substrate and passivation surface treatment, respectively.

Keywords: thin organic coating; galvanized steel; cup test; friction coefficient; corrosion resistance

---

Citation: Tomáš, M.; Németh, S.; Evin, E.; Hollý, F.; Kundracik, V.; Kulya, J.M.; Buber, M. Comparison of Friction Properties of GI Steel Plates with Various Surface Treatments. *Lubricants* 2024, 12, 198. https://doi.org/ 10.3390/lubricants12060198

Received: 26 April 2024
Revised: 24 May 2024
Accepted: 28 May 2024
Published: 31 May 2024

Copyright: © 2024 by the authors. Licensee MDPI, Basel, Switzerland. This article is an open access article distributed under the terms and conditions of the Creative Commons Attribution (CC BY) license (https:// creativecommons.org/licenses/by/ 4.0/).

1. Introduction

The formability of steel plates and the shape of the final parts made from them depend on factors such as the material (its mechanical properties and surface microgeometry) [1,2]; the forming die geometry (die clearance, radii of the punch and die) and microgeometry (roughness of contact surfaces) [3–5]; the technological parameters (temperature, strain rate, blank holding forces, contact pressure, etc.) [6–8]; the properties of the tool's material (hardness, chemical composition, structure) [9]; and the type and amount of lubricant used [10]. During the deep drawing process, the individual parameters change and their mutual interaction has an influence on the formability, which is largely influenced by the lubricant or lubrication strategy. Correct selection of the lubricant is therefore one of the fundamental factors that ensures either the quality of the deep drawing process or its result in the form of the stamped part itself [11,12].

The functions of the lubricant in the deep drawing process are usually of a different nature. In general, however, the role of the lubricant is to minimize the friction acting on the contact surfaces of the tool in order to achieve the greatest possible ductility and to take maximum advantage of all of the plastic properties of the material [13]. Lubricants must perform other functions, such as the protection of metal sheets against corrosion, easy application and the removal of surfaces when necessary [14]. Nowadays, novel tribological systems are being studied, such as using volatile medium-like liquid carbon dioxide (CO_2) and gaseous nitrogen, directly introduced into the friction zones during deep drawing. These alternatives aim to reduce environmental impacts and eliminate the need for post-forming cleaning processes [15]. Good adhesion and homogeneity of the lubricant on the surface of the sheet must also be ensured from application through to transport, storage and processing. Lubricants must be stable in the pressing process as well as non-hazardous and their application should be cost-effective [16].

One possible way to obtain these properties for a specific area of the stampings produced is the development of coatings based on a combination of a zinc coating with an applied permanent thin organic coating. Such a coating should be compatible with all Zn coatings and remain on the metal sheets during pressing [17,18]. The low coefficient of friction obtained should allow for more uniform reshaping in tooling and for maximum use of the mechanical properties for sharper lines of the stampings [19,20]. The dry lubricant on the steel plates should meet the requirement of eliminating coating flaking and the need for additional oil, thus keeping the forming presses (the contact surfaces of the blank holder, punch and die) clean and ready for the next step of production [21]. A typical permanent coating consists of a coating–forming material (resin) and different types of additives (forming additives and corrosion inhibitors) [22–24] enhancing conductivity and/or corrosion protection [25].

An example of such surface treatment on Zn, Al-Zn, Al-Zn-Si, Mg-Al-Zn and other substrates is a thin organic coating (TOC). The coating contains a limited amount of inorganic Cr^{3+} and Ti components [26]. After drying, they form a thin dry film with a thickness from 0.8 to 1.7 μm (i.e., 1.0 to 2.0 g/m^2). The coating serves as a permanent protection for galvanized sheet metal in transparent and colored versions [26–29]. Steel sheets with a thin organic coating currently provide an effective solution for applications where traditional passivated or oiled galvanized material is used, particularly in corrosion-sensitive conditions. The advantages and functional properties of the coating are excellent corrosion resistance and the absence of hexavalent chromium, which is RoHS-compliant based on Directive 2011/65/EU of the European Parliament and the Council [30]. The coating improves the formability due to the lower coefficient of friction "f", whose value of less than 0.15 is strictly required by sheet metal fabricators, especially in the white goods sector. When forming sheet metal with such surface protection, there is no need to add any pressing oils as it is also compatible with subsequent joining processes (gluing, welding...). A thin organic coating is also characterized by AFP—anti finger print, i.e., fingerprint resistance. In addition, these coated steel sheets are suitable for painting without any pre-treatment, for interior and exterior use, as well as for exposed and unexposed parts of products. This coating has a wide range of applications: construction—profiles, electronics—TV panels and Hi-Fi systems, and others—furniture, air conditioning, appliances, etc. [25–29].

The tribological behavior of TOCs is being tested by different methods to evaluate the friction coefficient in dry/wet contact conditions. The modified scratch test is well suited to serve as a first screening method to compare the tribological performance of different types of thin organic permanent coatings for forming applications [31,32]. Pin-on-disc/ball-on-disc tests are useful when determining the friction coefficient of organic/inorganic coatings [33,34]. While the mentioned tests belong to the basic testing methods performed in laboratory conditions, the bending under tension test allows for the modeling of the friction conditions on the edge of the die [31,35,36]. More complex tests, involving real conditions in deep drawing/stretching processes, are the cup test and biaxial stretching test. In [37], the authors tested the formability of organic coated steel sheet metal by using the cup test and biaxial stretching, focusing on the fracture behavior of the coating during forming operations. A biaxial tension state showed a greater effect on the coating compared to a mixed compression–tension state, even if the crack development was demonstrated to be similar. By using the cup test, it is possible to model the stress of the contact surfaces during sheet metal forming [38]. The cup test offers a more CF-process-relevant determination approach for the prediction of the technological characteristics of the deep drawing process [39,40].

The aim of this paper is to study the tribological and corrosion properties of a thin organic coating (TOC) based on hydroxyl resins dissolved in water and compare them with a standard Zn coating and a passivated one. All of these coatings were created on a standard drawing quality steel substrate.

2. Materials and Methods

2.1. Experimental Materials—Substrate and Coatings

This experimental study was carried out on 1.0 mm thick steel sheet specimens of deep-drawn grade DX56D + Z. Tested samples of hot-dip galvanized steel sheet (GI substrate) were used as the reference material with a Zn layer thickness of 80 g/m^2 (40/40 g/m^2). Further, for comparison, the same quality hot-dip Zn coating with an applied surface treatment called passivation (based on inorganic substances) and also the same Zn coating with the TOC in a white shade were used. It is a single-component product based on Ti and Cr^{3+}. All of these coatings were in the same condition "as delivered" [27].

Figure 1 depicts a selected region of the metallographic section of the sample with the TOC analyzed under an Olympus GX 71 light microscope (Olympus, Tokyo, Japan). The Zn layer thickness was from 5 to 7 μm on both the top and bottom of the steel substrate. The thickness of the organic coating was controlled during its application to the Zn substrate according to the titanium content of the coating and the thickness of the dry film on the surface of the Zn substrate.

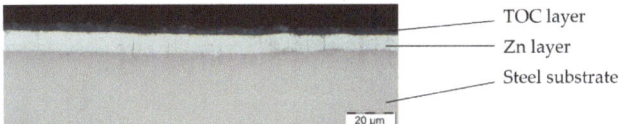

Figure 1. Cross-section of galvanized steel substrate with TOC layer.

The EDX/SEM analysis of the surface treatment (Figure 2b) confirmed a relatively stable chromium content in the range of 0.3 ± 0.1 wt%, which also corresponded with the required thickness of the coating applied. The uniform distribution of the pigment particles in the TOC after application to the steel sheet was also observed. The analysis was performed using a TESCAN VEGA 3 scanning electron microscope (TESCAN GROUP, a.s., Brno, Czech Republic) coupled with an OXFORD Instruments EDS analyzer.

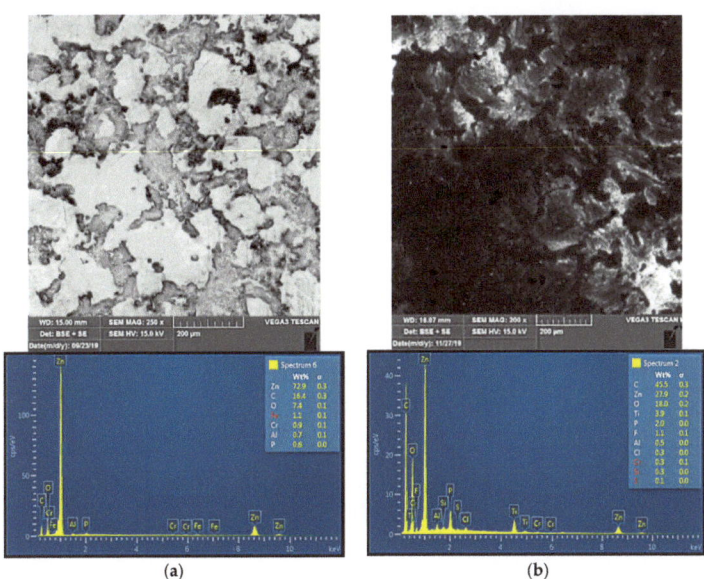

Figure 2. EDX analysis: (**a**) passivated GI coating; (**b**) GI coating with TOC.

The chemical composition of the substrate, which meets the prescribed limits of EN 10346 [41] (Table 1), was determined with an OBLF QSG 750 Analyzer (OBLF, Witten, DE, Germany) by using optical emission spectral analysis.

Table 1. Chemical composition of 1.0 mm thick substrate DX56D + Z (max. values) [wt%].

C	Mn	Si	P	S	Ti
0.12	0.6	0.5	0.1	0.045	0.3

2.2. Mechanical Properties of Substrate

The mechanical properties were determined by using a Zwick/Roell Z050 machine (Ulm, Germany) in the transversal direction (90°) to the rolling direction and meet the limits shown in Table 2. These properties were determined according to EN ISO 6892-1 [42], the normal anisotropy "r" was determined according to EN ISO 10113 [43] and the strain hardening exponent "n" was determined according to EN ISO 10275 [44] on standardized specimens (20 mm in width and 80 mm of original gauge length). The plastic strain ratio was determined at 20% of the engineering strain level and the strain hardening exponent was determined within the strain interval of 5% to uniform elongation A_g [42–44].

Table 2. Mechanical properties of 1.0 mm thick substrate DX56D + Z in transversal direction (90°).

	$R_{p0.2}$ (YS) [MPa]	R_m (UTS) [MPa]	A_{80} [%]	r [−]	n [−]
Measured	175 ± 1.3	301 ± 1.1	43.5 ± 0.9	2.30 ± 0.03	0.23 ± 0.005
Required	120–180	260–350	39% min.	1.9 min.	0.21 min.

$R_{p0.2}$ (YS) is proof strength; R_m (UTS) is ultimate tensile strength; A_{80} is percentage elongation after fracture; r is plastic strain ratio; n is strain hardening exponent.

2.3. Cup Test

To determine the lubrication properties (the value of the friction coefficient on the contact surfaces of the die and the blank holder), the well-known formability test—the cup test—was used, which simulates the stresses on the material in a real deep drawing process [38–40]. This test can be used to describe the most frequent contacts in the tribological system between the tool (punch, blank holder, die), friction material (sheet metal specimen/product) and intermediate material (lubricant, oil, foil or permanent special coating on sheet metal)—Figure 3 [12,38].

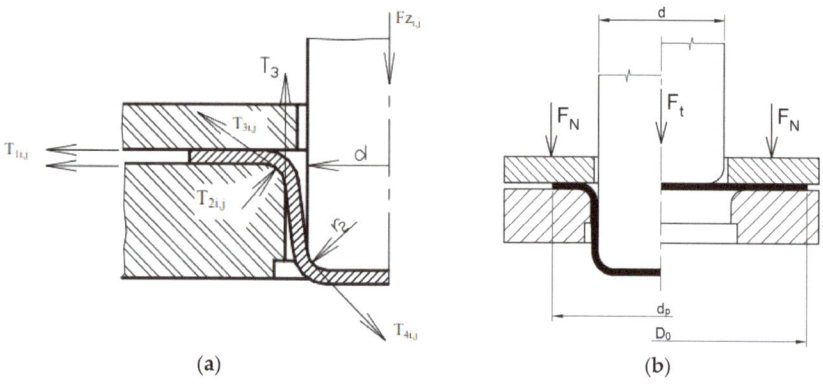

Figure 3. The principle of the cup test to determine the friction coefficient: (**a**) the friction forces during the cup test: $T_{1i,j}$ is the friction force under the blank holder, $T_{2i,j}$ is the normal force at the die radius, $T_{3i,j}$ is the friction force at the die radius and $T_{4i,j}$ is the friction force at the punch radius; (**b**) a schematic illustration of the cup test: F_N is the blank holder force and F_t is the drawing force [12,38].

The friction coefficient "f" of the tested specimens with different surface treatments was determined on the basis of the increase in the contact force and the subsequent results of the tensile forces. The coefficient of friction generally represents the ratio of the frictional force between two surfaces (press tool–sheet metal) and the load applied perpendicular to these surfaces. It should not be overlooked that the coefficient of friction depends not only on the condition of the surface (surface roughness, surface with or without lubricant) but also on other boundary conditions such as contact force, drawing speed, etc.

The cup test allows for the modeling of the tribological conditions at the contact surfaces of the tribo-system tool–lubricant–material, as they are carried out under semi-operational deep drawing conditions. Thus, this test is closer to the real deep drawing process and allows us to determine the friction coefficient under the blank holder. The friction coefficient "f" was determined for each type of surface (GI coating) specified as follows:

A—Reference GI coating (as supplied);
B—Passivated GI coating;
C—GI coating with the thin organic coating.

To determine the friction coefficient, a flat-bottomed punch with a diameter of ø 50 mm was used. Circular blanks with a constant diameter of 90 mm were cut out and subsequently subjected to plastic deformation on an Erichsen 145/60 hydraulic testing machine (Erichsen GmbH & Co. KG, Hemer, Germany). The blank holder force F_N represented loads of 10, 20, 30, 40 and 50 kN and the punch speed was kept constant (60 mm·min^{-1}) during the test. Ten cups were drawn for each level of the blank holder force. At these holding forces, the maximum drawing force F_t required to reshape the blank was recorded from the entire force course. Then, an average value of the drawing force and its standard deviation were calculated. After plotting F_t-F_N on a graph and interpolating the trend equation over these values, a linear relationship was obtained, where the angle under the curve represents the friction in the contact surfaces between the press tool and the sheet metal specimen. Thus, the evaluation method used a regression analysis where the slope of this linear trend line represents the value of the friction coefficient 'f' [12,40].

During the cup test (Figure 3), each blank holding force F_N corresponded to the maximum tensile force F_t—Figure 4a. A change in the blank holding force caused the maximum drawing force F_t to change. The graph shows that the relationship F_t—F_N is linearly dependent (curve 1, Figure 4b) and it can be approximated as follows [12,40]:

$$F_t = \text{Intercept} + \text{Slope} \cdot F_N, \qquad (1)$$

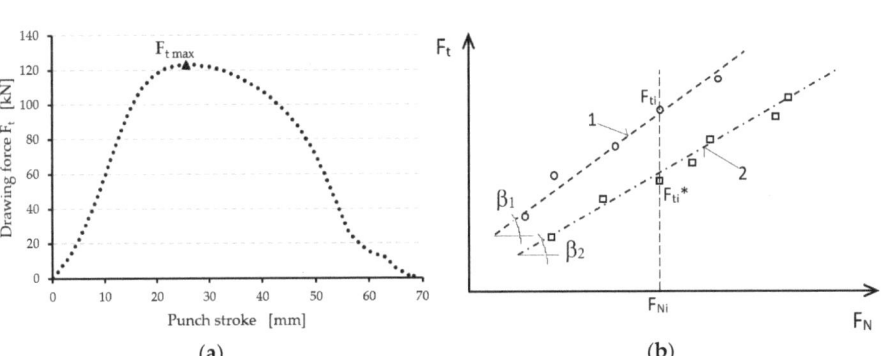

Figure 4. Evaluation of the cup test: (**a**) determination of the maximum drawing force F_t from the record of the force course; (**b**) determination of the friction coefficient for different lubricants [12,38,40].

The slope of the curve, given by the angle β_1, characterizes the friction conditions between the tool and the blank. Then, the friction coefficient is determined as follows:

$$f = \frac{\text{Slope}}{2} = \frac{\beta_i}{2} \qquad (2)$$

In the case of a lubricant with improved frictional properties (lower coefficient of friction), we obtain lower drawing forces $F_{Ti}*$ for the same blank holding force F_{Ni} and thus a new relationship $F_{Ti}*$—F_{Ni} (curve 2, Figure 4b). The slope of curve 2 (β_2) is less than the slope of curve 1 (β_1) and is proportional to the coefficient of friction between the tool and the blank.

2.4. Surface Microgeometry

Measuring the roughness parameters was carried out on a roughness measuring device, Hommel Tester T2000 (Hommelwerke, Schwenningen, Germany). Measurements of the roughness parameters Ra (arithmetic mean deviation of profile), RPc (peak count number) and Rz (maximum height of profile) were carried out on all sheet metal specimens from both sides of the blank (A side, B side) at five different points uniformly distributed on each specimen. Then, the average value of each evaluated parameter was calculated.

2.5. Corrosion Resistance according to ISO 9227

All tested specimens were placed in the SKBW 400 A-TR corrosion chamber (Liebisch GmbH & Co. KG, Bielefeld, Germany) and tested in an atmosphere of neutral salt spray (NSS) according to ISO 9227 [45]. A 5% sodium chloride solution (50 g/L $\pm$ 5 g/L of NaCl) was continuously sprayed onto the samples placed in the rack at an angle of $20° \pm 5°$ to the vertical direction using an air nozzle at an ambient temperature T = 35 $\pm$ 2 °C. The accelerated corrosion test was carried out to detect discontinuous layers, such as pores and other defects, in certain metallic, organic, anodic oxide and conversion specialty coatings. The size of all tested specimens was 210 × 290 mm and the corrosion resistance of the coatings was evaluated based on the time until the appearance of white corrosion, evaluated after 5, 24, 48, 72, 96, 120, 144, 216 and 288 h spent in the corrosion chamber.

3. Results

The average values of the surface microgeometry parameters calculated from the five measured values on each side of the specimens are presented in Table 3 for each coating used in the experiment. As it is shown in the table, the reference "as supplied" GI coating showed an Ra value of 1.153 $\pm$ 0.170 µm, an Rz value of 7.028 $\pm$ 0.878 µm and an RPc value of 50.7 $\pm$ 10.0 cm^{-1} with a low difference when measured on each side of the steel sheet. The same tendency was noted for the GI passivated surface, with average values of Ra = 1.050 $\pm$ 0.132 µm, Rz = 6.754 $\pm$ 0.453 µm and RPc = 49.2 $\pm$ 5.6 cm^{-1}, and for the GI coating with the TOC, with average values of Ra = 0.934 $\pm$ 0.101 µm, Rz = 5.957 $\pm$ 0.505 µm and RPc = 48.5 $\pm$ 5.681 cm^{-1}. It can be concluded that the applied organic coating follows the morphology of the substrate surface, and the applied dry film microlayer influences the microgeometry of the coating by decreasing the coating roughness parameters Ra and Rz, while maintaining RPc at almost the same value.

The results of the drawing force measurement in the cup test for each coating are shown in Table 4. When the GI coating was lubricated by the oil during the cup test, the drawing forces ranged from 60.3 to 69.8 kN (Figure 5), with an increasing blank holder force, while the standard deviation was from 0.9 to 1.3 kN. Thus, a linear trend line that approximates the average values of the drawing forces predicts a friction coefficient f_{GI} = 0.119 with a Pearson correlation coefficient R = 0.999. The same tendency but a lower friction coefficient was found when the GI coating after passivation was used (dry friction): the drawing forces increased from 52.1 to 59.2 kN, with a standard deviation from 0.7 to 1.0 kN. The friction coefficient was f_{GIpass} = 0.091, with a Pearson correlation coefficient

R = 0.999. The lowest drawing forces were measured when the GI coating with the TOC was tested: the drawing forces increased from 41.7 to 46.1 kN, with a standard deviation from 2.0 to 2.3 kN. Thus, the lowest friction coefficient was determined as $f_{GI + TOC} = 0.053$ with a Pearson correlation coefficient R = 0.995. For each coating tested, a very good correlation between drawing force and blank holder force was determined.

Table 3. The average values of the surface microgeometry parameters measured at 90° to the rolling direction.

Title 1	Title 2	Ra [μm]		Rz [μm]		RPc [cm^{-1}]	
		A Side	B Side	A Side	B Side	A Side	B Side
GI coating	Average	1.286	1.020	7.646	6.410	54.6	46.8
	St.dev	0.111	0.091	0.610	0.640	13.0	4.3
		1.153 ± 0.170		7.028 ± 0.878		50.7 ± 10.0	
Passivated GI coating	Average	1.162	0.938	7.132	6.376	53.0	45.4
	St.dev	0.058	0.066	0.141	0.290	3.9	4.3
		1.050 ± 0.132		6.754 ± 0.453		49.2 ± 5.6	
GI coating with TOC	Average	1.022	0.846	6.332	5.582	49.6	47.4
	St.dev	0.038	0.045	0.179	0.436	4.3	4.1
		0.934 ± 0.101		5.957 ± 0.505		48.5 ± 5.681	

Table 4. The deep drawing forces measured in the cup test for each blank holder force.

Blank Folder Force F_N [kN]	Deep Drawing Force Ft [kN]		
	GI	GI Passivated	GI + TOC
10	60.3 ± 0.9	52.1 ± 0.9	41.7 ± 2.0
20	62.7 ± 1.1	53.8 ± 0.7	43.1 ± 2.0
30	65.4 ± 1.3	55.6 ± 1.0	44.1 ± 2.3
40	67.5 ± 1.1	57.7 ± 0.8	44.9 ± 2.1
50	69.8 ± 0.9	59.2 ± 0.9	46.1 ± 2.0
Slope	0.238	0.182	0.105
Friction coefficient	0.119	0.091	0.053
R^2	0.9989	0.9985	0.9893

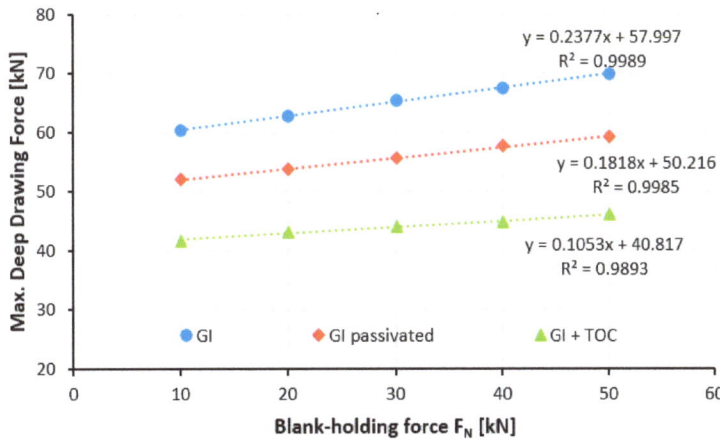

Figure 5. Regression analysis of measured drawing forces.

The adhesion of the coating to the substrate was evaluated after the cup test through visual inspection and microscopic analysis. No flaking or separation of the organic TOC from the substrate was observed on the surface of the cups (Figure 6; Canon EOS 100D 18 MPx with EFS 18–55 mm; Canon, Tokyo, Japan) at different F_N blank holding force loads. Even at the maximum loading force F_N = 50 kN, there was no adhesion of the coating or its elements to the drawing tool's contact surfaces (drawing die radius or blank holder area). Thus, good adhesion of the TOC was observed.

Figure 6. Adhesion of coating to GI substrate after cup test: (**a**) GI oiled; (**b**) GI passivated; (**c**) GI–TOC.

The results of the corrosion test after a selected number of hours in the corrosion chamber are shown in Figure 7. Based on the test results, it can be concluded that white corrosion appeared on the surface of the samples with the conventional GI coating without surface treatment during the first hours of exposure to a neutral atmosphere of salt solution. Surface treatment of the GI coating by means of standard passivation showed better corrosion resistance, i.e., the observation of white corrosion started after more than 24 h. The GI coating with the TOC showed the best corrosion resistance in relation to the compared coatings, with no observation of white corrosion for up to 144 h. Since the corrosion resistance limit for the TOC is set at 72 h (customer quality standard), the coated specimens still have a sufficient margin.

Figure 7. *Cont.*

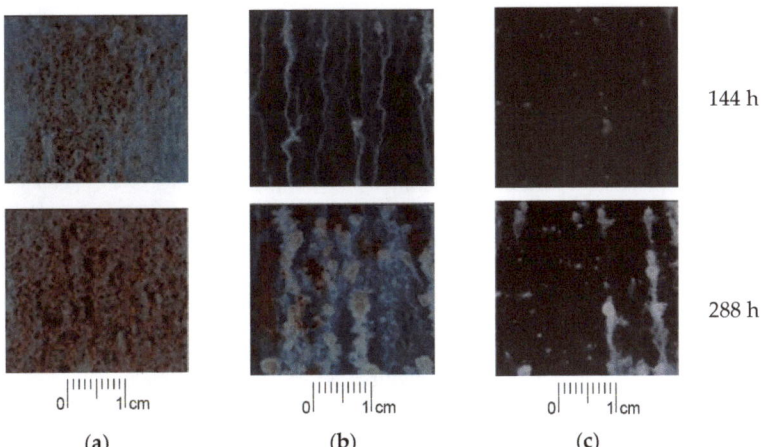

Figure 7. Kinetics of corrosion resistance of different coatings: (**a**) GI coating as reference one (as supplied); (**b**) passivated GI coating; (**c**) GI coating with TOC.

Figure 8 illustrates the comparison of time to white corrosion appearance between examined coatings.

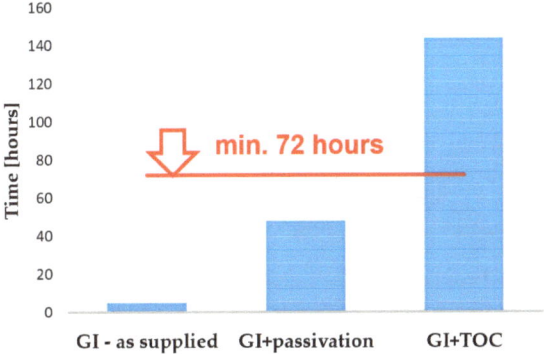

Figure 8. Comparison of time to white corrosion for examined coatings.

4. Discussion

From the perspective of steel sheet surface morphology, it can be concluded that the applied passivation and organic coating follow the morphology of the substrate surface, and the applied dry organic microlayer influences the microgeometry of the coating by decreasing the coating roughness parameters Ra and Rz. After passivation, the roughness of the passivated GI surface was Ra = 1.050 ± 0.132 µm and Rz = 6.754 ± 0.453 µm, while applying the organic coating decreased the roughness to Ra = 0.934 ± 0.101 µm and Rz = 5.957 ± 0.505. The surface morphology after phosphating depends on the deposition temperature and the concentration of the additives [46,47]; the self-lubricating coating can smooth the surface of the galvanized steel sheet and decrease the surface roughness of the galvanized steel sheet [48].

For all tested samples, GI as supplied, passivated GI and GI + TOC, the value of the friction coefficient determined by the cup test was f < 0.15 (the required parameter f_{max} for final sheet metal processors is 0.15). Thus, for samples without surface treatment (GI), the value 0.119 was reached when oil was used as a lubricant, while for dry friction (no oil used) samples with passivation (GI + passivation), it was 0.091, and for samples with the thin organic coating (GI + TOC), the value was 0.053.

The effect of phosphating on lowering the friction coefficient was proven by Narayanan [49]. Shih tested pre-phosphated DDQ and AHSS steels by using the bending under tension test and the results showed much a lower friction coefficient compared to without the pre-phosphated coating [50]. But, in a dry condition, phosphate coatings usually do not have inherent low-friction properties and they have no practical effect on the friction coefficient [21,51]. Carlsson [31,52] evaluated the friction coefficient of organic permanent coatings by using different basic tests. They achieved the lowest friction coefficient for a pure organic coating in the modified scratch test and pin-on-disc test within the range from 0.06 to 0.11, while in the bending under tension test, the friction coefficient varied from 0.16 to 0.17. Liu et al. [48] also studied self-lubricating coatings from the view of the influence of surface roughness on the friction coefficient and wear. Their results showed that within the range from 0.857 to 1.629 μm of the Ra value, the friction coefficient varied from 0.073 to 0.108 when measured by the reciprocating friction and wear tester. It may be concluded that when comparing the friction coefficient of 0.053 measured by the authors for the TOC (based on hydroxyl resins dissolved in water) by using the cup test, this value is comparable to the results mentioned. Thus, the positive effect of the TOC on lowering the friction coefficient in the deep drawing process can be elucidated.

The corrosion test showed the enhanced properties of the TOC applied on the GI steel compared to the passivated surface. The results are supported by other authors. Podjuklova et al. [53] proved the enhanced short-term corrosion protection of a transparent coating system containing particles of zinc compounds applied on steel sheets for enameling. Sarli et al. [54] applied polyurethane-based polymeric films on pretreated electrogalvanized steel which provided very effective protection against corrosion as a result of their excellent barrier properties. Gao et al. [55] studied acrylic coatings composed of modified MoS_2 nanosheets successfully prepared on the surface of galvanized sheets. Their results showed that modified MoS_2 nanosheets reduced the number of cracks in the coatings and made the coatings more compact. Furthermore, the coating's corrosion resistance was significantly enhanced. Li et al. [56] clarified the roles of titanium-containing materials in improving coatings when studying organic–inorganic hybrid coatings. As Kuznetsov analyzed [57], appropriately selected inhibitors will form a durable protective layer on the surface, preventing the formation of corrosion.

5. Conclusions

Based on this study's findings, it can be concluded that the tested sheet metal specimens with the TOC based on hydroxyl resins dissolved in water showed the most suitable lubrication properties due to the continuous coating layer compared to the other tested specimens. The lowest friction coefficient when measured by using the cup test in dry conditions resulted in lower drawing forces.

The dependence of the deep drawing force on the blank holding force is linear, and thanks to the special TOC surface treatment, the coefficient of friction was reduced from 0.119 to 0.053, which is more than half.

Good adhesion of the TOC was observed at each level of the blank holding force. No elements of the coating adhered to the drawing tool's contact surfaces.

As proved by the corrosion test, GI + TOC showed enhanced corrosion properties. The time to white corrosion appearance was greater than 72 h, as required by the customer, or more than twice that when compared to the GI passivated surface.

Thus, using GI + TOC based on hydroxyl resins dissolved in water, Ti and Cr^{3+} without the addition of pressing oils as a lubricant has several advantages—cost savings, an ecological clean process, savings on pressing tools and equipment, a decreased maintenance time and also much better corrosion resistance. Potential industrial applications of the studied TOC lie in a wide range of industries (construction, electronics, air conditioning, appliances, etc.) where metal sheet processing using stamping operations is involved.

Author Contributions: Conceptualization, M.T. and S.N.; methodology, S.N., F.H. and V.K.; validation, S.N., F.H. and V.K.; formal analysis, J.M.K. and M.B.; data curation, S.N., F.H. and V.K.;

writing—original draft preparation, M.T. and S.N.; writing—review and editing, M.T. and S.N.; visualization, J.M.K. and M.B.; supervision, E.E.; project administration, E.E.; funding acquisition, E.E. All authors have read and agreed to the published version of the manuscript.

Funding: This study was funded by Vedecká grantová agentúra MŠVVaŠ SR a SAV, grant number VEGA 1-0238-23.

Data Availability Statement: The data are only available on request due to restrictions provided by the funder.

Acknowledgments: This study was accomplished under the grant project VEGA 1-0238-23 "Implementation of CAx systems and virtual engineering techniques in the redesign of car-body parts for deformation zones".

Conflicts of Interest: Authors Stanislav Németh, František Hollý and Vladimír Kundracik are employed by the company U.S. Steel Košice s.r.o. The remaining authors declare that the research was conducted in the absence of any commercial or financial relationships that could be construed as a potential conflict of interest.

References

1. Čada, R. Formability of Deep-Drawing Steel Sheets. In Proceedings of the 5th European Conference on Advanced Materials and Processes and Applications (EUROMAT 97): Materials, Functionality Design: Volume 4-Characterization and Production/Design, Maastricht, The Netherlands, 21–23 April 1997; Netherlands Society for Materials Science: Maastricht, The Netherlands, 1997; pp. 463–466, ISBN 90-803513-4-2.
2. Liewald, M.; Wagne, S.; Becker, D. Influence of surface topography on the tribological behaviour of aluminium alloy 5182 with EDT surface. *Tribol. Lett.* **2010**, *39*, 135–142. [CrossRef]
3. Do, T.T.; Minh, P.S.; Le, N. Effect of Tool Geometry Parameters on the Formability of a Camera Cover in the Deep Drawing Process. *Materials* **2021**, *14*, 3993. [CrossRef] [PubMed]
4. Hazrati, J.; Stein, P.; Kramer, P.; van den Boogaard, A.H. Tool Texturing for Deep Drawing Applications. *IOP Conf. Ser. Mater. Sci. Eng.* **2018**, *418*, 012095. [CrossRef]
5. Šugárová, J.; Šugár, P.; Frnčík, M.; Necpal, M.; Moravčíková, J.; Kusý, M. The influence of the tool surface texture on friction and the surface layers properties of formed component. *Adv. Sci. Technol. Res. J.* **2018**, *12*, 181–193. [CrossRef] [PubMed]
6. Hou, J.; Deng, P.; Wang, S.; Xu, H.; Shi, Y. Study on formability and microstructure evolution of hot deep drawing manufactured 7005 aluminum alloy sheet metal. *Mater. Today Commun.* **2023**, *36*, 106794. [CrossRef]
7. Mihaliková, M.; Zgodavová, K.; Bober, P.; Špegárová, A. The Performance of CR180IF and DP600 Laser Welded Steel Sheets under Different Strain Rates. *Materials* **2021**, *14*, 1553. [CrossRef] [PubMed]
8. Feng, Y.; Hong, Z.; Gao, Y.; Lu, R.; Wang, Y.; Tan, J. Optimization of variable blank holder force in deep drawing based on support vector regression model and trust region. *Int. J. Adv. Manuf. Technol.* **2019**, *105*, 4265–4278. [CrossRef]
9. Kirkhorn, L.; Bushlya, V.; Andersson, M.; Stahl, J.-E. The influence of tool microstructure on friction in sheet metal forming. *Wear* **2013**, *302*, 1268–1278. [CrossRef]
10. Thipprakmas, S.; Sriborwornmongkol, J.; Jankree, R.; Phanitwong, W. Application of an Oleophobic Coating to Improve Formability in the Deep-Drawing Process. *Lubricants* **2023**, *11*, 104. [CrossRef]
11. Lange, K. *Handbook of Metal Forming*, 2nd ed.; Society of Manufacturing Engineers: Millersville, PA, USA, 1995.
12. Hrivnak, A.; Evin, E. *Formability of Steel Sheets*, 1st ed.; Elfa: Košice, Slovakia, 2004.
13. Chen, D.; Zhao, C.; Chen, X.; Li, H.; Zhang, X. Research on the active pressurized forced lubrication deep drawing process and evaluation of the lubrication effect. *Int. J. Adv. Manuf. Technol.* **2022**, *120*, 2815–2826. [CrossRef]
14. Bay, N.; Azushima, A.; Groche, P.; Ishibashi, I.; Merklein, M.; Morishita, M.; Nakamura, T.; Schmid, S.; Yoshida, M. Environmentally benign tribo-systems for metal forming. *CIRP Ann.* **2010**, *59*, 760–780. [CrossRef]
15. Reichardt, G.; Henn, M.; Reichle, P.; Umlauf, G.; Riedmüller, K.; Weber, R.; Barz, J.; Liewald, M.; Graf, T.; Tovar, G.E.M. Friction and wear behavior of deep drawing tools using volatile lubricants injected through laser-drilled micro-holes. *JOM* **2022**, *74*, 826–836. [CrossRef]
16. Jivan, R.; Eskandarzade, M.; Bewsher, S.; Leighton, M.; Mohammadpour, M.; Saremi-Yarahmadi, S. Application of solid lubricant for enhanced frictional efficiency of deep drawing process. *Proceedings of the Institution of Mechanical Engineers. Part C J. Mech. Eng. Sci.* **2022**, *236*, 624–634. [CrossRef]
17. European Commission. *Directorate-General for Research and Innovation*; Ferrari, V., Nicolle, R., Eds.; New surface treatment to improve adhesion of organic coatings and corrosion—In situ methods for the characterisation of Zn polymer coatings—Final report; Publications Office: Luxembourg, 2002.
18. European Commission. *Directorate-General for Research and Innovation*; Hardy, Y., Wormuth, R., Barranco Asensio, V., Eds.; New chromium-free thin organic coatings for Z, ZA and ZF—Final report; Publications Office: Luxembourg, 2003.
19. Trzepiecinski, T. A Study of the Coefficient of Friction in Steel Sheets Forming. *Metals* **2019**, *9*, 988. [CrossRef]

20. Padmanabhan, R.; Oliveira, M.C.; Alves, J.L.; Menezes, L.F. Influence of Process Parameters on the Deep Drawing of Stainless Steel. *Finite Elem. Anal. Des.* **2007**, *43*, 1062–1067. [CrossRef]
21. Zhao, H.; Cao, L.; Wan, Y.; Yang, S.; Gao, J.; Pu, J. Enhanced lubricity of zinc phosphate coating by stearic acid. *Lubr. Sci.* **2018**, *30*, 331–337. [CrossRef]
22. Bexell, U.; Carlsson, P.; Olsson, A. Tribological characterisation of an organic coating by the use of ToF-SIMS. *Appl. Surf. Sci.* **2003**, *203*, 596–599. [CrossRef]
23. Ma, I.A.W.; Ammar, S.; Kumar, S.S.A.; Ramesh, K.; Ramesh, A. A concise review on corrosion inhibitors: Types, mechanisms and electrochemical evaluation studies. *J. Coat. Technol. Res.* **2022**, *19*, 241–268. [CrossRef]
24. Lessa, R.C.d.S. Synthetic Organic Molecules as Metallic Corrosion Inhibitors: General Aspects and Trends. *Organics* **2023**, *4*, 232–250. [CrossRef]
25. Bammel, B.D.; Comoford, J.; Donaldson, G.T.; McGee, J.D.; Smith, T.S., II; Zimmerman, J. Novel Non-Chrome Thin Organic Hybrid Coating for Coil Steels. In Proceedings of the 38th Annual Waterborne Symposium, New Orleans, LA, USA, 28 February–4 March 2011; pp. 44–58.
26. Roos, O. Brugal® thin organic coatings: Effective and gainful alternative to traditional methods of protection of steels from corrosion. *Met. Sci. Heat Treat.* **2011**, *53*, 350–352. [CrossRef]
27. Thin Organic Coating. Available online: https://www.usske.sk/en/products/hot-dip-galvanized/thin-organic-coating (accessed on 3 March 2024).
28. Multiface®. Available online: https://www.voestalpine.com/surface-treatments/en/Products/Surface-treatments/multiface-R (accessed on 3 March 2024).
29. Easyfilm®. Available online: https://industry.arcelormittal.com/catalogue/E80/EN (accessed on 3 March 2024).
30. Directive 2011/65/EU of the European Parliament and of the Council of 8 June 2011 on the Restriction of the Use of Certain Hazardous Substances in Electrical and Electronic Equipment. Available online: http://data.europa.eu/eli/dir/2011/65/oj (accessed on 1 April 2024).
31. Carlsson, P.; Bexel, U.; Olsson, M. Friction and wear mechanisms of thin organic permanent coatings deposited on hot-dip coated steel. *Wear* **2001**, *247*, 88–99. [CrossRef]
32. Barletta, M.; Gisario, A.; Puopolo, M.; Vesco, S. Scratch, wear and corrosion resistant organic inorganic hybrid materials for metals protection and barrier. *Mater. Des.* **2015**, *69*, 130–140. [CrossRef]
33. Tsai, P.-Y.; Chen, T.-E.; Lee, Y.-L. Development and Characterization of Anticorrosion and Antifriction Properties for High Performance Polyurethane/Graphene Composite Coatings. *Coatings* **2018**, *8*, 250. [CrossRef]
34. Ying, L.; Wu, Y.; Nie, C.; Wu, C.; Wang, G. Improvement of the Tribological Properties and Corrosion Resistance of Epoxy–PTFE Composite Coating by Nanoparticle Modification. *Coatings* **2021**, *11*, 10. [CrossRef]
35. Trzepiecinski, T.; Lemu, H.G. Effect of Lubrication on Friction in Bending under Tension Test-Experimental and Numerical Approach. *Metals* **2020**, *10*, 544. [CrossRef]
36. Evin, E.; Daneshjo, N.; Mareš, A.; Tomáš, M.; Petrovčiková, K. Experimental Assessment of Friction Coefficient in Deep Drawing and Its Verification by Numerical Simulation. *Appl. Sci.* **2021**, *11*, 2756. [CrossRef]
37. Heinzel, H.; Ramezani, M.; Neitzert, T. Experimental Investigation of the Formability of Organic Coated Steel Sheet Metal. *Procedia Manuf.* **2015**, *1*, 854–865. [CrossRef]
38. Evin, E.; Tomáš, M. Influence of Friction on the Formability of Fe-Zn-Coated IF Steels for Car Body Parts. *Lubricants* **2022**, *10*, 297. [CrossRef]
39. Xia, J.; Zhao, J.; Dou, S.; Shen, X. A Novel Method for Friction Coefficient Calculation in Metal Sheet Forming of Axis-Symmetric Deep Drawing Parts. *Symmetry* **2022**, *14*, 414. [CrossRef]
40. Evin, E.; Németh, S.; Výrostek, M. Evaluation of Friction Coefficient of Stamping. *Acta Mech. Slovaca* **2014**, *18*, 20–27. [CrossRef]
41. EN 10346:2015; Continuously hot-dip coated steel flat products—Technical delivery conditions. CEN-CENELEC Management Centre: Brussels, Belgium, 2015.
42. ISO 6892-1:2019; Metallic Materials—Tensile Testing—Part 1: Method of Test at Room Temperature. International Organization for Standardization: Geneva, Switzerland, 2019.
43. ISO 10113:2020; Metallic Materials—Sheet and Strip—Determination of Plastic Strain Ratio. International Organization for Standardization: Geneva, Switzerland, 2020.
44. ISO 10275:2020; Metallic Materials—Sheet and Strip—Determination of Tensile Strain Hardening Exponent. International Organization for Standardization: Geneva, Switzerland, 2020.
45. ISO 9227:2022; Corrosion Tests in Artificial Atmospheres—Salt Spray Test. International Organization for Standardization: Geneva, Switzerland, 2022.
46. Popic, J.P.; Jegdic, B.V.; Bajat, J.B.; Veljovic, D.; Stevanovic, S.I.; Miskovic-Stankovic, V.B. The effect of deposition temperature on the surface coverage and morphology of iron-phosphate coatings on low carbon steel. *Appl. Surf. Sci.* **2011**, *257*, 10855–10862. [CrossRef]
47. Hara, A.; Kazimierczak, H.; Bigos, A.; Świątek, Z.; Ozga, P. Effect of different organic additives on surface morphology and microstructure of Zn-Mo coatings electrodeposited from citrate baths. *Arch. Metall. Mater.* **2019**, *64*, 207–220. [CrossRef]
48. Liu, X.; Yu, W.; Zhang, Q.; Jiang, S. Influence of Surface Roughness of Galvanized Steel Sheet on Self-lubricated Coating. *J. Iron Steel Res. Int.* **2014**, *21*, 342–347. [CrossRef]

49. Sankara Narayanan, T.S.N. Surface Pretreatment by Phosphate Conversion Coatings—A Review. *Rev. Adv. Mater. Sci.* **2005**, *9*, 130–177.
50. Shih, H. Friction and Die Wear in Stamping Prephospated Advanced High Strength Steels. *SAE Int. J. Mater. Manf.* **2016**, *9*, 481–487. [CrossRef]
51. Saffarzade, P.; Ali Amadeh, A.; Agahi, N. Study of tribological and friction behavior of magnesium phosphate coating and comparison with traditional zinc phosphate coating under dry and lubricated conditions. *Tribol. Int.* **2020**, *144*, 106122. [CrossRef]
52. Carlsson, P.; Bexel, U.; Olsson, M. Tribological behaviour of thin organic permanent coatings deposited on hot-dip coated steel sheet—A laboratory study. *Surf. Coat. Technol.* **2000**, *132*, 169–180. [CrossRef]
53. Podjuklová, J.; Laník, T.; Hrabovská, K.; Bártek, V.; Suchánková, K.; Kopaňáková, S.; Šrubař, P.; Dvorský, R. Study on thin organic coatings for short-term anticorrosive protection of metallurgical materials production. In Proceedings of the 20th Anniversary International Conference on Metallurgy and materials, Brno, Czech Republic, 18–20 May 2011.
54. Sarli, A.R.D.; Elsner, C.I.; Tomachuk, C.R. Characterization and Corrosion Resistance of Galvanized Steel/Passivation Composite/Polyurethane Paint Systems. *Curr. J. Appl. Sci. Technol.* **2013**, *4*, 853–878. [CrossRef]
55. Gao, F.; Du, A.; Ma, R.; Lv, C.; Yang, H.; Fan, Y.; Zhao, X.; Wu, J.; Cao, X. Improved corrosion resistance of acrylic coatings prepared with modified MoS2 nanosheets. *Colloids Surf. A Physicochem. Eng. Asp.* **2020**, *587*, 124318. [CrossRef]
56. Li, H.; Sun, L.; Li, W. Application of organosilanes in titanium-containing organic–inorganic hybrid coatings. *J. Mater. Sci.* **2022**, *57*, 13845–13870. [CrossRef]
57. Kuznetsov, Y.I.; Redkina, G.V. Thin Protective Coatings on Metals Formed by Organic Corrosion Inhibitors in Neutral Media. *Coatings* **2022**, *12*, 149. [CrossRef]

Disclaimer/Publisher's Note: The statements, opinions and data contained in all publications are solely those of the individual author(s) and contributor(s) and not of MDPI and/or the editor(s). MDPI and/or the editor(s) disclaim responsibility for any injury to people or property resulting from any ideas, methods, instructions or products referred to in the content.

Article

Optimization of Sustainable Production Processes in C45 Steel Machining Using a Confocal Chromatic Sensor

Jozef Jurko [1], Katarína Paľová [2], Peter Michalík [3] and Martin Kondrát [1,*]

[1] Department of Industrial Engineering and Informatics, Faculty of Manufacturing Technologies, Technical University of Košice, Bayerova 1, 08001 Prešov, Slovakia; jozef.jurko@tuke.sk

[2] Department of Biomedical Engineering and Measurement, Faculty of Mechanical Engineering Technical University of Košice, Letná 9, 04200 Košice, Slovakia; katarina.palova@student.tuke.sk

[3] Department of Automotive and Manufacturing Technologies, Faculty of Manufacturing Technologies, Technical University of Košice, Štúrova 31, 08001 Prešov, Slovakia; peter.michalik@tuke.sk

* Correspondence: martin.kondrat@tuke.sk; Tel.: +421-55-602-6420

Abstract: Metal machining production faces a myriad of demands encompassing ecology, automation, product control, and cost reduction. Within this framework, an exploration into employing a direct inspection of the machined area within the work zone of a given machine through a confocal chromatic sensor was undertaken. In the turning process, parameters including cutting speed (A), feed (B), depth of cut (C), workpiece length from clamping (D), and cutting edge radius (E) were designated as input variables. Roundness deviation (Rd) and tool face wear (KM) parameters were identified as output factors for assessing process performance. The experimental phase adhered to the Taguchi Orthogonal Array L27. Confirmatory tests revealed that optimizing process parameters according to the Taguchi method could enhance the turning performance of C45 steel. ANOVA results underscored the significant impact of cutting speed (A), feed (B), depth of cut (C), and workpiece length from clamping (D) on turning performance concerning Rd and KM. Furthermore, initial regression models were formulated to forecast roundness variation and tool face wear. The proposed parameters were found to not only influence the machined surface but also affect confocal sensor measurements. Consequently, we advocate for the adoption of these optimal cutting conditions in product production to bolster turning performance when machining C45 steel.

Keywords: confocal chromatic sensor (CCHS); turning; steel C45; roundness deviation; tool face wear; machined surface; Taguchi method; ANOVA

Citation: Jurko, J.; Paľová, K.; Michalík, P.; Kondrát, M. Optimization of Sustainable Production Processes in C45 Steel Machining Using a Confocal Chromatic Sensor. *Lubricants* **2024**, *12*, 99. https://doi.org/10.3390/lubricants12030099

Received: 13 January 2024
Revised: 5 March 2024
Accepted: 14 March 2024
Published: 16 March 2024

Copyright: © 2024 by the authors. Licensee MDPI, Basel, Switzerland. This article is an open access article distributed under the terms and conditions of the Creative Commons Attribution (CC BY) license (https://creativecommons.org/licenses/by/4.0/).

1. Introduction

Laser scanning technology has been well established for several years, with its efficacy in product inspection demonstrated across various research endeavors. While the utilization of scanners in inspecting machined surfaces within the working zone of machines post process completion offers numerous benefits, it also presents limitations [1]. Sustainable and efficient manufacturing serves as the impetus behind the burgeoning Zero Defect Manufacturing (ZDM) trend. ZDM aims not only to detect defective products but also to predict and prevent defects [2]. As concluded by the authors of [3], dry machining proves optimal for clean manufacturing. Laser technology currently finds critical applications in the non-contact inspection of product parameters, with researchers exploring the integration of laser sensors in measuring parameters within the cutting zone during machining. According to [4], CCHS represents a high-precision measuring device, with data analysis stability outweighing mere accuracy. Reflectivity of the beam from the object to be measured and light source fluctuation pose challenges, as noted by [5]. Other researchers [6] delved into the utilization of CCHS in three-dimensional product surface measurement. Meanwhile, in their exploration of product measurement, Ref. [7] employed a high-speed 3D camera system. The authors of [8] investigated the application of CCHS for assessing the surface properties of product layers in [9]. However, the machining process itself can be significantly influenced by factors such as surface topography and microhardness [10–12], surface roughness [10,11,13], residual stress [14–16], and microstructure [17]. In their analysis in [18], the authors provide insights and solutions for machining hardened steels with regard to surface integrity. Specifically, they explore the impact of hardening on surface roughness, cylindricity, and roundness after turning C45 steel. Research findings [19] indicate a detrimental effect of hardening on surface quality. Furthermore, surface treatment via machining methods can alter parameters associated with surface integrity [20]. The investigation and analysis of the surface quality parameters of milled surfaces were conducted using a confocal laser scanning microscope by the authors of [21]. Mechanical hardening is described in detail in studies by the authors of [22,23]. Additionally, other researchers have endeavored to identify the influence of material properties on the surface hardening of materials such as Al 6061-T6 [24] and Hadfield steels [25–31]. The issue of surface hardening in AISI 304 steel is addressed in [32]. The authors of [33] examine samples of surface roughness after turning. The influence of surface texture on surface roughness is investigated in [34], while Ref. [35] analyzes the impact of the tip radius of measuring contacts on surface roughness parameters. Additionally, Ref. [36] asserts results that surface roughness significantly affects machined product performance. Investigations into factors affecting the machined surface of metallic materials are carried out by the authors of [37,38]. The authors of [39] observe the deterioration of machined surfaces in Ti6Al4V alloys due to thermal softening using the HSM method. Similarly, results from [40] indicate a trade-off relationship between significant process factors' influence on material removal rate and surface roughness. Regarding EDM technology, Ref. [41] investigates the performance of machined stainless steel. Changes in machined surfaces affect quality parameter attainment in milling due to thermal deformation [42], in turning due to stress and strain prediction [43], in turning due to cutting tool modification [44] and cutting tool damage [45], and in milling Ti alloys due to elastic deformation [46]. The investigation of the effect of tool wear on surface roughness increase is addressed by [47], while [48] demonstrates through research that low-temperature cooling can enhance machined surface quality while minimizing tool wear. The determination of tool wear in the machining of duplex stainless steel is presented in papers [49,50], with the issue of tool wear and surface topography also discussed in [51–55]. Additional researchers [56] conduct experimental investigations into factors affecting stainless steel turning and assess wear parameters on the cutting tool's face and back surface. Additionally, surface integrity evaluation parameters encompass shape and position deviations. In this regard, the application of CCHS proves highly beneficial, particularly considering its comprehensive data and swift data acquisition. Nonetheless, the challenge remains in efficiently processing acquired data and accurately interpreting them for theoretical and practical needs, as per the research results in [1]. The authors of [57] utilize CCHS to measure roundness deviation and demonstrate, through experimental results, the method's steady and reliable evaluation of roundness deviations. Several authors have explored different machinability characteristics of C45 steel through varying heat treatment methods [58], employing coated tools in milling [59], utilizing process media [60], implementing ultrasonics in turning [61], and modifying cutting inserts [62]. Furthermore, the authors of [63–65] investigate various parameters of machined surfaces using non-contact methods.

Our primary research focus revolves around controlling factors affecting machined surfaces through non-contact methods, utilizing laser sensors directly within a machine's working zone. We aim to optimize input parameters and analyze adverse phenomena on machined surfaces.

In this investigation, we explored the feasibility of directly measuring roundness deviation within a machine's working zone using a confocal chromatic sensor during the turning of C45 steel. In the turning process, parameters such as cutting speed (A), feed (B), depth of cut (C), workpiece length from clamping (D), and cutting edge radius (E) were chosen as input parameters. Roundness deviation and tool face wear (KM) parameters, following ISO 3685 standards, were proposed as output factors for evaluating process performance, measured using a microscope. To measure roundness deviation, we developed new-generation CCHS sensors (CL P070 from Keyence—Mechelen, Belguim). To analyze the effect of cutting insert face wear on roundness deviation under defined cutting conditions using Taguchi Design in Minitab 21.4.2 software, we designed the Orthogonal Array L27 experiment matrix. All research tests were conducted under dry turning conditions, with a key assumption being that machine tool vibrations align with prescribed values certified for specific machine tools (as outlined in Table 1). Currently, several companies in our region manufacture various C45 steel products for the engineering industry.

Table 1. Experimental data.

M—Machine	Leadwell T5 CNC machine tool equipped with a FANUC Oi-MATE-TC control system. Maximum radial sweep certified by the manufacturer is 0.030 mm, and maximum axial sweep is 0.020 mm.
T—Cutting insert Tool holder Working insert tool geometry	Cutting tool clamped with a tool holder marked SSDCN1212 F09 from Dormer Pramet Ltd., Šumperk, Czech Republic. Cutting insert SCMT 09T308E-FM, T9325 made of sintered carbide, with specific geometry parameters (nose angle ε_r = 90°, main cutting edge setting angle κ_r = 45°, clearance angle major α = 7°, and nose radius r_ε = 0.8 mm).
W—Workpiece material and dimensions	Workpiece material C45 steel (1.0503). Test specimen dimensions: diameter (d) = 40 mm, length (L) = 150 mm. The chemical composition of the steel is given in Table 2 and was verified prior to the start of the research. Table 3 shows the main properties of the tested steel.
F—Fixture for tool and object	Specimen: A round bar clamped in a chuck. Tool: Clamped in the cutter head.
Machining conditions	Dry machining. Machining method—turning.
Mobile Measuring System (MMS) is composed of CCHS sensor	PLC Siemens-1511C, Communication module KEYENCE-CL3000, Amplifier KEYENCE-CLP070N, Communication module KEYENCE-DL-PN1 and sensor CL P070 by Keyence. Measurement range: 75 mm to 130 mm. Reference distance: 100 mm. Resolutions: ±1 mm. Spot diameter: 600 mm. Linearity: ±0.15% of F.S. (IL-100: ±20 mm). Repeatability: 4 mm. For the research purposes, a bespoke holder tailored for the CCHS sensor was meticulously designed and subsequently fabricated utilizing advanced 3D printing technology.

Table 2. Chemical composition of C45 steel (EN 10083-2-91) *.

Steel C45	(%)
C	0.50
Mn	0.80
Si	0.37
Cr	0.22
Ni	0.28
Cu	0.18
P	0.035
S	0.032

* Chemical composition verified through analysis of the sample.

Table 3. Verified properties of C45 steel products.

Steel C45	Values
Yield stress Re (MPa)	202
Tensile strength Rm (MPa)	650
Density (g/cm^3)	7.85
Hardness HB	max. 220
Elastic modulus (GPa)	81
Flexural strength (MPA)	606
Thermal conductivity (W/mK)	50

2. Description, Implementation, and Experimental Results

2.1. Experimental Design

In this section, we outline the experimental setup and methodology employed in our study. We proposed to utilize the experimental data within the framework of the technology system (TS) depicted in Figure 1, along with the mobile measurement system (MMS) previously introduced by the authors [1], illustrated in Figure 2 and described in detail in Table 1. The experimental parameters considered in this study are enumerated in Table 1. The range for each controlled parameter was determined based on prior experiments and information obtained from manufacturers of C45 steel products, as summarized in Table 2. Additionally, adhering to recommendations from the tool manufacturer, a specific cutting insert was selected for machining C45 steel, as detailed in Table 1.

Figure 1. Technological System (TS). The TS comprises the machine (M), tool (T), workpiece (W), and fixture (F).

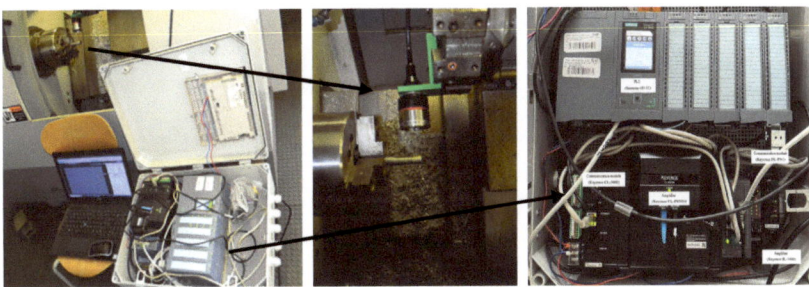

Figure 2. Mobile Measuring System (MMS).

The cutting process conditions are outlined in Table 4, with cutting speed designated as parameter A, feed as parameter B, depth of cut as parameter C, workpiece length from clamping as parameter D, and cutting edge radius as parameter E. The assessment of the output factor, specifically the roundness deviation of the circular bar, was conducted using CCHS. The experimental matrix and resultant output factors, represented as average values from the measurements, are detailed in Tables 5 and 6. Roundness deviation of the machined surface was directly measured and analyzed within the working zone of the machine tool utilizing CCHS. This measurement was performed post-machining on a 240 mm length workpiece without unclamping, employing MMS (mobile measurement system) for CCHS. Rd was assessed at three points along the machined length (60 mm) of the circular bar, specifically at distances of 5.0 mm, 30.0 mm, and 55.0 mm from the clamping point (refer to Figure 3). KM face wear was gauged following a 240 mm traversal, adhering to ISO 3685 [66] standards, as illustrated in Figure 4. Rd values for specific local locations were determined as averages. Resultant average values of Rd and cutting insert face wear were derived from 12 replicate measurements, with the exclusion of minimum and maximum outliers. Each experiment employed a cutting insert with a new cutting edge. Literature reviews [55,56] highlight the importance of surface integrity research incorporating new technologies, necessitating further investigation. The exploration of roundness deviation control using CCHS on a reference specimen of C45 steel serves as a foundation for other materials. A systematic examination of process parameter effects on the machined surface underscores the potential to influence output factors towards desired values by parameter adjustments. Enhanced machined surface integrity is achievable through the optimization of input parameters such as cutting speed, feed, depth of cut, workpiece clamping length, and cutting edge radius. The alignment of cutting conditions with CCHS requirements is imperative. These adjustable input parameters during turning render them conducive to automation and intelligent processing.

Table 4. The primary machining conditions.

Symbol	Process Parameters Units	Levels		
		1	2	3
A	Cutting speed (m/min)	90	180	270
B	Feed (mm/rev.)	0.1	0.2	0.3
C	Depth of cut (mm)	0.1	0.4	0.8
D	Workpiece length from clamping (mm)	5.0	30.0	55.0
E	Cutting edge radius (mm)	0.003 + 0.0005	0.005 + 0.0005	0.008 + 0.0005

Table 5. Experimental matrix, design, and experimental results.

Number of Exp.-RUN	Controllable Process Parameter					Experimental Results			KM (mm)
	A	B	C	D	E	Rd (mm)			
						Average Rd	STDVP	STDV ERROR SE (AVERAGE)	
1	1	1	1	1	1	0.0260	0.001	0.0003	0.1420
2	1	1	1	1	2	0.0184	0.001	0.0005	0.1120
3	1	1	1	1	3	0.0220	0.001	0.0006	0.1180
4	1	2	2	2	1	0.0440	0.001	0.0006	0.1850
5	1	2	2	2	2	0.0440	0.002	0.0009	0.1740
6	1	2	2	2	3	0.0410	0.001	0.0006	0.1670
7	1	3	3	3	1	0.0860	0.005	0.0024	0.1820
8	1	3	3	3	2	0.0720	0.003	0.0015	0.1810
9	1	3	3	3	3	0.0640	0.003	0.0012	0.1960
10	2	1	2	3	1	0.0520	0.003	0.0012	0.1730
11	2	1	2	3	2	0.0450	0.002	0.0010	0.1710
12	2	1	2	3	3	0.0460	0.013	0.0059	0.1690
13	2	2	3	1	1	0.0640	0.008	0.0035	0.1860
14	2	2	3	1	2	0.0420	0.012	0.0054	0.1820
15	2	2	3	1	3	0.0390	0.011	0.0051	0.1670
16	2	3	1	2	1	0.0340	0.005	0.0025	0.1750
17	2	3	1	2	2	0.0360	0.007	0.0030	0.1700
18	2	3	1	2	3	0.0320	0.003	0.0012	0.1710
19	3	1	3	2	1	0.0370	0.010	0.0043	0.1880
20	3	1	3	2	2	0.0410	0.001	0.0006	0.1770
21	3	1	3	2	3	0.0390	0.003	0.0013	0.1860
22	3	2	1	3	1	0.0300	0.007	0.0031	0.1970
23	3	2	1	3	2	0.0370	0.008	0.0034	0.1820
24	3	2	1	3	3	0.0440	0.010	0.0045	0.1820
25	3	3	2	1	1	0.0290	0.001	0.0004	0.1820
26	3	3	2	1	2	0.0370	0.002	0.0007	0.1920
27	3	3	2	1	3	0.0180	0.002	0.0009	0.1810

Table 6. Calculated S/N ratios.

Number of Exp.-RUN	S/N Ratios of Results	
	Rd (dB)	KM (dB)
1	31.7005	16.9542
2	34.7036	19.0156
3	33.1515	18.5624
4	27.1309	14.6566
5	27.1309	15.1890
6	27.7443	15.5457
7	21.3100	14.7986
8	22.8534	14.8464
9	23.8764	14.1549
10	25.6799	15.2391
11	26.9357	15.3401
12	26.7448	15.4423
13	23.8764	14.6097
14	27.5350	14.7986
15	28.1787	15.5457
16	29.3704	15.1392
17	28.8739	15.3910
18	29.8970	15.3401
19	28.6360	14.5168
20	27.7443	15.0405
21	28.1787	14.6097
22	30.4576	14.1107
23	28.6360	14.7986
24	27.1309	14.7986
25	30.7520	14.7986
26	28.6360	14.3340
27	34.8945	14.8464

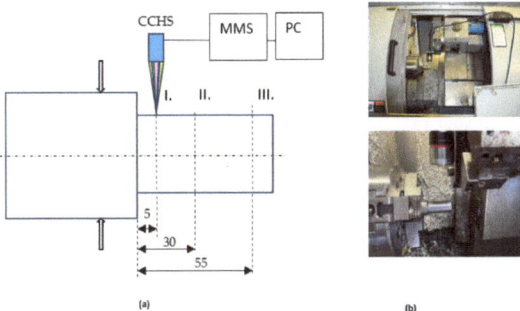

Figure 3. The measurement layout within the Leadwell T5 CNC machine zone. (**a**) Measurement scheme utilizing CCHS (confocal chromatic sensor), MMS (Mobile Measuring System), and PC (Personal Computer). (**b**) Visualization of the working area of the CNC machine.

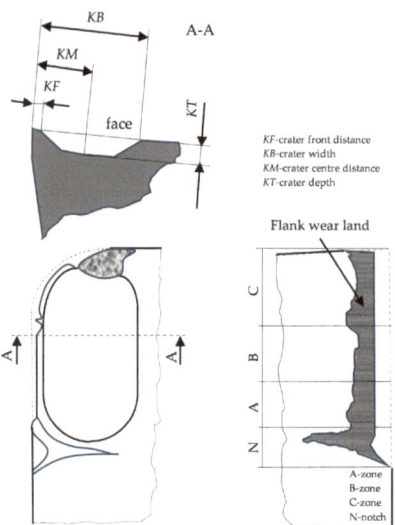

Figure 4. The cutting tool wear parameters as per ISO 3685 [66].

The face wear on the cutting insert was assessed in accordance with ISO standard 3685, as depicted in Figure 4. and analyzed by using a Carl Zeiss Primotech D/A ESD microscope – Carl Zeiss Jena GmbH, Jenna, Germany as shown in Figure 5.

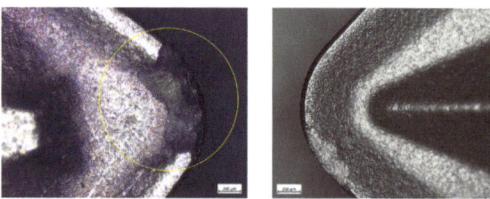

Figure 5. The measurement of wear on the cutting insert face employing a Carl Zeiss Primotech D/A ESD microscope. Magnification 5× The yellow circle characterizes the face wear of tool.

2.2. Optimization of Turning Input Parameters Using Taguchi Method with S/N Ratio Analysis

An experiment matrix employing Orthogonal Array L27 was designed following Taguchi principles using Minitab 21.4.2 software—Coventry, United Kingdom. Five parameters were set at three levels each, resulting in a total of 27 runs.

Researchers [67,68] have noted that Taguchi's methodology [69] considers factors such as resource wastage, warranty costs, customer complaints, and repair expenses, all of which influence product quality. Taguchi's approach is particularly beneficial for optimizing products or processes while simultaneously reducing experimentation time and research costs. Orthogonal arrays, coupled with signal-to-noise ratio analysis, are employed to gauge output parameter quality. Various models incorporating Taguchi methods have been applied in research, with parameters' significance determined through ANOVA. These models include L8OA [70–72], L18OA [73], L27OA [74,75], L27OA with ANOVA [58,76,77], L18OA with ANOVA [78], L9OA with ANOVA [37,60,72], and the L8OA model combined with Response Surface Methodology (RSM) and ANOVA [38,79,80]. Additionally, the GRA (Grey Relation Analysis) method with ANOVA has been proposed [74,81–84], along with other variations by different researchers [85].

Genichi Taguchi [69] introduced a loss function in data processing to represent the disparity between experimental and target values, subsequently converted into signal-to-noise (S/N) ratios. The S/N ratio, defined as the ratio of mean value to standard deviation, categorizes into "Larger is Better," "Medium is Better," and "Smaller is Better," based on response requirements. This study focuses on the output factors of roundness deviation and wear on the cutting insert's face surface, both adopting the "Smaller is Better" methodology, aligned with practical and C45 steel machinability requirements. Equation (1) was utilized for S/N ratio calculation, with the results tabulated in Table 6. Taguchi analysis, mean S/N ratio graphs, and ANOVA were conducted using Minitab 21.4.2 software.

The impact severity of each input parameter on output factors is depicted in Figures 6 and 7. Cutting conditions and cutting insert geometry influence chip evacuation from the cutting zone [37,43,44,78]. Continuous cutting processes yield regular machined surfaces devoid of Plastically Deformed Material (PDM) elements. Conversely, discontinuous cutting may produce irregular chips with PDM, adversely affecting surface quality and increasing roundness deviation (see Figure 6). Increasing Rd may signify elevated cutting insert wear, notably on the KM face, further exacerbating surface quality due to augmented PDM elements.

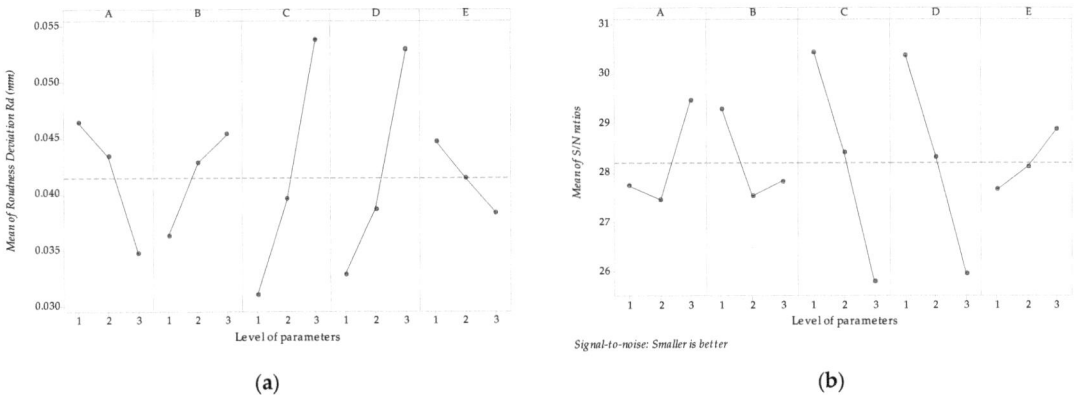

Figure 6. Main effects plots: (**a**) illustrates the effects of input factors on roundness deviation, while (**b**) displays the mean S/N ratios corresponding to roundness deviation.

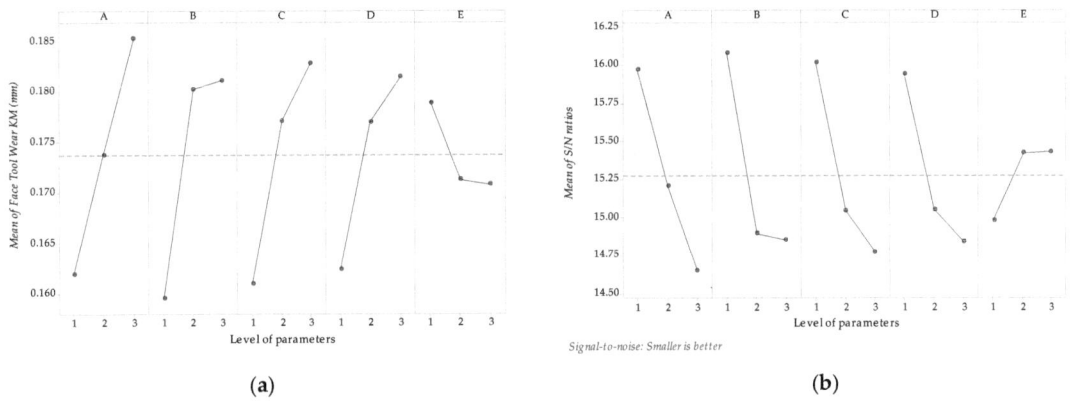

Figure 7. Main effects plots: (**a**) the effects of input factors on the tool wear of the face, and (**b**) the mean S/N ratios corresponding to the tool wear of the face.

The S/N ratio (Equation (1)) serves as a quantitative tool, with Rd proposed as a qualitative machined surface factor, ideally minimized. The S/N ratio is as follows:

$$S/N = -10\log\left[\frac{1}{n}\left(y_1^2 + y_2^2 + \ldots + y_n^2\right)\right] \quad (1)$$

where

S/N represents parameter values (unit dB)
and
$y_1, y_2, \ldots, y_n$ are the observed output values for the test condition repeated n times.

Figures 6a and 7a depict the relationship between mean values of roundness deviation and the face wear of the cutting insert. Cutting speed emerges as the most influential parameter impacting roundness deviation (as evident in Figure 6a), similarly affecting tool wear at the cutting insert face (as observed in Figure 7a). Increased feed, depth of cut, and distance of the workpiece from the fixture also exert a negative effect on Rd. With rising roundness deviation, surface quality diminishes, accompanied by increased tool wear. Elevated cutting speed mitigates roundness deviation but heightens the risk of tool damage and wear. Optimal regions lie between cutting speed levels 1 and 2, as illustrated in Figures 6b and 7b. Figures 6a and 7a highlight the optimal levels of individual parameters to achieve the desired output factors in terms of roundness deviation (Rd) and the face wear of the cutting inserts, respectively. These optimal levels correspond to A3B1C1D1E3 for Rd and A1B1C1D1E3 for the face wear of the cutting inserts, as corroborated by the S/N ratio values in Figures 6b and 7b.

Lower roundness deviation (Rd) values can enhance the operational reliability of the studied functional surface of a product. This improvement stems from increased friction between the tool and workpiece, resulting in higher cutting temperatures with increasing cutting speed. Elevated temperatures in the machining zone cause thermal softening of the workpiece, reducing smeared materials on the machined surface and consequently minimizing roundness deviation [77,85,86]. As depicted in Figure 8, Rd values escalate with rising cutting speed due to heightened tool resistance against the workpiece, particularly as feed increases, leading to the formation of more built-up edges (BUE) on the tool face. This phenomenon induces surface deterioration and consequently elevates Rd values. Moreover, an increasing trend in Rd values accompanies higher depths of cut. These trends align with the existing literature findings on machining difficult-to-machine materials [87]. Precision in determining the local measurement spot on the specimen significantly influences Rd.

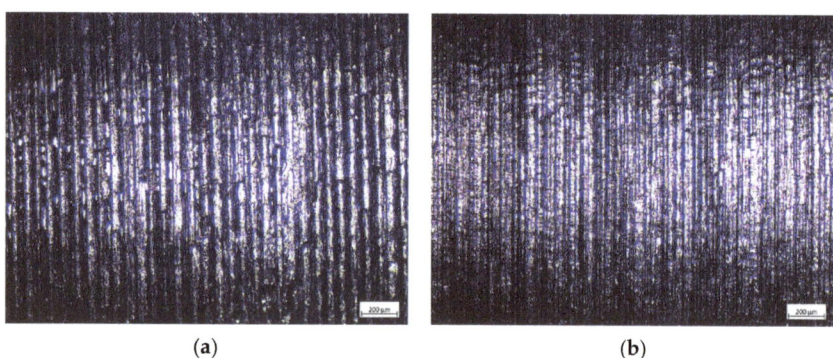

(a) (b)

Figure 8. Visualization of machined surface. (**a**) Initial parameter settings at A = 180 m/min, B = 0.2 mm/rev., C = 0.4 mm, D = 30 mm, and E = 0.005 mm, (A2B2C2D2E2). Magnification 5×. (**b**) Taguchi optimum settings at A = 270 m/min, B = 0.1 mm/rev., C = 0.1 mm, D = 5.0 mm, and E = 0.008 mm, (A3B1C1D1E3). Magnification 5×.

Simultaneously, this measurement process provides insights into the cutting tool's compliance with the specified technical requirements of the product and identifies any undesirable wear. Variations in Rd magnitude may indicate changes in the cutting tool. Consequently, this measurement method facilitates the rapid and accurate identification of local spots on the functional surface and enables the monitoring of changes in cutting tool wear rate. However, the mean values of output factors are contingent on specific conditions, necessitating test repetition for other materials.

The S/N ratio response factor table for Rd is presented in Table 7. Figure 7b illustrates the S/N ratio chart generated using Minitab software. A higher S/N ratio indicates minimal deviation between the desired and measured outputs. As depicted in Figure 9b, the highest average S/N ratio values obtained for Rd are 270 m/min, 0.1 mm/rev., 0.1 mm, 5.0 mm, and 0.008 mm. Therefore, the assumed optimal process parameters for achieving low Rd using Taguchi's method are 270 m/min, 0.1 mm/rev., 0.1 mm, 5.0 mm, and 0.008 mm. These optimal combinations are highlighted in bold in Table 6 for clarity, with the corresponding levels identified. This predicted optimal combination is represented as A3B1C1D1E3 for roundness deviation.

Table 7. Mean S/N ratio response table for roundness deviation.

Symbol	Process Parameters and Units	Mean S/N Ratios				
		Level 1	Level 2	Level 3	Max–Min	Rank
A	Cutting speed (m/min)	27.73	27.45	**29.45**	2.00	3
B	Feed (mm/rev.)	**29.28**	27.54	27.83	1.74	4
C	Depth of cut (mm)	**30.44**	28.41	25.80	4.64	1
D	Workpiece length from clamping (mm)	**30.38**	28.30	25.96	4.42	2
E	Cutting edge radius (mm)	27.66	28.12	**28.87**	1.21	5

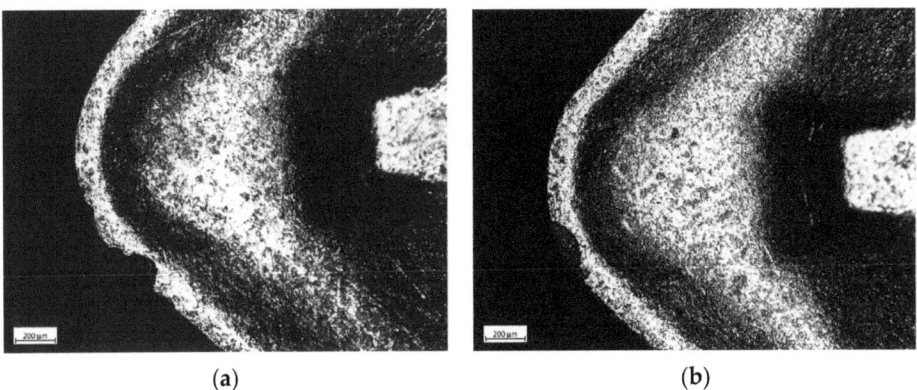

Figure 9. Wear display on cutting insert face. (**a**) Initial parameter settings at A = 180 m/min, B = 0.2 mm/rev., C = 0.4 mm, D = 30 mm, and E = 0.005 mm, (A2B2C2D2E2). Magnification 5×. (**b**) Taguchi optimum settings at A = 90 m/min, B = 0.1 mm/rev., C = 0.1 mm, D = 5.0 mm, and E = 0.008 mm, (A1B1C1D1E3). Magnification 5×.

Table 8 presents the generated S/N ratio factor table for face wear on the cutting insert. Mean values of the S/N ratio for face wear are depicted in Figure 7b. Based on the data from Figure 7b and Table 8, the estimated optimal process parameters for minimizing face wear are 90 m/min, 0.1 mm/rev., 0.1 mm, 5.0 mm, 0.008 mm. This predicted optimal combination is represented as A1B1C1D1E3 for face wear on the cutting insert.

Table 8. Mean S/N ratio response table for tool face wear.

Symbol	Process Parameters and Units	Mean S/N Ratio				
		Level 1	Level 2	Level 3	Max–Min	Rank
A	Cutting speed (m/min)	**15.97**	15.21	14.65	1.32	1
B	Feed (mm/rev.)	**16.08**	14.89	14.85	1.23	3
C	Depth of cut (mm)	**16.01**	15.04	14.77	1.24	2
D	Workpiece length from clamping (mm)	**15.94**	15.05	14.84	1.10	4
E	Cutting edge radius (mm)	14.98	15.42	**15.43**	0.45	5

The S/N ratio response table for Rd is also presented in Table 7, while Figure 6b illustrates the mean S/N ratio chart obtained through Minitab software—Coventry, United Kingdom. Again, a higher S/N ratio signifies minimal deviation between the desired and measured outputs. As indicated in Figure 6b, the highest average S/N ratio values obtained for Rd are A3 = 270 m/min, B1 = 0.1 mm/rev., C1 = 0.1 mm, D1 = 5.0 mm, and E3 = 0.008 mm. Thus, the assumed optimal process parameters for achieving low roundness deviation using Taguchi's method are represented as A3B1C1D1E3, with the corresponding level values highlighted in bold in Table 7. Table 8 presents the obtained S/N ratio response table for face wear on the cutting insert. The average S/N ratio values for the face wear of the cutting insert are illustrated in Figure 7b. From Figure 7a, it can be observed that the estimated optimum process parameters for achieving low wear on the cutting insert face are A1 = 90 m/min, B1 = 0.1 mm/rev., C1 = 0.1 mm, D1 = 5.0 mm, and E3 = 0.008 mm.

Confirmation Test

To validate the optimality of the predicted optimal settings according to Taguchi, confirmatory tests must be conducted. The predicted signal-to-noise (S/N) ratio (e) was utilized to estimate and verify the response under the predicted optimal cutting settings, calculated using Equation (2).

$$\varepsilon_{predicted} = \varepsilon_{tm} + \sum_{i=1}^{p}(\varepsilon_o - \varepsilon_{tm}) \qquad (2)$$

where

ε_{tm} is the total mean S/N ratio;
ε_o is the mean S/N ratio at the optimal level;
p is the number of input process parameters.
Confirmatory tests are essential to validate the predicted optimal cutting settings.

After predicting the optimum cutting settings, confirmation experiments were conducted, and the results are presented in Tables 9 and 10 for Rd and KM, respectively. The predicted optimal cutting settings for both Rd and KM led to improved process performance outcomes. Tables 9 and 10 demonstrate that the S/N ratios under the predicted and optimal cutting settings closely align for both Rd and KM. The enhancement in the S/N ratio at the optimal cutting settings for Rd and KM amounted to 4.95 dB and 2.01 dB, respectively, compared to the original parameter settings shown in Tables 9 and 10. From the confirmation experiments, it was observed that the predicted optimal cutting settings by Taguchi yielded favorable results compared to the initial parameter settings, particularly with regard to the reduction in Rd and KM. Specifically, the reductions in Rd and KM compared to the initial parameter settings were found to be 15.29% and 12.5%, respectively. Hence, the predicted optimal cutting settings by Taguchi are deemed to be the optimum conditions for achieving low Rd and low KM in machining C45 steel under the given conditions. From Figures 8 and 9, it is evident that the optimal cutting settings as per Taguchi led to low Rd and KM values. Figure 10 demonstrates the reduced impact of removed material and fewer smeared particles on the machined surface under the optimum cutting conditions according to Taguchi compared to the initial setup. Similarly, less face wear (smaller pitting) was observed at the optimal cutting setting by Taguchi compared to the initial setting.

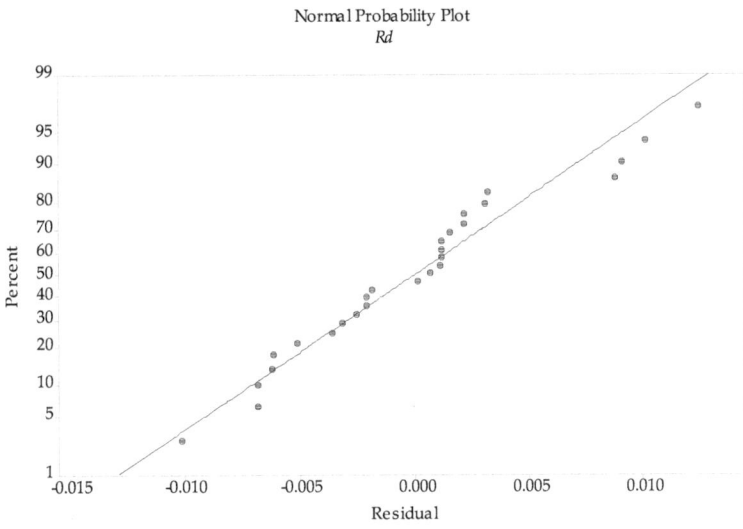

Figure 10. Normal probability plot of the residuals for roundness deviation.

Table 9. Confirmation test results for roundness deviation.

	Initial Process Parameter	Optimal Process Parameters	
		Prediction	Experiment
Level	A2B2C2D2E2	A3B1C1D1E3	A3B1C1D1E3
Roundness deviation (mm)	0.041		0.024
S/N ratio (dB)	27.44	35.56	32.39
Improvement in S/N ratio (dB)	4.95		
Percentage reduction in tool face wear	15.29%		

Table 10. Confirmation test results for tool face wear.

	Initial Process Parameter	Optimal Process Parameters	
		Prediction	Experiment
Level	A2B2C2D2E2	A1B1C1D1E3	A1B1C1D1E3
Tool face wear (mm)	0.198		0.157
S/N ratio (dB)	14.07	15.97	16.08
Improvement in S/N ratio (dB)	2.01		
Percentage reduction in tool face wear	12.5%		

3. Optimization of Input Parameters by ANOVA, Regression Analysis, and Modeling

ANOVA identifies the process parameters most influencing performance characteristics. Table 11 presents the analysis of variance for the output factor of roundness deviation (Rd). It indicates that cutting speed, feed, depth of cut, and workpiece length from clamping significantly affect roundness deviation (with p-values less than 0.05 at a 95% confidence interval). However, the significance of the cutting edge radius of the curvature parameter regarding roundness deviation was not demonstrated. Rd is notably influenced by depth of cut (37.51%), workpiece length from clamping (30.29%), cutting speed (10.50%), and feed

(6.21%), while cutting edge radius exhibits the least significance (2.85%). Table 12 presents the analysis of variance for the output factor of cutting insert face wear. It indicates that cutting speed, feed, depth of cut, and length of workpiece from clamping significantly affect *KM* cutting insert face wear (with *p*-values less than 0.05 at a 95% confidence level). However, the significance of the cutting edge radius parameter for *KM* was not demonstrated. *KM* is significantly influenced by feed (25.43%), cutting speed (23.27%), depth of cut (21.83%), and workpiece length from clamping (16.89%), with cutting edge radius exhibiting the least effect (3.56%).

Table 11. Analysis of variance for roundness deviation.

Source	DF	Adj SS	Adj MS	F-Value	*p*-Value	Contribution	Remarks
A	2	0.000665	0.000332	6.65	0.008	10.50%	Significant
B	2	0.000393	0.000197	3.94	0.041	6.21%	Significant
C	2	0.002375	0.001187	23.76	0.000	37.51%	Significant
D	2	0.001917	0.000959	19.18	0.000	30.29%	Significant
E	2	0.000181	0.000090	1.81	0.196	2.85%	Insignificant
Error	16	0.000800	0.000050			12.63%	
Total	26	0.006330				100.00%	

Table 12. Analysis of variance for face wear.

Source	DF	Adj SS	Adj MS	F-Value	*p*-Value	Contribution	Remarks
A	2	0.002450	0.001225	20.63	0.003	23.27%	Significant
B	2	0.002678	0.001339	22.54	0.001	25.43%	Significant
C	2	0.002298	0.001149	19.34	0.005	21.83%	Significant
D	2	0.001778	0.000889	14.97	0.008	16.89%	Significant
E	2	0.000374	0.000187	3.15	0.070	3.56%	Insignificant
Error	16	0.000950	0.000059			9.03%	
Total	26	0.010528				100.00%	

In this study, Minitab 21.4.2 software was utilized to develop predictive mathematical models for the dependent variable of roundness deviation, considering cutting speed (A), feed (B), depth of cut (C), workpiece length from clamping (D), and cutting edge radius (E), through linear regression analysis. No transformation was applied to each response. The prediction equation obtained from the regression analysis for *Rd* (3) and *KM* (4) is provided below.

Regression analysis model for roundness deviation versus A, B, C, D, and E with regression Equation (3):

$$Rd = 0.00764 - 0.00586 \, A + 0.00453 \, B + 0.01137 \, C + 0.01003 \, D - 0.00317 \, E, \quad (3)$$

For *Rd*, an R^2 value of 83.81% was calculated.

Regression analysis model for face wear versus A, B, C, D, and E with regression Equation (4):

$$KM = 0.0961 + 0.01167 \, A + 0.01078 \, B + 0.01089 \, C + 0.00850 \, D - 0.00406 \, E, \quad (4)$$

For *KM*, an R^2 value of 81.64% was calculated.

The fitness of the developed models was validated using the coefficient of determination R^2 [88,89]. The coefficient of determination ranges from zero to one, where a value closer to one indicates a strong agreement between the dependent and independent variables. For instance, an R^2 value of 95% signifies that 95% of the variability in new observations has been estimated. In this study, the regression models for *Rd* and *KM* achieved high R^2 values, namely 83.81% and 81.64%, respectively.

A graph of the residuals was employed to assess the significance of coefficients in the predicted model. The linearity of the residual graph indicates that the residual errors in the model are normally distributed, and the coefficients are significant. Figure 10 displays the residuals for roundness deviation. It illustrates that the residuals closely align with a straight line, indicating the significance of the coefficients in the developed model. Similarly, Figure 11 depicts the residuals for tool face wear. The residuals also fall near a straight line, suggesting the significance of the coefficients in the developed model for KM. To validate the developed models, confirmatory tests were conducted, and the results are presented in Table 11. The test results, taken randomly from the orthogonal array L27, demonstrate good agreement between the predicted and experimental results across a range of parameters. The response was found to be favorable for machining different difficult-to-cut materials [74,75,81].

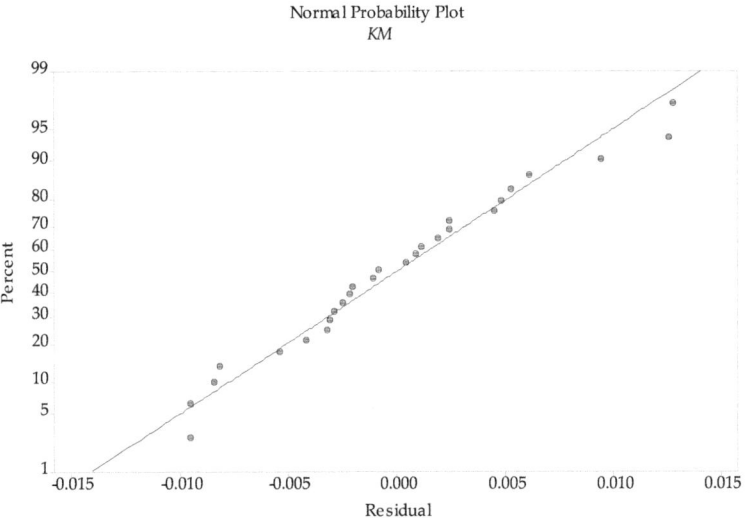

Figure 11. Normal probability plot of the residuals for tool face wear.

Confirmation Test

Confirmation tests of the control factors were conducted following the Taguchi method and regression equations for both the optimum level and randomly selected levels. Table 13 provides a comparison between the test results and the predicted values obtained using the Taguchi method and regression equations (Equations (3) and (4)). The proximity of the predicted values to the experimental values indicates a high level of agreement between them. To ensure reliable statistics in the analyses, the error values must be kept below 20% [88]. An examination of the results in terms of percentages reveals that the errors for the output factors of Rd and KM did not exceed the 20% threshold. Consequently, the results of the confirmatory test demonstrate a successful optimization process. According to the authors of [89,90], similar results were obtained during material machining processes, affirming the effectiveness of the Taguchi optimization method in enhancing the machining performance of C45 steel under the specified process parameters.

This validation process was conducted under both practical conditions, addressing the need for controlling the factors of machined surfaces, and under laboratory conditions, ensuring the reliability and robustness of the optimization procedure.

Table 13. Confirmation results for the developed models.

Run	Experimental		Predicted		Residuals		Error	
	Rd (mm)	KM (mm)	Rd (mm)	KM (mm)	Rd (mm)	KM (mm)	Rd (%)	KM (%)
1	0.026	0.112	0.025	0.122	0.001	−0.01	3.846	8.929
6	0.041	0.167	0.040	0.172	0.001	−0.005	2.439	2.994
10	0.052	0.173	0.051	0.176	0.001	−0.003	1.923	1.734
11	0.045	0.169	0.048	0.168	−0.003	0.001	6.667	0.592
16	0.034	0.167	0.037	0.175	−0.003	−0.008	8.824	4.79
19	0.037	0.188	0.042	0.189	−0.005	−0.001	13.513	0.532
23	0.037	0.182	0.036	0.184	0.001	−0.002	2.703	1.099
25	0.029	0.182	0.031	0.191	−0.002	−0.009	6.896	4.945

4. Conclusions

The discrepancies between the theoretical and actual values of the machined surface factors stem from the deformation process occurring ahead of the cutting edge of the tool. This process is primarily influenced by the properties of the material being machined and the prevailing working conditions [91–93]. The authors of [1,94] presented both positive and negative experiences regarding the use of CCHS in measuring roundness deviation on machined surfaces after turning C45 steel. Furthermore, this research uncovered several concomitant phenomena, including the following:

- Plastic deformation occurred at localized sites on the machined surface, as illustrated in Figure 12. These observations were captured using scanning electron microscopy with a JEOL JSM 7000F autoemission nozzle—JEOL Ltd., Hertfordshire, England, United Kingdom. The results of this analysis warrant further investigation.

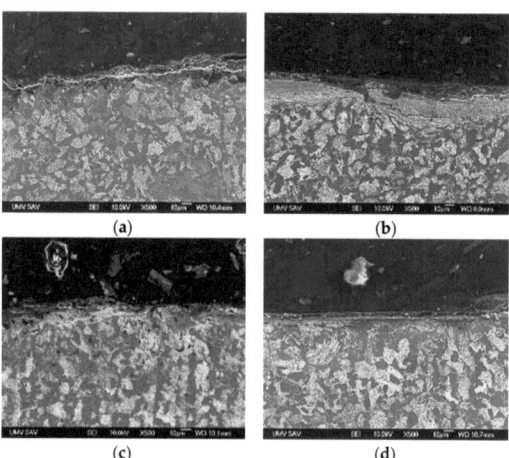

Figure 12. Examples of machined surfaces after turning C45 steel. (**a**) Deformation of the machined surface not measurable. Perlite and ferrite. Nital. Magnification 500×, A = 180 m/min, B = 0.2 mm/rev., C = 0.4 mm, D = 30 mm, E = 0.005 mm. (**b**) Local plastic deformation of machined surface ~34 µm. Perlite and ferrite. Nital. Magnification 500×, A = 270 m/min, B = 0.1 mm/rev., C = 0.1 mm, D = 5.0 mm, E = 0.008 mm. (**c**) Local plastic deformation of machined surface. Traces of volume extraction of the machined surface. Perlite and ferrite. Nital. Magnification 500×, A = 180 m/min, B = 0.2 mm/rev., C = 0.4 mm, D = 30 mm, E = 0.005 mm. (**d**) Local plastic deformation of machined surface ~12 µm. Perlite and ferrite. Nital. Magnification 500×, A = 90 m/min, B = 0.1 mm/rev., C = 0.1 mm, D = 5.0 mm, E = 0.008 mm.

The research presented in this paper offers several key conclusions based on the analysis conducted. The confocal chromatic sensor (CCHS) proved to be suitable for measuring the roundness deviation of machined surfaces, particularly on C45 steel. The data collected using CCHS can serve as a reference for further research, potentially extending to the comparison of other types of steels. Roundness deviation values were effectively measured and evaluated within the range of 0.002 to 0.08 mm using CCHS, aligning with the specified limit of 0.1 mm. The observed wear value on the cutting insert face reached up to 0.3 mm under the research conditions.

Key research findings include the following:

- The Taguchi method identified an optimal combination of cutting conditions (A = 270 m/min, B = 0.1 mm/rev., C = 0.1 mm, D = 5.0 mm, and E = 0.008 mm) resulting in a 53% reduction in roundness deviation.
- Similarly, Taguchi's method determined optimal cutting conditions (A = 90 m/min, B = 0.1 mm/rev., C = 0.1 mm, D = 5.0 mm, and E = 0.008 mm), leading to a 31% reduction in the face wear of the cutting insert.
- ANOVA analysis revealed that depth of cut had the most significant influence on roundness deviation (37.51%), followed by workpiece distance from clamping, cutting speed, and feed. Feed was found to be the most significant factor influencing tool cutting insert wear, with a percentage influence of 25.43%.

These results underscore the effectiveness of the determined optimal cutting settings in reducing roundness deviation and cutting insert face wear during the machining of C45 steel. The findings also highlight the potential of these optimal settings to positively impact roundness deviation.

Practical implications relevant for customers include the following:

- Fine-tuning conditions and controlling factors for laser sensor use on machined surfaces of C45 steel.
- The identification of negative phenomena on machined surfaces after turning C45 steel.
- The consideration of non-contact laser sensor methods for measuring the roundness deviation of machined surfaces in order to implement optimal cutting settings and enhance the quality of turning C45 steel within the specified range.

Author Contributions: Conceptualization, J.J., K.P., and P.M.; methodology, J.J. and P.M.; calculation, J.J.; validation, J.J. and K.P.; formal analysis, M.K.; investigation, J.J.; resources, P.M.; data curation, M.K; writing—original draft preparation, J.J.; writing—review and editing, M.K.; visualization, M.K.; supervision, P.M. and J.J.; project administration, M.K.; funding acquisition, J.J. All authors have read and agreed to the published version of the manuscript.

Funding: This work was supported by the project AMTRteam (Advanced Manufacturing Technology Research Team), awarded by the Ministry of Education, Science, Research and Sport of the Slovak Republic.

Data Availability Statement: Data are contained within the article.

Acknowledgments: As the authors of this article, we would like to thank the support of the project AMTRteam (Advanced Manufacturing Technology Research Team) awarded by the Ministry of Education, Science, Research and Sport of the Slovak Republic.

Conflicts of Interest: The authors declare no conflicts of interest.

References

1. Jurko, J.; Miškiv-Pavlík, M.; Hladký, V.; Lazorík, P.; Michalík, P.; Petruška, I. Measurement of the Machined Surface Diameter by a Laser Triangulation Sensor and Optimalization of Turning Conditions Based on the Diameter Deviation and Tool Wear by GRA and ANOVA. *Appl. Sci.* **2022**, *12*, 5266. [CrossRef]
2. Powell, D.; Magnanini, M.C.; Colledani, M.; Myklebust, O. Advancing Zero Defect Manufacturing: A State-of-the-Art Perspective and Future Research Directions. *Comput. Ind.* **2022**, *136*, 103596. [CrossRef]
3. Bagaber, S.A.; Yusoff, A.R. Multi-objective optimization of cutting parameters to minimize power consumption in dry turning of stainless steel 316. *J. Clean. Prod.* **2017**, *157*, 30–46. [CrossRef]

4. Chen, C.; Leach, R.; Wang, J.; Liu, X.; Jiang, X.; Lu, W. Locally adaptive thresholding centroid localization in confocal microscopy. *Opt. Lett.* **2021**, *46*, 1616–1619. [CrossRef] [PubMed]
5. Wang, Z.; Wang, T.; Yang, Y.; Yang, Y.; Mi, X.; Wang, J. Precise Two-Dimensional Tilt Measurement Sensor with Double-Cylindrical Mirror Structure and Modified Mean-Shift Algorithm for a Confocal Microscopy System. *Sensors* **2022**, *22*, 6794. [CrossRef]
6. Saeidi Aminabadi, S.; Jafari-Tabrizi, A.; Gruber, D.P.; Berger-Weber, G.; Friesenbichler, W. An Automatic, Contactless, High-Precision, High-Speed Measurement System to Provide In-Line, As-Molded Three-Dimensional Measurements of a Curved-Shape Injection-Molded Part. *Technologies* **2022**, *10*, 95. [CrossRef]
7. Liu, Y.; Zhang, Q.; Liu, Y.; Yu, X.; Hou, Y.; Chen, W. High-Speed 3D Shape Measurement Using a Rotary Mechanical Projector. *Opt. Express* **2021**, *29*, 7885–7903. [CrossRef]
8. Yang, Y.; Dong, Z.; Meng, Y.; Shao, C. Data-Driven Intelligent 3D Surface Measurement in Smart Manufacturing: Review and Outlook. *Machines* **2021**, *9*, 13. [CrossRef]
9. Ruan, Y.; Xue, M.; Teng, J.; Wu, Y.; Shi, M. Horizontal Oxidation Diffusion Behavior of MEMS-Based Tungsten-Rhenium Thin Film Thermocouples. *Materials* **2022**, *15*, 5071. [CrossRef]
10. Andrea, L.M.; James, W.M.; Zhirong, L.; Alistair, S.; Jose, A.R.; Dragos, A.A.; Mark, C.H.; Adam, T.C. Surface integrity in metal machining-Part II: Functional performance. *Int. J. Mach. Tools Manuf.* **2021**, *164*, 103718. [CrossRef]
11. Liu, G.; Huang, C.; Zhao, B.; Wang, W.; Sun, S. Effect of Machined Surface Integrity on Fatigue Performance of Metal Workpiece: A Review. *Chin. J. Mech. Eng.* **2021**, *34*, 118. [CrossRef]
12. Felho, C.; Varga, G. Theoretical Roughness Modeling of Hard Turned Surfaces Considering Tool Wear. *Machines* **2022**, *10*, 188. [CrossRef]
13. Kovacı, H.; Bozkurt, Y.; Yetim, A.; Aslan, M.; Çelik, A. The effect of surface plastic deformation produced by shot peening on corrosion behavior of a low-alloy steel. *Surf. Coat. Technol.* **2019**, *360*, 78–86. [CrossRef]
14. Böhm, M.; Kowalski, M.; Niesłony, A. Influence of the Elastoplastic Strain on Fatigue Durability Determined with the Use of the Spectral Method. *Materials* **2020**, *13*, 423. [CrossRef]
15. Xiong, Y.; Yue, Y.; He, T.; Lu, Y.; Ren, F.; Cao, W. Effect of Rolling Temperature on Microstructure Evolution and Mechanical Properties of AISI316LN Austenitic Stainless Steel. *Materials* **2018**, *11*, 1557. [CrossRef] [PubMed]
16. Kiran, B.; Nagaraju, D. Simulation and experimental study on the effect of surface residual stresses in dry orthogonal micro turning sample. *Aust. J. Mech. Eng.* **2022**, *21*, 1396–1408. [CrossRef]
17. Yu, C.; Shiue, R.-K.; Chen, C.; Tsay, L.-W. Effect of Low-Temperature Sensitization on Hydrogen Embrittlement of 301 Stainless Steel. *Metals* **2017**, *7*, 58. [CrossRef]
18. Wisley, F.S.; Julius, S.; Leonardo, R.R.D.S.; Álisson, R.M.; Jawahir, I.S. A review of surface integrity in machining of hardened steels. *J. Manuf. Process* **2020**, *58*, 136–162. [CrossRef]
19. Moravčíková, J.; Moravčík, R.; Palcut, M. Effect of Heat Treatment on the Resulting Dimensional Characteristics of the C45 Carbon Steel after Turning. *Metals* **2022**, *12*, 1899. [CrossRef]
20. Chung, Y.-H.; Chen, T.-C.; Lee, H.-B.; Tsay, L.-W. Effect of Micro-Shot Peening on the Fatigue Performance of AISI 304 Stainless Steel. *Metals* **2021**, *11*, 1408. [CrossRef]
21. Groeb, M.; Hagelüken, L.; Groeb, J.; Ensinger, W. Experimental Analysis of Ductile Cutting Regime in Face Milling of Sintered Silicon Carbide. *Materials* **2022**, *15*, 2409. [CrossRef]
22. Peng, C.-H.; Hou, P.-Y.; Lin, W.-S.; Shen, P.-K.; Huang, H.-H.; Yeh, J.-W.; Yen, H.-W.; Huang, C.-Y.; Tsai, C.-W. Investigation of Microstructure and Wear Properties of Precipitates-Strengthened Cu-Ni-Si-Fe Alloy. *Materials* **2023**, *16*, 1193. [CrossRef]
23. Zhu, P.; Zhao, Y.; Agarwal, S.; Henry, J.; Zinkle, S.J. Toward accurate evaluation of bulk hardness from nanoindentation testing at low indent depths. *Mater. Des.* **2022**, *213*, 110317. [CrossRef]
24. Clayton, J.D.; Casem, D.T.; Lloyd, J.T.; Retzlaff, E.H. Toward Material Property Extraction from Dynamic Spherical Indentation Experiments on Hardening Polycrystalline Metals. *Metals* **2023**, *13*, 276. [CrossRef]
25. Luo, Q.; Kitchen, M. Microhardness, Indentation Size Effect and Real Hardness of Plastically Deformed Austenitic Hadfield Steel. *Materials* **2023**, *16*, 1117. [CrossRef]
26. Nix, W.D.; Gao, H. Indentation size effects in crystalline materials: A law for strain gradient plasticity. *J. Mech. Phys. Solids* **1998**, *46*, 411–425. [CrossRef]
27. Sarangi, S.S.; Lavakumar, A.; Singh, P.K.; Katiyar, P.K.; Ray, R.K. Indentation size effect in steels with different carbon contents and microstructures. *Mater. Sci. Technol.* **2023**, *39*, 338–346. [CrossRef]
28. Song, P.; Yabuuchi, K.; Spaetig, P. Insights into hardening, plastically deformed zone and geometrically necessary dislocations of two ion-irradiated FeCrAl (Zr)-ODS ferritic steels: A combined experimental and simulation study. *Acta Mater.* **2022**, *234*, 117991. [CrossRef]
29. Broitman, E. Indentation Hardness Measurements at Macro-, Micro-, and Nanoscale: A Critical Overview. *Tribol. Lett.* **2017**, *65*, 23. [CrossRef]
30. Balos, S.; Rajnovic, D.; Sidjanin, L.; Cekic, O.E.; Moraca, S.; Trivkovic, M.; Dedic, M. Vickers hardness indentation size effect in selective laser melted MSI maraging steel. *J. Mech. Eng. Sci.* **2021**, *235*, 1724–1730. [CrossRef]
31. Das, A.; Altstadt, E.; Kaden, C.; Kapoor, G.; Akhmadaliev, S.; Bergner, F. Nanoindentation Response of Ion-Irradiated Fe, Fe-Cr Alloys and Ferritic-Martensitic Steel Eurofer 97: The Effect of Ion Energy. *Front. Mater.* **2022**, *8*, 811851. [CrossRef]

32. Amanov, A.; Karimbaev, R.; Maleki, E.; Okan, U.; Young-Sik, P.; Amanov, T. Effect of combined shot peening and ultrasonic nanocrystal surface modification processes on the fatigue performance of AISI 304. *Surf. Coat. Technol.* **2019**, *358*, 695–705. [CrossRef]
33. Baleani, A.; Paone, N.; Gladines, J.; Vanlanduit, S. Design and Metrological Analysis of a Backlit Vision System for Surface Roughness Measurements of Turned Parts. *Sensors* **2023**, *23*, 1584. [CrossRef]
34. Jayabarathi, S.B.; Ratnam, M.M. Comparison of Correlation between 3D Surface Roughness and Laser Speckle Pattern for Experimental Setup Using He-Ne as Laser Source and Laser Pointer as Laser Source. *Sensors* **2022**, *22*, 6003. [CrossRef] [PubMed]
35. dos Santos Motta Neto, W.; Leal, J.E.S.; Arantes, L.J.; Arencibia, R.V. The Effect of Stylus Tip Radius on Ra, Rq, Rp, Rv, and Rt Parameters in Turned and Milled Samples. *Int. J. Adv. Manuf. Technol.* **2018**, *99*, 1979–1992. [CrossRef]
36. Maruda, R.W.; Krolczyk, G.M.; Wojciechowski, S.; Powalka, B.; Klos, S.; Szczotkarz, N.; Matuszak, M.; Khanna, N. Evaluation of turning with different cooling-lubricating techniques in terms of surface integrity and tribologic properties. *Tribol. Int.* **2020**, *148*, 106334. [CrossRef]
37. Yıldırım, Ç.V.; Kıvak, T.; Sarıkaya, M.; Şirin, Ş. Evaluation of tool wear, surface roughness/topography and chip morphology when machining of Ni-based alloy 625 under MQL, cryogenic cooling and CryoMQL. *J. Mater. Res. Technol.* **2020**, *9*, 2079–2092. [CrossRef]
38. Leksycki, K.; Feldshtein, E.; Lisowicz, J.; Chudy, R.; Mrugalski, R. Cutting Forces and Chip Shaping When Finish Turning of 17-4 PH Stainless Steel under Dry, Wet, and MQL Machining Conditions. *Metals* **2020**, *10*, 1187. [CrossRef]
39. Yadav, S.P.; Pawade, R.S. Manufacturing Methods Induced Property Variations in Ti6Al4V Using High-Speed Machining and Additive Manufacturing (AM). *Metals* **2023**, *13*, 287. [CrossRef]
40. Naeim, N.; AbouEleaz, M.A.; Elkaseer, A. Experimental Investigation of Surface Roughness and Material Removal Rate in Wire EDM of Stainless Steel 304. *Materials* **2023**, *16*, 1022. [CrossRef]
41. Abu Qudeiri, J.E.; Saleh, A.; Ziout, A.; Mourad, A.-H.I.; Abidi, M.H.; Elkaseer, A. Advanced Electric Discharge Machining of Stainless Steels: Assessment of the State of the Art, Gaps and Future Prospect. *Materials* **2019**, *12*, 907. [CrossRef] [PubMed]
42. Nguyen, D.-K.; Huang, H.-C.; Feng, T.-C. Prediction of Thermal Deformation and Real-Time Error Compensation of a CNC Milling Machine in Cutting Processes. *Machines* **2023**, *11*, 248. [CrossRef]
43. Chodór, J.; Kukiełka, L.; Chomka, G.; Bohdal, Ł.; Patyk, R.; Kowalik, M.; Trzepieciński, T.; Radchenko, A.M. Using the FEM Method in the Prediction of Stress and Deformation in the Processing Zone of an Elastic/Visco-Plastic Material during Diamond Sliding Burnishing. *Appl. Sci.* **2023**, *13*, 1963. [CrossRef]
44. Tagiuri, Z.A.M.; Dao, T.-M.; Samuel, A.M.; Songmene, V. Numerical Prediction of the Performance of Chamfered and Sharp Cutting Tools during Orthogonal Cutting of AISI 1045 Steel. *Processes* **2022**, *10*, 2171. [CrossRef]
45. Wang, R.; Yang, D.; Wang, W.; Wei, F.; Lu, Y.; Li, Y. Tool Wear in Nickel-Based Superalloy Machining: An Overview. *Processes* **2022**, *10*, 2380. [CrossRef]
46. Hailong, M.; Aijun, T.; Shubo, X.; Tong, L. Finite Element Simulation of Bending Thin-Walled Parts and Optimization of Cutting Parameters. *Metals* **2023**, *13*, 115. [CrossRef]
47. Dyl, T. The Designation Degree of Tool Wear after Machining of the Surface Layer of Duplex Stainless Steel. *Materials* **2021**, *14*, 6425. [CrossRef]
48. Wang, W.; Wang, B.; Liu, B.; Gao, H.; Wei, Z. Machinability and chip morphology evolution of hardened stainless steel using liquid nitrogen cryogenic. *Int. J. Adv. Manuf. Technol.* **2022**, *125*, 967–987. [CrossRef]
49. Królczyk, G.M.; Niesłony, P.; Legutko, S. Determination of tool life and research wear during duplex stainless steel turning. *Arch. Civ. Mech. Eng.* **2015**, *15*, 347–354. [CrossRef]
50. Królczyk, G.M.; Niesłony, P.; Legutko, S.; Hloch, S.; Samardzic, I. Investigation of selected surface integrity features of duplex stainless steel after turning. *Metalurgija* **2015**, *54*, 91–94.
51. Sarıkaya, M.; Gupta, M.K.; Tomaz, I.; Pimenov, D.Y.; Kuntoğlu, M.; Khanna, N.; Yıldırım, Ç.V.; Krolczyk, G.M. A state-of-the-art review on tool wear and surface integrity characteristics in machining of superalloys. *CIRP J. Manuf. Sci. Technol.* **2021**, *35*, 624–658. [CrossRef]
52. Dzierwa, A.; Markopoulos, A.P. Influence of Ball-Burnishing Process on Surface Topography Parameters and Tribological Properties of Hardened Steel. *Machines* **2019**, *7*, 11. [CrossRef]
53. Khanna, N.; Airao, J.; Gupta, M.K.; Song, Q.; Liu, Z.; Mia, M.; Maruda, R.; Krolczyk, G. Optimization of Power Consumption Associated with Surface Roughness in Ultrasonic Assisted Turning of Nimonic-90 Using Hybrid Particle Swarm-Simplex Method. *Materials* **2019**, *12*, 3418. [CrossRef]
54. Sousa, V.F.C.; Silva, F.J.G.; Lopes, H.; Casais, R.C.B.; Baptista, A.; Pinto, G.; Alexandre, R. Wear Behavior and Machining Performance of TiAlSiN-Coated Tools Obtained by dc MS and HiPIMS: A Comparative Study. *Materials* **2021**, *14*, 5122. [CrossRef] [PubMed]
55. Wojciechowski, S.; Królczyk, G.M.; Maruda, R.W. Advances in Hard–to–Cut Materials: Manufacturing, Properties, Process Mechanics and Evaluation of Surface Integrity. *Materials* **2020**, *13*, 612. [CrossRef]
56. Cardoso, L.G.; Madeira, D.S.; Ricomini, T.E.P.A.; Miranda, R.A.; Brito, T.G.; Paiva, E.J. Optimization of machining parameters using response surface methodology with desirability function in turning duplex stainless steel UNS S32760. *Int. J. Adv. Manuf. Technol.* **2021**, *117*, 1633–1644. [CrossRef]

57. Cao, Z.-M.; Wu, Y.; Han, J. Roundness deviation evaluation method based on statistical analysis of local least square circles. *Meas. Sci. Technol.* **2017**, *28*, 10. [CrossRef]
58. Kumar, S.; Riyaz, A.M.; Marulaiah, L.; Manjunatha, L. Investigation of machinability characteristics on C45 steel alloy while turning with untreated and cryotreated M2 HSS cutting tools. *ARPN J. Eng. Appl. Sci.* **2019**, *14*, 307–317. [CrossRef]
59. Sivaprakash, E.; Aswin, S.; Dhanaruban, D.; Dinesh, G.; Inbamathi, M. Machining Character Analysis of Coated and Uncoated End Mill on Heat Treated C45 Steel. *Int. J. Res. Appl. Sci. Eng. Technol.* **2022**, *10*, 2708–2713. [CrossRef]
60. Usca, Ü.A. The Effect of Cellulose Nanocrystal-Based Nanofluid on Milling Performance: An Investigation of Dillimax 690T. *Polymers* **2023**, *15*, 4521. [CrossRef]
61. Tuan, B.; Hai, N.; Kien, L.; Hai, N. Investigation of surface topography in ultrasonic-assisted turning of C45 carbon steel. *Jpn. J. Appl. Phys.* **2023**, *63*, 016501. [CrossRef]
62. SreeramaReddy, T.V.; Sornakumar, T.; Venkatarama Reddy, M.; Venkatram, R. Machinability of C45 steel with deep cryogenic treated tungsten carbide cutting tool inserts. *Int. J. Refract. Met. Hard Mater.* **2009**, *27*, 181–185. [CrossRef]
63. Huang, G.; Bai, J.; Feng, F.; Zeng, L.; Feng, P.; Li, X. A Hybrid Strategy for Profile Measurement of Micro Gear Teeth. *Micromachines* **2023**, *14*, 1729. [CrossRef]
64. Lishchenko, N.; O'Donnell, G.E.; Culleton, M. Contactless Method for Measurement of Surface Roughness Based on a Chromatic Confocal Sensor. *Machines* **2023**, *11*, 836. [CrossRef]
65. Cheng, F.; Fu, S.; Chen, Z. Surface Texture Measurement on Complex Geometry Using Dual-Scan Positioning Strategy. *Appl. Sci.* **2020**, *10*, 8418. [CrossRef]
66. ISO 3685:1993 (E); Tool-Life Testing with Single-Point Turning Tools. International Organization for Standardization: Geneva, Switzerland, 1993.
67. Belavendram, N. *Quality by Design: Taguchi Techniques for Industrial Experimentation*; Prentice Hall: London, UK, 1995; ISBN 9780131863620.
68. Antony, J.; Kaye, M. The Taguchi Approach to Industrial Experimentation. In *Experimental Quality*; Springer: Boston, MA, USA, 2000. [CrossRef]
69. Taguchi, G. *System of Experimental Design: Engineering Methods to Optimize Quality and Minimize Costs*; UNIPUB/Kraus International Publications: White Plains, NY, USA, 1987; p. 1189. ISBN 9780941243001.
70. Sap, E.; Usca, Ü.A.; Gupta, M.K.; Kuntoğlu, M.; Sarıkaya, M.; Pimenov, D.Y.; Mia, M. Parametric Optimization for Improving the Machining Process of Cu/Mo-SiCP Composites Produced by Powder Metallurgy. *Materials* **2021**, *14*, 1921. [CrossRef] [PubMed]
71. Vora, J.; Chaudhari, R.; Patel, C.; Pimenov, D.Y.; Patel, V.K.; Giasin, K.; Sharma, S. Experimental Investigations and Pareto Optimization of Fiber Laser Cutting Process of Ti6Al4V. *Metals* **2021**, *11*, 1461. [CrossRef]
72. Selvam, M.D.; Senthil, P. Investigation on the effect of turning operation on surface roughness of hardened C45 carbon steel. *Aust. J. Mech. Eng.* **2016**, *14*, 131–137. [CrossRef]
73. Singh, M.; Garg, H.K.; Maharana, S.; Yadav, A.; Singh, R.; Maharana, P.; Nguyen, T.V.T.; Yadav, S.; Loganathan, M.K. An Experimental Investigation on the Material Removal Rate and Surface Roughness of a Hybrid Aluminum Metal Matrix Composite (Al6061/SiC/Gr). *Metals* **2021**, *11*, 1449. [CrossRef]
74. Prakash, K.S.; Gopal, P.; Karthik, S. Multi-objective optimization using Taguchi based grey relational analysis in turning of Rock dust reinforced Aluminum MMC. *Measurement* **2020**, *157*, 107664. [CrossRef]
75. Akhtar, M.N.; Sathish, T.; Mohanavel, V.; Afzal, A.; Arul, K.; Ravichandran, M.; Rahim, I.A.; Alhady, S.S.N.; Bakar, E.A.; Saleh, B. Optimization of Process Parameters in CNC Turning of Aluminum 7075 Alloy Using L27 Array-Based Taguchi Method. *Materials* **2021**, *14*, 4470. [CrossRef] [PubMed]
76. Krolczyk, J.B.; Maruda, R.W.; Krolczyk, G.M.; Wojciechowski, S.; Gupta, M.K.; Korkmaz, M.E. Investigations on surface induced tribological characteristics in MQCL assisted machining of duplex stainless steel. *J. Mater. Res. Technol.* **2022**, *18*, 2754–2769. [CrossRef]
77. Fratila, D.; Caizar, C. Application of Taguchi method to selection of optimal lubrication and cutting conditions in face milling of AlMg3. *J. Clean. Prod.* **2011**, *19*, 640–645. [CrossRef]
78. Lubis, S.M.; Darmawan'Adianto, S. Effect of cutting speed on temperature cutting tools and surface roughness of AISI 4340 steel. *IOP Conf. Ser. Mater. Sci. Eng.* **2019**, *508*, 012053. [CrossRef]
79. Gunjal, S.U.; Patil, N.G. Experimental Investigations into Turning of Hardened AISI 4340 Steel using Vegetable based Cutting Fluids under Minimum Quantity Lubrication. *Procedia Manuf.* **2018**, *20*, 18–23. [CrossRef]
80. Fnides, M. Optimization and Mathematical Modelling of Surface Roughness Criteria and Material Removal Rate when Milling C45 Steel using RSM and Desirability Approach. *J. Mech. Eng.* **2023**, *20*, 173–197. [CrossRef]
81. Jamil, M.; Khan, A.M.; He, N.; Li, L.; Zhao, W.; Sarfraz, S. Multi-response optimisation of machining aluminium-6061 under eco-friendly electrostatic minimum quantity lubrication environment. *Int. J. Mach. Mach. Mater.* **2019**, *21*, 459–479. [CrossRef]
82. Mia, M.; Gupta, M.K.; Lozano, J.A.; Carou, D.; Pimenov, D.Y.; Królczyk, G.; Khan, A.M.; Dhar, N.R. Multi-objective optimization and life cycle assessment of eco-friendly cryogenic N2 assisted turning of Ti-6Al-4V. *J. Clean. Prod.* **2019**, *210*, 121–133. [CrossRef]
83. Pu, Y.; Zhao, Y.; Meng, J.; Zhao, G.; Zhang, H.; Liu, Q. Process Parameters Optimization Using Taguchi-Based Grey Relational Analysis in Laser-Assisted Machining of Si_3N_4. *Materials* **2021**, *14*, 529. [CrossRef]
84. Mufarrih, A.; Istiqlaliyah, H.; Ilha, M. Optimization of Roundness, MRR and Surface Roughness on Turning Process using Taguchi-GRA. *J. Phys. Conf. Ser.* **2019**, *1179*, 012099. [CrossRef]

85. Jin, L.; Wang, G.; Deng, J.; Li, Z.; Zhu, M.; Wang, R. A New Model for Cleaning Small Cuttings in Extended-Reach Drilling Based on Dimensional Analysis. *Appl. Sci.* **2023**, *13*, 12118. [CrossRef]
86. Tian, P.; He, L.; Zhou, T.; Du, F.; Zou, Z.; Zhou, X. Experimental characterization of the performance of MQL-assisted turning of solution heat-treated and aged Inconel 718 alloy. *Int. J. Adv. Manuf. Technol.* **2023**, *125*, 3839–3851. [CrossRef]
87. Thakur, A.; Gangopadhyay, S. State-of-the-art in surface integrity in machining of nickel-based super alloys. *Int. J. Mach. Tool. Manuf.* **2016**, *100*, 25–54. [CrossRef]
88. Kıvak, T. Optimization of surface roughness and flank wear using the Taguchi method in milling of Hadfield steel with PVD and CVD coated inserts. *Measurement* **2014**, *50*, 19–28. [CrossRef]
89. Debnath, S.; Reddy, M.M.; Yi, Q.S. Influence of cutting fluid conditions and cutting parameters on surface roughness and tool wear in turning process using Taguchi method. *Measurement* **2016**, *78*, 111–119. [CrossRef]
90. Maruda, R.W.; Krolczyk, G.M.; Feldshtein, E.; Nieslony, P.; Tyliszczak, B.; Pusavec, F. Tool wear characterizations in finish turning of AISI 1045 carbon steel for MQCL conditions. *Wear* **2017**, *372–373*, 54–67. [CrossRef]
91. Klocke, F. *Fertigungsverfahren 1*; Springer Science and Business Media LLC: Dordrecht, The Netherlands, 2018; ISBN 978-3-662-54206-4. [CrossRef]
92. Hou, Z.; Yuan, Y.; Chen, Y.; Jiang, E.; Wang, H.; Zhang, X. A Review of the Settling Law of Drill Cuttings in Drilling Fluids. *Processes* **2023**, *11*, 3165. [CrossRef]
93. Yin, W.-H.; Yue, H.; Wang, X. A Study on the Depositional Law of Road Cutting in the Tengger Desert. *Appl. Sci.* **2023**, *13*, 11967. [CrossRef]
94. Jurko, J.; Miškiv-Pavlík, M.; Husár, J.; Michalik, P. Turned Surface Monitoring Using a Confocal Sensor and the Tool Wear Process Optimization. *Processes* **2022**, *10*, 2599. [CrossRef]

Disclaimer/Publisher's Note: The statements, opinions and data contained in all publications are solely those of the individual author(s) and contributor(s) and not of MDPI and/or the editor(s). MDPI and/or the editor(s) disclaim responsibility for any injury to people or property resulting from any ideas, methods, instructions or products referred to in the content.

Review

Application and Prospect of Wear Simulation Based on ABAQUS: A Review

Liang Yan [1], Linyi Guan [1], Di Wang [1] and Dingding Xiang [1,2,*]

[1] School of Mechanical Engineering and Automation, Foshan Graduate School of Innovation, Northeastern University, Shenyang 110819, China; yl839025195@163.com (L.Y.); skdsj159357@126.com (L.G.); cc28252023@163.com (D.W.)

[2] State Key Laboratory of Solid Lubrication, Lanzhou Institute of Chemical Physics, Chinese Academy of Sciences, Lanzhou 730000, China

* Correspondence: xiangdd@mail.neu.edu.cn

Abstract: The finite element method(FEM) is a powerful tool for studying friction and wear. Compared to experimental methods, it has outstanding advantages, such as saving financial costs and time. In addition, it has been widely used in friction and wear research. This paper discusses the application of the FEM in the study of friction and wear in terms of the finite element modeling methods, factors affecting wear behavior, wear theory, and the practical application of the method. Finally, the latest progress of finite element simulation wear research is summarized, and the future research direction is proposed.

Keywords: friction and wear; wear theory; finite element method; UMESHMOTION

1. Introduction

Wear, an important factor affecting the service life and reliability of mechanical parts, is one of the most common topics in tribology, which is defined as the progressive loss of material due to the relative movement between the surfaces in contact [1]. Wear can make components lose the correct shape and size, leading to vibration, noise, and other undesirable effects. In addition, excessive wear may also cause early failure of parts, resulting in mechanical failure, economic losses, and safety hazards [2–6]. Therefore, the study of friction and wear behavior of materials is of great significance for improving machine performance and economic development.

The experimental method is the primary method in wear research. However, the cost of human, material, and financial resources required by the experimental method is high [1], and the experimental conditions are harsh. Wear simulation is an alternative technique for predicting wear characteristics based on experimental material properties. Among the simulation methods, the FEM is the most popular because of its wide applicability. FEM is used to simulate the wear characteristics under different conditions, which can effectively predict the wear characteristics. FEM is a powerful tool for wear prediction and parametric studies, which, compared to physical experiments, can provide a cost-effective solution for optimizing friction systems to reduce wear.

ABAQUS, one of the mainstream simulation software, has outstanding advantages in the nonlinear behavior of materials, such as plastic deformation, contact, and friction. ABAQUS can model the wear behavior of materials and consider the impact of wear on structural performance through coupling analysis. In addition, ABAQUS has powerful computing power and provides users with various subroutines. Based on the above advantages, ABAQUS has been widely applied in the field of wear research [7].

The purpose of this article is to review the latest methods and progress based on ABAQUS in friction and wear research. This article is divided into four parts: the establishment of finite element models, factors affecting wear behavior, wear theory, and application

Citation: Yan, L.; Guan, L.; Wang, D.; Xiang, D. Application and Prospect of Wear Simulation Based on ABAQUS: A Review. *Lubricants* **2024**, *12*, 57. https://doi.org/10.3390/lubricants12020057

Received: 26 December 2023
Revised: 31 January 2024
Accepted: 5 February 2024
Published: 16 February 2024

Copyright: © 2024 by the authors. Licensee MDPI, Basel, Switzerland. This article is an open access article distributed under the terms and conditions of the Creative Commons Attribution (CC BY) license (https://creativecommons.org/licenses/by/4.0/).

of wear simulation. On this basis, the future development direction of finite element friction and wear simulation research is proposed. The overall structure of this paper is shown in Figure 1.

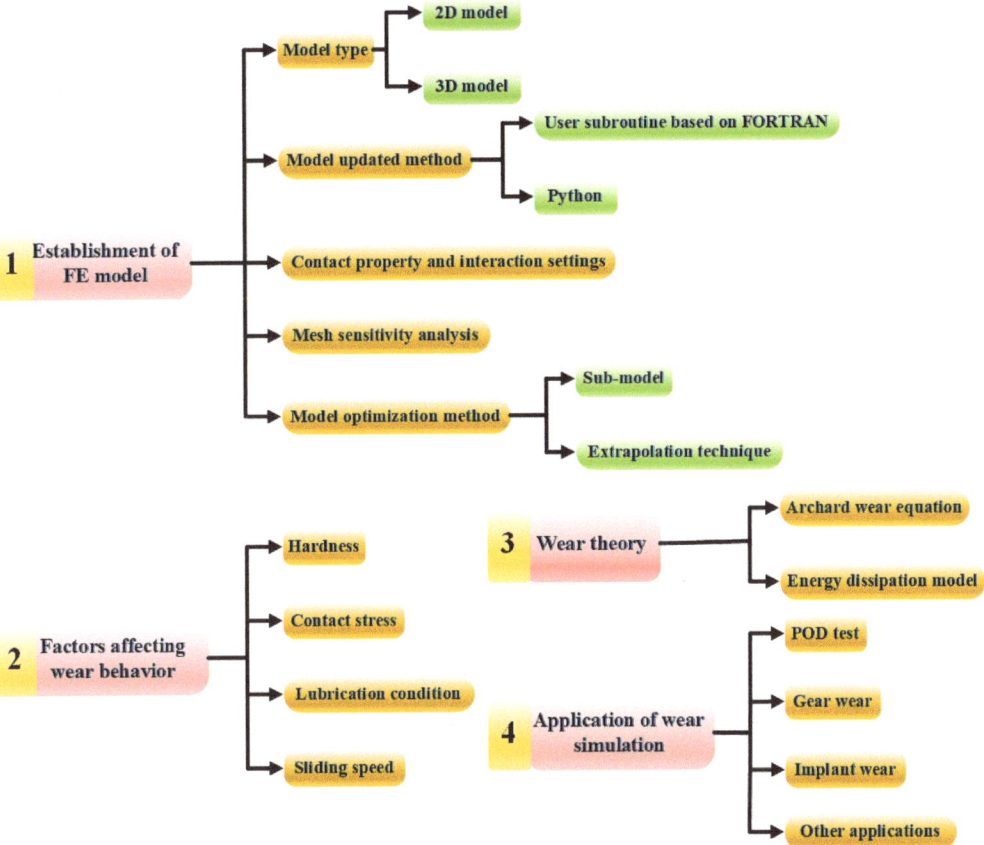

Figure 1. Schematic illustration of the main content in this review.

2. Finite Element Model

Establishing a finite element model is the first and crucial step in simulation. Factors such as the type of model, mesh division, and selection of element types directly affect the accuracy of the final results and the length of calculation time. On the premise of ensuring the accuracy of the calculation results, time cost is the biggest issue that scholars have focused on. A good finite element (FE) model can not only obtain accurate results but also have a short calculation time. The authors have summarized the research methods used in 34 articles related to wear simulation, as shown in Table 1, and elaborated on model types, model updating methods, and model optimization methods. Figure 2 displays the frequency of commonly used research methods, including model updating methods and wear theories, as observed in 34 research papers.

Table 1. Wear simulation research method statistics.

References	Model Type	Model Updating Method	Wear Theory	Model Optimization Method	Application
[8]	2D	UMESHMOTION	Energy dissipation model	Extrapolation technique	Fretting wear
[9]	2D	UMESHMOTION	Archard's wear law	Extrapolation technique	Service life prediction
[10]	2D/3D	Nope	Nope	Sub-model	Wear profile prediction
[11]	2D	UMESHMOTION	Energy dissipation model + Damage-coupled elastic–plastic constitutive model	Extrapolation technique	Service life prediction
[12]	3D	UMESHMOTION	Archard's wear law	Extrapolation technique	Tribocorrosion
[13]	3D	Nope	Energy dissipation model	Nope	Wear mechanism auxiliary analysis
[14]	3D	UMESHMOTION	Archard's wear law	Extrapolation technique	Wear profile prediction
[15]	3D	UMESHMOTION	Archard's wear law	Sub-model	Fretting wear
[16]	3D	UMESHMOTION	Energy dissipation model	Extrapolation technique	Fretting wear
[17]	3D	UMESHMOTION	Archard's wear law	Extrapolation technique	POD tribometer wear prediction
[18]	3D	UMESHMOTION	Archard's wear law	Extrapolation technique	Tire tread wear
[19]	2D	UMESHMOTION	Archard's wear law	Nope	Casing wear
[20]	2D	UMESHMOTION	Archard's wear law	Mesh and increment size optimization	Fretting wear
[2]	3D	UMESHMOTION	Power hardening law +Archard's wear law	Extrapolation technique	Thermo-mechanical wear
[21]	3D	UMESHMOTION	Electrochemical equation + Archard's wear law	Nope	Corrosive wear
[22]	3D	Python	Archard's wear law	Nope	Orthopedic implant wear
[23]	3D	Python	Archard's wear law	Nope	Orthopedic implant wear
[24]	2D/3D	Python	Usui's tool wear model	Nope	Tool wear
[25]	3D	Python	Archard's wear law	Extrapolation technique	POD tribometer wear prediction
[26,27]	3D	UMESHMOTION	Archard's wear law	Sub-model	Orthopedic implant wear
[28]	3D	UMESHMOTION	Archard's wear law	Sub-model	Orthopedic implant wear
[29]	2D	UMESHMOTION	Archard's wear law	Extrapolation technique	Pin wear prediction

Table 1. Cont.

References	Model Type	Model Updating Method	Wear Theory	Model Optimization Method	Application
[30]	3D	UMESHMOTION	Archard's wear law	Extrapolation technique	POD tribometer wear prediction
[31]	3D	UMESHMOTION	Archard's wear law	Extrapolation technique	Dry sliding wear prediction
[1]	2D	UMESHMOTION	Archard's wear law	Extrapolation technique	POD tribometer wear prediction
[32]	3D	UMESHMOTION	Archard's wear law	Nope	Wear profile prediction
[33]	2D	UMESHMOTION	Archard's wear law	Extrapolation technique	POD tribometer wear prediction
[34]	3D	UMESHMOTION	Archard's wear law	Nope	Gear wear prediction
[35]	3D	UMESHMOTION	Archard's wear law + shape functions and Newton–Raphson formulation	Nope	POD tribometer wear prediction
[36]	2D	UMESHMOTION	Energy dissipation model	Nope	Fretting wear
[37]	3D	UMESHMOTION	Energy dissipation model	Nope	Wear profile evolution
[38]	3D	UMESHMOTION	Energy dissipation model	Extrapolation technique	Wear simulation in automotive bush chain
[39]	3D	UMESHMOTION	Energy dissipation model	Extrapolation technique	Fretting wear

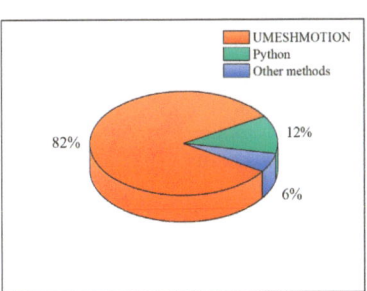
(a) Model updating methods

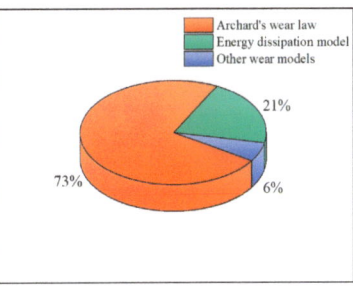
(b) Wear theories

Figure 2. Comparison chart of the frequency of commonly used research methods.

2.1. Model Type

2.1.1. 2D Model

Finite element models can be divided into two categories: 2D and 3D models. The types of FE models selected vary depending on the needs of practical problems. The 2D model is suitable for cases where there is no concern about the overall wear profile. Using this approach, the number of elements, nodes, degrees of freedom, and boundary conditions is reduced, thereby improving computational efficiency. In the field of friction and wear research, the pin-on-disk (POD) test is a typical example. The 2D and 3D FE models of the POD test are shown in Figure 3.

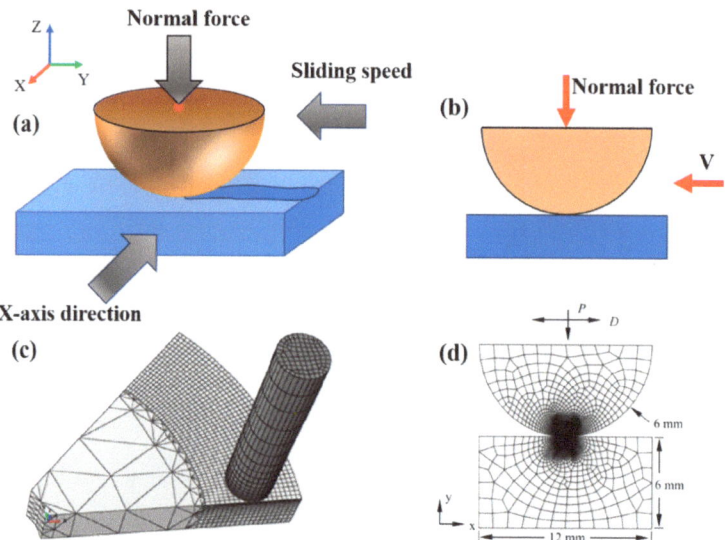

Figure 3. (**a**) 3D solid model, (**b**) 2D solid model, (**c**) 3D mesh model [30], and (**d**) 2D mesh model [40].

However, as shown in Table 1, a large number of studies used 3D FE models instead of 2D models. This is because, in practical engineering applications, many models are very complex, such as orthopedic implants, gears, and cutting tools. The 2D model cannot truly reflect its true structure and working conditions. Moreover, due to the simplification of conditions, the 2D model may cause significant computational errors. In this case, the 2D model is not applicable.

2.1.2. 3D Model

3D FE models have the advantage of more accurate calculation results. However, 3D FE models are not economical because of their complex geometric structure and long calculation time [15,17,41]. Hence, 3D FE models are often used in situations where the structure or the stress situation is complex and when the model cannot be simplified.

Sadeghi and Ahmadi [13] applied a 3D FE model to study the Hertz circular and line contacts. Compared to the results of fretting wear tests, the numerical results are well-confirmed. Bastola et al. [14] proposed a generalized 3D FEM to obtain the wear between the contacting components. The results were validated by the POD test.

From the above literature, it can be seen that the calculation results of the 3D FE models are more accurate. However, as mentioned at the beginning, the problem of long calculation time becomes more serious in 3D wear simulation. To solve this problem, scholars have proposed many solutions. Hegadekatte et al. [41] proposed an incremental implementation of Archard's wear law, which greatly improved the computing efficiency. Bae et al. [15] used the sub-model method to reduce the complexity of the model and the significant computational time required for finite element analysis (FEA). Bose and Penchaliah [17] introduced a numerical wear simulation approximation technique based on the FEM to solve the problem. In addition, as shown in Table 1, the main model optimization methods include a substructure method and extrapolation technique, which will be discussed in Section 2.5.

2.2. Model Update Method

Wear is an accumulation process. The FEM is employed to simulate wear formation by updating the mesh model with moving nodes, enabling the determination of wear depth and volume for wear assessment. In ABAQUS-based wear simulation, two primary methods are utilized to update the mesh model: user subroutine based on programming languages, FORTRAN and Python. This section will elaborate on these two methods.

2.2.1. User Subroutine Based on FORTRAN

ABAQUS provides users with various subroutines, with the most commonly used in wear simulation research being the UMESHMOTION subroutine, which can be used to simulate the movement of nodes in an FE model [42]. During the wear process, as the number of wear cycles increases, material loss also increases, and the actual contact situation also changes. In order to simulate the actual wear process and obtain correct calculation results, it is necessary to continuously update node and mesh information during wear simulation. The UMESHMOTION subroutine provides the conditions for this process [43,44]. The process of wear simulation is shown in Figure 4. The determination of the initial parameters is a key step to ensure the accuracy and efficiency of the FE model, particularly for the parameter "ΔN"(step sizes). To tackle this problem, McColl proposed a method that has been widely applied in wear simulation [14,20].

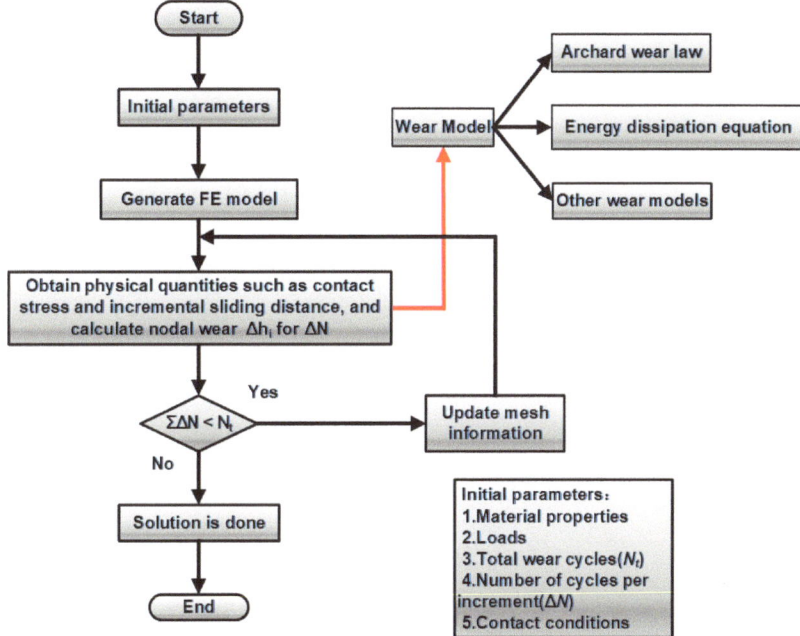

Figure 4. Finite element model updating process.

In addition, the UMESHMOTION subroutine is always combined with the ALE (Arbitrary-Lagrangian-Eulerian) technique. The ALE technique is an adaptive mesh method in ABAQUS [42]. By incorporating the attributes of unadulterated Lagrangian and Eulerian examination, it permits the mesh to shift independently of the material [1]. This technique can automatically adjust the mesh according to the degree of distortion in the analysis of materials with large deformation or loss, ensuring high quality of the mesh and improving calculation accuracy and model stability. In wear simulation, UMESHMOTION is used to move the contact node by the amount of local wear increment. As the nodes move, the contact mesh model is updated, and preparations are made for the node update after the next wear increment. The movement of nodes may cause mesh distortion, thereby affecting the FEM results. This mesh distortion can be prevented through the ALE technique [1,9,45].

The majority of examples in Table 1 used the Archard wear equation to calculate the wear depth, which does not mean that the UMESHMOTION subroutine can only be bound to Archard's wear equation. The UMESHMOTION subroutine only serves the purpose of moving nodes. The distance and mode of movement of nodes are determined by the wear model used. The Archard wear equation is only one of the methods for calculating the distance of node movement. The UMESHMOTION subroutine can also be used in conjunction with other theoretical models. Chemical corrosion is an important factor affecting wear [46], Fallahnezhad et al. [21] combined Archard's wear law with chemical equations to study fretting corrosion wear of CoCr. Temperature is an important factor affecting wear [47,48]; Gan et al. [2] took into account the effect of temperature on wear by combining heat transfer analysis with Archard's wear law. After verification through POD experiments, comparative analysis shows that friction heat and plasticity have a substantial effect on the progression of wear. Li et al. [49] proposed a wear equation, which combines the hydrodynamic lubrication wear with the thermochemical erosion to study the wear of artillery barrels under hydrodynamic friction. In addition, as shown in Figure 2, the energy dissipation model can also be used in conjunction with the UMESHMOTION subroutine to simulate wear.

2.2.2. Python

Except for the UMESHMOTION subroutine, Python is also an important way to update node information. By using Python scripts, the ABAQUS/CAE graphical user interface can be bypassed and the ABAQUS kernel can be directly operated to modify the finite element model and related parameters [42]. Figure 4 shows that the calculation process is similar to the UMESHMOTION subroutine. However, as shown in Table 1 and Figure 2, most studies apply the UMESHMOTION subroutine instead of Python. Compared to the UMESHMOTION subroutine, Python has lower accuracy in FEA. This is because the UMESHMOTION subroutine is usually used in conjunction with the ALE technique, resulting in good mesh quality. However, when using Python scripts, the mesh quality is poor, leading to increased stress concentration and convergence issues [50].

2.3. Contact Property and Interaction Settings

The setting of contact attributes and interactions is one of the important links to ensure the correctness of the solution. In ABAQUS, the contact surface interaction is established using the contact pair approach, which employs the master–slave algorithm to implement the contact constraints [20]. In addition, surface-to-surface contact discretization is utilized instead of node-to-surface contact discretization. When the contact geometry is well depicted, the surface-to-surface discretization produces more accurate stress and pressure outcomes [14]. Arbitrary separation, sliding, and rotation of the contact surfaces are allowed by the finite-sliding contact tracking approach [42]. The definitions of tangential and normal contact properties are needed. For the tangential behavior, constant penalty friction formulation is utilized with the coefficient of friction tested in the experiment. For normal behavior, "Hard" contact pressure-overclosure is applied. In addition, two constraint enforcement methods, augmented Lagrange and penalty, are selected [14]. Compared with augmented Lagrange, a lower error in the maximum contact pressure can be obtained by the penalty method [9]. However, the comparable pressure distribution throughout the contact region can be obtained by the augmented Lagrange method [14]. Both constraint enforcement methods can be applied in simulation.

2.4. Mesh Sensitivity Analysis

Meshing is also one of the important links to ensure the accuracy of the results. In general, a fine mesh is needed in the contact area. However, as the elements increase, the calculation cost will also increase, and the accuracy does not necessarily increase [14]. Therefore, mesh sensitivity analysis is of the essence, which plays a role in determining the size and number of elements [20]. Based on the literature survey, the Hertz formula is always used to validate the worn model [9,14,20,51].

The Hertz contact pressure distribution varies with 'x' as:

$$p(x) = p_0 \sqrt{1 - \frac{x^2}{a^2}} \tag{1}$$

where a and p_0 are the half-width of the contact region and the maximum contact pressure, respectively, given by the following formulas:

$$a = \left(\frac{4PR}{\pi E^*}\right)^{1/2} \tag{2}$$

$$p_0 = \left(\frac{PE^*}{\pi R}\right)^{1/2} \tag{3}$$

where P is the applied normal load and E^* is the composite modulus of two contacting bodies. E^* and R are given by:

$$E^* = \left(\frac{1-\left(v^f\right)^2}{E^f} + \frac{1-(v^c)^2}{E^c}\right)^{-1} \tag{4}$$

$$R = \left(\frac{1}{R^f} + \frac{1}{R^c}\right)^{-1} \tag{5}$$

where v^f, v^c and E^f, E^c are the Poisson's ratios and the elastic modulus of flat and cylindrical bodies, respectively. R^f and R^c are the radii of the contacting surfaces.

2.5. Model Optimization Method

The effectiveness and computational efficiency of a model are the two most concerned issues in FEA. With regard to ensuring the effectiveness of the model, the computational efficiency of the finite element model is the primary concern of scholars because faster calculation speed will bring better economic benefits. Currently, common methods for solving computational time problems include the sub-model method and extrapolation method. This section will elaborate on the two aspects mentioned above.

2.5.1. Sub-Model

The main idea of the sub-model method is to combine the global coarse model of the entire system with the local fine model of key regions, to minimize computational costs and provide accurate numerical results. When conducting finite element analysis on large and complex structures, the order of the equation and the computer resources required for calculation can be reduced, resulting in an improvement of the solving efficiency. Therefore, the sub-model method is suitable for wear analysis of large and complex structures. The program for the wear sub-model mainly consists of three steps [52], which are shown in Figure 5a. The first step is to determine the boundary conditions required for creating a detailed local model based on the initial rough FE global model. After checking the convergence of the quantity to be transmitted, the next step is to create the local model with the appropriately defined mesh. Eventually, the local model, to which boundary conditions from the global model are applied, is used to conduct the wear simulation. The research of Currelli et al. [52,53] on the substructure method showed that the sub-model method can greatly reduce the computational cost of FE wear simulation. In particular, in the biomedical field, complex finite element models result in high time costs [26]. Substructure technology provides an effective solution.

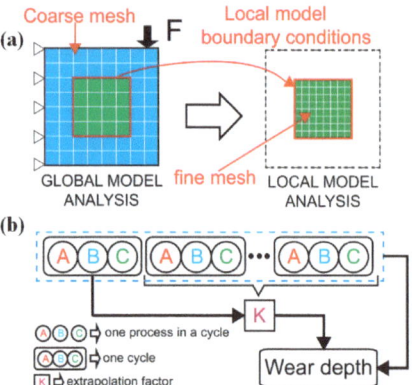

Figure 5. Schematic diagrams of the (**a**) sub-model method and (**b**) extrapolation technique.

Using total hip arthroplasty (THA) as an example, Shankar et al. [26,27] applied this method to study the contact pressure and wear in hip joint prostheses made of metal, ceramic, and polycrystalline diamond materials, as well as the wear behavior of silicon nitride and Ti6Al4V alloy under the influence of five different biological lubricants during various gait activities. Prasad and Ramkumar [28] employed the sub-model method to investigate the wear performance of ceramic hip joint implants under dynamic edge loading conditions. The sub-model technique improves the computational efficiency of FEA in the above studies.

2.5.2. Extrapolation Technique

The extrapolation technique is one of the most commonly used methods, with the main idea involving the assumption that the state of multiple wear cycles is the same as the state of one cycle [52]. An extrapolation factor is introduced to calculate the wear depth after several wear cycles. As shown in Figure 5b and using Bose's study as an example [9], each FEM wear cycle consists of 4 steps. In each cycle, the pin moves a sliding distance increment of 2 mm, which is called step size. Increasing the step size will lead to a decrease in the stability of the wear simulation. Therefore, in order to simulate a sliding distance of 200 m without using extrapolation techniques, 100,000 FEM cycles and 400,000 steps are required. Assuming a constant contact pressure in the extrapolation process, using an extrapolation factor of 100, the computer only needs 1000 FEM cycles and 4000 steps to simulate a sliding distance of 200 m, thus improving the computational efficiency of the finite element model and reducing time costs. The extrapolation factor depends on the applied load [1]. However, it should be noted that a large extrapolation factor can affect the stability and accuracy of the model [9,29–31], while a small extrapolation factor will result in higher utilization of computer resources. Even if an appropriate extrapolation size is selected at the beginning of the simulation, different sizes of extrapolation factors may be required at different stages of the simulation to maximize resource utilization [52]. Therefore, the key to the extrapolation technique lies in selecting the appropriate extrapolation factor to balance computational efficiency and simulation accuracy. According to current studies, there is no uniform method for determining the extrapolation factor.

Bose and Ramkumar [1] improved the extrapolation technique. They found that in the wear model of the POD model, point contact begins, and the contact stress is high. As the contact area increases, the contact pressure gradually decreases, and using a constant extrapolation factor will result in a large error. After several wear cycles, the contact pressure begins to stabilize, and at this moment, using extrapolation techniques will have a better effect. On this basis, they proposed the linear extrapolation technique. In comparison to the constant extrapolation technique, this method uses a small extrapolation factor at the beginning of the cycle, and as the number of wear cycles increases, the extrapolation factor also increases, demonstrating better efficiency and accuracy in wear simulation. The extrapolation method can be used in general wear simulation.

3. Factors Affecting Wear Behavior

Many factors affect the wear behavior, and the influence of the influencing factors on the wear is often not a single effect; there is an interaction between the influencing factors. This section will cover four aspects: material hardness, contact stress, lubrication, and sliding speed.

3.1. Hardness

Hardness is an important factor affecting the material wear rate. In general, the higher the material hardness, the stronger its surface wear resistance. When two material surfaces come into contact, they are subjected to stress and friction from the other surface. If the surface hardness is high, it often exhibits strong resistance to scratches and indentation, and the degree of wear caused by interaction between contacting surfaces is also small, thereby reducing the wear rate. On the contrary, if the surface hardness of the material is low, the surface is susceptible to scratches and indentation, leading to microstructure damage and material loss, resulting in a higher wear rate. The research of Rigney [54] shows that severe wear occurs when the hardness ratio range of the tested pin to disc is less than or equal to 1.0, and mild wear occurs when the hardness ratio is greater than 1.0. Lemm et al. [55] studied the effect of hardness in an AISI Type O1 steel-on-steel fretting contact. As shown in Figure 6a, the wear rate initially increases and then decreases with increasing hardness.

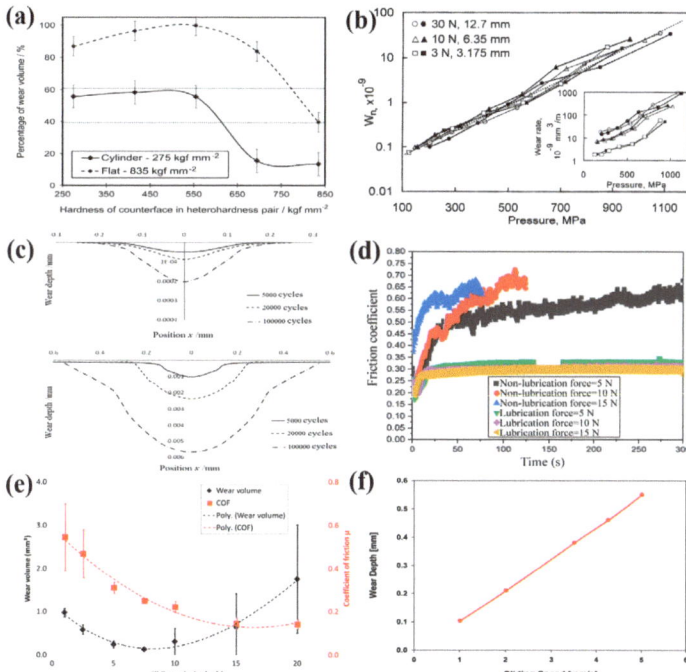

Figure 6. (**a**) Distribution of the wear volume between the specimens (fraction of the total net wear volume) [55], (**b**) variation in the normalized incremental wear rate (wear rate per unit contact area) with contact pressure at different loads and ball diameters [56], (**c**) simulated worn profiles of the flat part: lubricated contact and dry contact [57], (**d**) friction coefficient obtained from different lubrication conditions [58], (**e**) wear coefficient and coefficient of friction (mean values) plotted against the frictional power [59], and (**f**) variation of wear depth with the change in sliding speed [60].

3.2. Contact Stress

Contact stress has a greater impact on wear compared to other factors, and higher contact stress increases shear forces between contacting surfaces, thereby exacerbating wear. When the contact stress exceeds the strength limit of the material, it can cause plastic deformation and peeling on the surface of the material, further exacerbating wear. Ravikiran and Jahamir [56] studied the effects of contact pressure and load on the wear of alumina. As shown in Figure 6b, the result indicates that the wear rate increases with the increase of contact pressure, regardless of the applied load used in the test. Guo et al. [61] studied the effect of contact stress on the cyclic wear behavior of ceramic restorations. The result shows that the high contact stress promotes the veneer porcelain to enter a severe wear stage. On the contrary, lower contact stress can easily keep the veneered porcelain in a stable wear stage, thereby delaying the arrival of a severe wear stage. The contact area affects wear by affecting contact stress. Under the same load conditions, the larger the contact area, the smaller the contact stress, thereby reducing wear.

3.3. Lubrication Condition

Lubrication is a common method to reduce wear as lubricants can form a thin film between the surfaces of two objects in contact, avoiding direct contact, and thus reducing wear [62]. Lubricants can also absorb and dissipate the heat generated by friction between two contact objects, reducing the impact of temperature on material properties and thereby reducing material wear. Qin et al. [57] compared the simulated geometric shape of lubricated frictional contact and the effect of lubricants on friction and wear with experimental results. The result shown in Figure 6c indicates that oil lubrication can effectively reduce microwear and friction in low flexible contact while having little effect on high flexible contact. Zhao et al. [63] proposed that under mixed lubrication conditions, increasing the viscosity of lubricants can reduce the degree of wear on rough surfaces as it can reduce the direct contact area of rough surfaces. Cao et al. [58] established a wear model for all-metal progressive cavity pumps (AMPCP) based on Archard's wear theory and studied the effects of factors such as rotational speed, lubrication, and clearance on wear behavior. As shown in Figure 6d, lubrication can significantly reduce wear and prolong the service life of AMPCP.

3.4. Sliding Speed

The impact of sliding speed on wear varies in different situations. Khader et al. [59] established a wear model for dry sliding contact between silicon nitride and nickel-based alloys. As shown in Figure 6e, the wear rate decreases with an increase in sliding speed, reaching a minimum value. Beyond this threshold, wear escalates as the sliding speed increases. The same behavior was also observed in the relationship between wear and friction. Okonkwo et al. [64] conducted a study on the influence of sliding speed on the wear of steel and tool steel pairs. The results show that at all sliding speeds, adhesive wear dominates on the surface of tool steel, with the highest amount of adhesion occurring at the slowest sliding speed. Moreover, temperature has a significant impact on wear, and a slight increase in contact interface temperature can lead to a significant change in wear rate. Arjmandi et al. [60] conducted parameterized research on the developed wear model to investigate the impact of key parameters on wear rate. The result shown in Figure 6f indicates that under higher normal loads and sliding speeds, the wear rate of three-dimensional woven textiles increases, but the change in friction coefficient has little effect. Chowdhury et al. [65] studied the effects of sliding speed and normal load on the friction and wear performance of aluminum. The study showed that the friction coefficient decreases with the increase of sliding speed and normal load, while the wear rate increases with the increase of sliding speed and normal load. Maintaining sliding speed and normal load at an appropriate low level helps to reduce wear.

4. Wear Theory

Wear is always one of the most complicated and difficult problems in tribology because there are numerous factors that affect wear, such as load, speed, temperature, surface roughness, mechanical properties, and microscopic flaws. Currently, there are over 300 wear-related equations [66], most of which are tailored to a specific situation and cannot be applied to other situations. These theories can be divided into two categories: mechanistic and phenomenological [14]. The mechanistic models aim to explore the potential mechanism, but multiple wear mechanisms occur simultaneously in a practical matter. Furthermore, the mechanistic models are also limited to specific length scales. The phenomenological models are easier, presented in mathematical equations based on the contact mechanics variables [67]. In the field of wear simulation based on FEM, the phenomenological type is preferred because it can be localized and provides an acceptable level of accuracy and reliability in tracking the wear process [67]. In this section, two commonly used wear equations are presented.

4.1. Archard's Wear Law

Archard's wear law is probably the most famous theory of wear analysis. This equation relates volume loss, normal load, sliding distance, and material hardness through a dimensionless coefficient, which lays the foundation for the establishment of wear simulation models. Archard deduced the relationship between the material wear volume (V), sliding distance (S), dimensionless wear coefficient (k), normal load (F_N), and hardness of the softer material (H) [41,60] in the two contacting materials, expressed as [68]:

$$V = k \cdot \frac{F_N}{H} \cdot S \quad (6)$$

It can be seen from the formula that the wear volume is proportional to the normal load, inversely proportional to the material hardness, and proportional to the sliding distance.

In the field of wear simulation, in order to simulate the evolution of the contact surface with the wear cycle, the wear depth at each contact node needs to be determined. Therefore, Equation (6) needs to be deformed. For very small contact areas (ΔA), the relationship between the incremental wear depth (dh) and the incremental slip distance (dS) can be expressed by dividing the left and right sides of Equation (6) by the contact area (ΔA):

$$\frac{dV}{\Delta A} = k \cdot \frac{F_N}{H \cdot \Delta A} \cdot dS \quad (7)$$

In this formula, $\frac{dV}{\Delta A}$ is the wear increment dh, $\frac{F_N}{\Delta A}$ is the local contact stress p, and $\frac{k}{H}$ is replaced by k_D; then, the above formula can be expressed as:

$$dh = k_D \cdot p \cdot dS \quad (8)$$

where k_D is dimensional local wear coefficient, which is determined by multiple factors including contact size [69], temperature [66], sliding condition [70], and pressure [71]. There is no one explicit approach to obtain the value of k_D during the wear process [14,20]. In wear simulation, the wear coefficient is always considered as a constant [8] and obtained from experimental results. The equation is given by:

$$k_D = \frac{V}{S \cdot F_N} \quad (9)$$

Archard's wear equation provides a theoretical basis for node updates in wear simulation, and as shown in Table 1 and Figure 2, and it is now widely used [67,72–76]. However, as an empirical formula obtained from experiments, it has a weak theoretical foundation and cannot fully reflect actual wear behavior. The disadvantages are shown below [77–80]:
1. Fatigue, corrosion, oxidation, and other wear mechanisms are ignored.

2. The effects of temperature and lubrication on wear are not considered.
3. The wear coefficient is set to constant in the simulation.
4. The effect of transverse shear stress is not taken into account.

In view of the above shortcomings, many scholars have improved Archard's wear theory. Most improvement methods are coupled with the addition of related factors to the analysis. For example, chemical corrosion [21], temperature [2,81,82], and Archard's wear law are combined for coupling analysis.

For the effect of temperature, Yin et al. [83] proposed a comprehensive modeling approach to predict the thermomechanical tribological behaviors. In view of the lack of consideration of transverse shear stress, Wang [84] used the POD test and hip simulator wear test results to establish a wear model dependent on cross shear and contact pressure. However, for conventional non-crosslinked polyethylene, this model is not suitable, because applying the model without crosslinking would result in the prediction of infinite wear. Using a POD test, Kang et al. [85] showed that cross-shear ratio (CS) has significant effects on wear coefficient and established a wear model including cross-shear effect. In addition, Kang et al. [86] incorporated cross-shear motion and wear factors related to contact pressure into Archard's law to establish an independent computational wear model to predict hip polyethylene wear. Goreham-Voss et al. [7] proposed an improved model considering the influence of transverse shear motion on the main molecular orientation of polyethylene joint surfaces. Shu et al. [50] developed an improved Archard's wear theory by considering CS for wear prediction of total knee replacements.

4.2. Energy Dissipation Model

The energy-based wear theory was proposed by Fouvry et al. [87], which relates material volume loss with dissipated interfacial shear energy and considers the influence of interfacial shear work as an important wear parameter. The energy dissipation method is expressed as follows:

$$V = \alpha \sum E_d \tag{10}$$

where V, α, and E_d represent the wear volume, energy wear coefficient, and accumulated dissipated energy, respectively. The factors affecting the value of α are the same as those determining the wear coefficient in Archard's wear equation [67]. E_d is given by:

$$E_d = Q \cdot S \tag{11}$$

where Q and S represent the shear force and the sliding distance. Based on Coulomb's friction law, E_d is given by:

$$E_d = \mu \cdot P \cdot S \tag{12}$$

where μ and P represent the coefficient of friction and the normal load. According to the equations from (10) to (12), the local wear depth for each wear cycle can be described as [39]:

$$dh(x) = \alpha \cdot q(x) \cdot ds(x) \tag{13}$$

where q is local shear stress and ds is the local sliding distance.

In the numerical simulation study of fretting wear, the energy wear model is considered better than the Archard-based approach, because it allows for the use of a single wear coefficient across a variety of fretting load-stroke combinations, encompassing both partial slip and gross sliding regimes [11,88,89]. Changes in the friction coefficient can be considered during the wear process [39]. Moreover, as shown in Equations (6) and (10), the wear coefficient in Archard's wear equation equals the wear coefficient in the energy method divided by the coefficient of friction. Therefore, once one of them is known, both wear equations can be utilized in wear simulation [67]. Nowadays, many scholars use the energy dissipation model for wear simulation. Shen et al. [11] established a coupled damage elastoplastic constitutive model and developed a method to predict fretting fatigue life. The study used the energy dissipation model to simulate the evolution of contact geometry. Li et al. [36] introduced the friction coefficient as a function of fretting cycle numbers in numerical simulations, combining the energy consumption model with an adaptive mesh method to establish a wear model considering variable friction coefficients. Zhang et al. [37] proposed a finite element model for thread surface wear based on the energy dissipation model, simulating the phenomenon of self-loosening of bolted connections under transverse loads. Tandler et al. [38] simulated wear in automotive chain drive systems after high mileage using the energy dissipation model, and a comparison of the simulation data with experimental data demonstrated the effectiveness of the established model. Imran et al. [39] utilized the energy wear theory to establish a 3D FE model for simulating fretting wear in steel wire ropes used in coal mining processes. The influence of contact parameters on the fretting wear process during fretting cycles was examined.

5. Application of Wear Simulation

Wear simulation provides a dynamic representation of the changes in various parameters throughout the wear process. By leveraging the visualization capabilities of ABAQUS, the wear process can be vividly illustrated. This allows researchers to analyze wear behavior by referencing both the computational results and the actual conditions under which wear occurs. Utilizing wear simulation technology for analyzing wear mechanisms offers several distinct advantages:

(1) Using wear simulation before physical experiments allows for early assessment of wear mechanisms, enabling product optimization and reducing the need for extensive testing and improvements.
(2) Compared to physical experiments, wear simulation is a cost-effective, fast, and adaptable method. It allows for multiple calculations with adjustable parameters based on real working conditions.
(3) Wear simulation, compared to physical experiments, provides an intuitive visualization of the distribution of contact stress, displacement, temperature, and more, making it easier for researchers to analyze.

As can be seen from Tables 1 and 2, wear simulation is widely used in wear research, which can be mainly divided into three aspects: service life prediction, wear profile prediction, and wear mechanism auxiliary analysis. This section will elucidate the application of wear simulation in three areas: POD test, gear wear, and wear of orthopedic implants.

Table 2. Articles related to wear simulation applications.

Reference	Application Field	Year	Aim
[90]	Service life prediction	2023	The study proposed an approach based on FEM to predict the electrical contact resistance endurance of AgNi10 alloy.
[91]	Service life prediction	2015	The study predicted thrust bearing run-out, with the intention of using linear and non-linear wear models to predict bearing failure/life.
[4]	Service life prediction	2021	The study introduced a combined 3D wear and fatigue numerical method for fretting issues in ultra-high-strength steel wires.
[92]	Service life prediction	2022	The study analyzed the friction and wear conditions of dynamic and static metal wires inside the metal rubber.
[3]	Service life prediction Wear mechanism auxiliary analysis	2018	The study established a 3D FM model to simulate the failure process of self-lubricating spherical plain bearings under swinging wear conditions.

Table 2. Cont.

Reference	Application Field	Year	Aim
[5]	Service life prediction Wear mechanism auxiliary analysis	2022	The study investigated the fretting fatigue mechanism of WC-12Co coating through experiments and simulations.
[93]	Wear profile prediction	2023	The study aimed to predict the wear of a tenon connection structure by FEM.
[28]	Wear profile prediction	2023	A new fundamental FEM model was developed to predict wear for ceramic hip replacement bearings.
[94]	Wear mechanism auxiliary analysis	2009	The study examined the impact of normal load and attack angle of a conical indenter on wear mechanisms.
[88]	Wear mechanism auxiliary analysis	2023	The study aimed to explore the wear mechanism of Inconel 690 alloy and 403 stainless-steel anti-vibration strips.
[95]	Wear mechanism auxiliary analysis	2022	The impact of adding 3 wt.% of Y on the wear characteristics of ZK60 extruded alloy was studied.
[96]	Wear mechanism auxiliary analysis	2022	The study investigated the influence of loading frequency on fatigue performance and uncovered the wear mechanisms of bolted joints.

5.1. POD Test

Fretting wear, resulting in component failures and financial loss, is defined as a surface degradation process when small amplitude oscillatory sliding occurs between two contacting surfaces [69]. To estimate fretting wear, the POD test, which is an essential approach to estimating the wear performance of a specific material, is always applied in fretting wear research. However, as previously mentioned, the cost of the test is high. The FEM, which has been widely used in POD simulation, has a distinct advantage in reducing cost [97]. This section will discuss the application of POD simulation in fretting wear.

The POD simulation in fretting wear is a typical case, as it shows the basic architecture of wear simulation, as shown in Figure 7. First, the FE model is established according to the test conditions. Then, the FE model is validated by the Hertz formula to ensure accuracy [9,98]. The outputs such as wear depth, wear rate, and wear profile are obtained in the final step. To track the wear process in fretting wear, two wear equations, Archard's wear law and the energy dissipation model [67], are implemented through the UMESHMOTION subroutine. Since the FE model of the POD test is simple, the sub-model method is rarely applied. On the contrary, the number of wear cycles is generally large, and the extrapolation technique is always utilized to enhance the efficiency [1,14].

Wear profile prediction and wear mechanism analysis are the primary applications. McColl et al. [20] presented a 2D FE model based on Archard's wear equation for fretting wear simulation. The study showed that measured and predicted wear profiles are well confirmed under the low normal load situation. However, under the high normal load condition, the results were overestimated. The same conclusion was obtained by several other studies [17,30,33]. McColl attributed this phenomenon to the fact that changes in the wear coefficient were not taken into account [20]. Despite this drawback, the depth of wear and the changing trend of the wear profile are generally consistent with the experimental results. The energy dissipation model is considered superior to the Archard wear equation [8,88]. Li et al. [97] presented a method based on the energy law to study the fretting wear of the double rough surfaces. The model was validated by Hertz's theory and experiments. Zhang et al. [98] compared the significance of the Hertzian assumption to that of a rounded punch-on-flat in terms of fretting behavior. The model was validated by Hertz's theory and experiments. Cai et al. [99] applied the energy dissipation model to simulate fretting wear under the ball on flat contacting conditions. The research demonstrated that the combined effect of the normal load and amplitude had an impact on the contact pressure and shear stress, which in turn influenced the kinetic behavior, wear behavior, and evolution of worn surfaces. Li et al. [8] studied the fretting wear performance of the Inconel 718 alloy-based energy wear approach. In addition, Bastola et al [14] showed that there is no FE method for describing the wear of two 3D bodies simultaneously using adaptive mesh. They presented a method to predict wear on both contact surfaces.

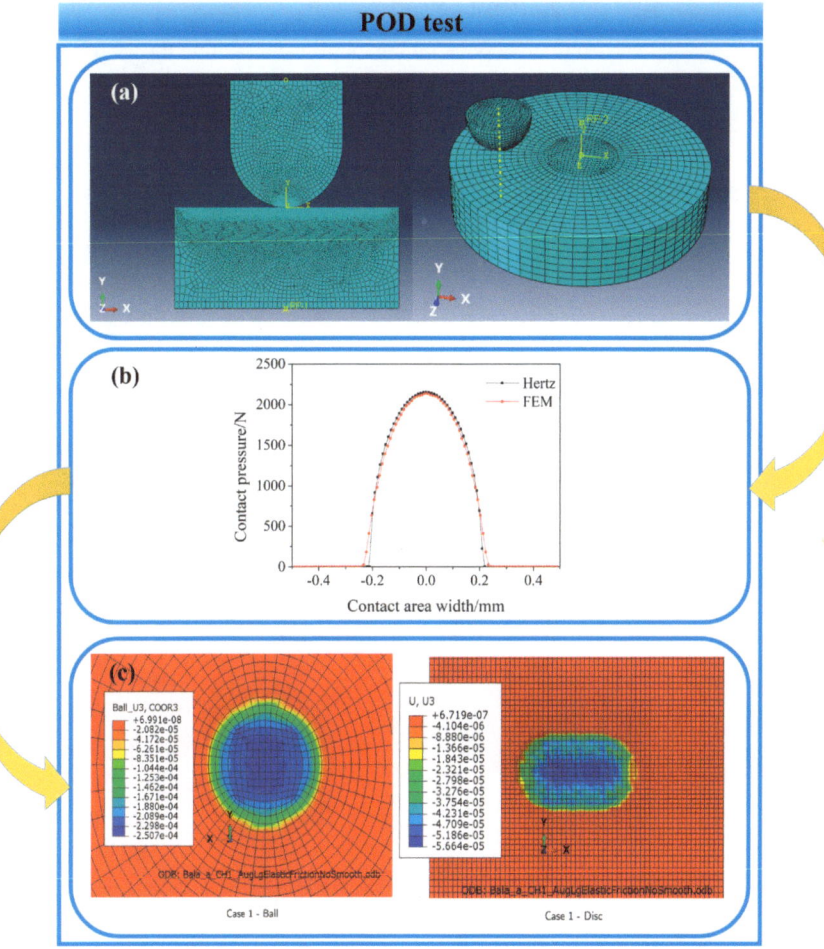

Figure 7. The basic architecture of wear simulation: (**a**) FE modeling, (**b**) model validation [99], and (**c**) the results of simulation [14].

5.2. Gear Wear

Gears, integral to mechanical transmission, facilitate power and motion transfer between parallel and non-parallel axes. With their precise transmission ratio, extensive power range, high efficiency, and smooth operation, they find widespread use in diverse power transmission domains. Wear, a primary cause of gear failure, occurs on the tooth contact surface due to friction between meshing gears, excessive load, and unsuitable working conditions. Excessive wear can distort the gear tooth shape, resulting in increased noise, vibration, and reduced transmission efficiency. The reduction in the contact area between meshing gears exacerbates gear force conditions, increases contact stress, and accelerates other gear failure modes. Hence, gear wear research holds significant importance.

Numerous studies about gear wear in existing research are predicated on Archard's wear theory [100–105]. As shown in Figure 8b, Xue et al. [103] used Archard's wear law to calculate the slip distance of an aero-engine's involute spline coupling. The typical simulation flow in gear wear simulation studies is shown in Figure 8a. For wear simulation, the ongoing gear meshing process is broken down into discrete steps, and the points of contact on the tooth profile are considered as the elastic contact of two cylinders based on Hertz's contact theory [106,107]. Subsequently, the contact pressure is determined, and the wear at each discrete point on the tooth surface within the wear cycle is computed.

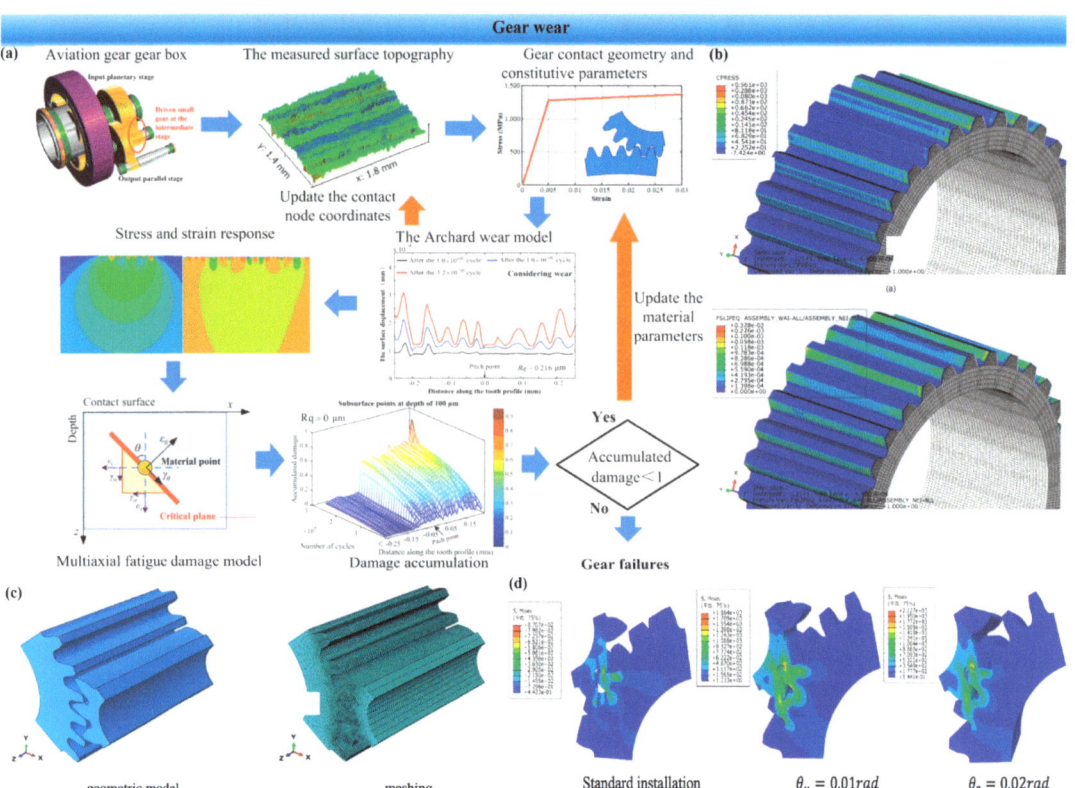

Figure 8. (**a**) The technical diagram of the simulation process [104], (**b**) FE simulation map: contact stress distribution of the initial model and slide distance distribution of the initial model [103], (**c**) FE model of gears [108], and (**d**) contact stress of gears [108].

In gear wear calculations, factors such as positional tolerances and installation errors can modify the contact path, leading to accelerated tooth surface wear [109]. Therefore, gear tolerance modeling is crucial. Tooth Contact Analysis (TCA) is conducted to determine the contact path and meshing state of the gear pair. However, traditional TCA can lead to high nonlinearity when errors exist in all six degrees of freedom, reducing efficiency. To mitigate this, the Small Displacement Torsor (SDT) concept has been introduced in TCA [110,111]. Sun et al. [108] conducted an analysis of the wear law of the tooth surface by integrating Archard's wear law with the SDT theory, tolerance theory, and TCA. Their approach involved the use of modeling and numerical simulation methods (Figure 8c). It was found that standard assembly of the gear pair results in an evenly distributed contact ellipse with a complete shape. However, installation errors can cause the contact area to become skewed, resulting in an incomplete contact ellipse and a smaller contact area, which leads to edge contact. As shown in Figure 8d, installation error results in a doubling of the contact surface stress. The results of their research can serve as a guide for the design and installation of a small module gear drive system.

5.3. Implant Wear

Joint replacement technology, a pivotal treatment for orthopedic diseases, has revitalized patients immobilized by conditions like arthritis and fractures. With the maturation of this technology and a growing demand, surgical cases are on the rise. Despite its benefits, the technology has limitations, with implant wear being a primary concern. Excessive wear can lead to implant-bone loosening, patient discomfort, increased risk of revision surgery, and potential local inflammation due to wear debris [22,71,112–115]. Hence, implant wear research is crucial. Wear simulation technology offers a novel approach to study implant friction wear, outperforming physical experiments in certain aspects. This discussion will further illustrate this using total knee arthroplasty (TKA) as an example.

TKA is a surgical technique that replaces the worn-out native knee joint. The TKA schematic is shown in Figure 9b. Specifically, the articular surface consisting of cartilage, meniscus, and cartilage is substituted with an Ultra-High Molecular Weight Polyethylene (UHMWPE) insert in a metal backing [116]. In this field, wear prediction is still the primary application. Innocenti et al. [116] devised and validated a finite element model to anticipate wear in polyethylene-based TKA. Zhang et al. [117] studied how internal-external rotation and anterior-posterior translation affect the wear of knee implants. It was shown that both internal-external rotation and anterior-posterior movement were vital factors that influenced the contact mechanism and wear of total knee implants. Kang et al. [118] employed finite element analysis under gait cycle loading conditions to examine the effects of various surface characteristics on the femoral component's weight loss, wear depth, and kinematics in TKA. Furthermore, Koh et al. [119] mitigated wear in personalized TKA through the design, optimization, and parameterization of a 3D finite element model, corroborated by experimental wear test outcomes.

The design of joint surface curvature is crucial in TKA for tibiofemoral kinematics and contact mechanics. Yet, the effects of this curvature on various designs remain underexplored [120]. Mukhtar et al. [120] optimized and personalized the design parameters of a knee implant using the Taguchi method. The constructed model and simulation results are shown in Figure 9a,d. Koh et al. [121] performed computational simulations to contrast wear performance between conventional and patient-specific TKA under gait loading conditions. They found that different TKA designs result in kinematic variations, with contact pressure and area not directly influencing wear performance, as shown in Figure 9c. Notably, conforming individualized TKA exhibited the highest volume wear and wear rate, with a 29% increase in volume wear compared to internally rotated center individualized TKA.

UHMWPE is frequently employed in knee joint replacements. However, research indicates that Polyether ether ketone (PEEK) and Carbon Fiber Reinforced PEEK (CFR-PEEK) could serve as alternatives. The wear particles of CFR-PEEK exhibit no cytotoxicity, suggesting minimal adverse tissue reactions. Koh et al. [122] constructed a finite element model using tomography and magnetic resonance imaging techniques, investigating the biomechanical implications of UHMWPE and CFR-PEEK on mobile bearing TKA. The findings revealed a significant reduction in wear volume and the depth of CFR-PEEK compared to UHMWPE, while PEEK showed an increase (Figure 9e). This underscores the potential of CFR-PEEK as a promising substitute for UHMWPE in tibial implants. Nonetheless, comprehensive orthopedic research is warranted for newly introduced biomaterials to ascertain their threshold conditions and appropriate applications.

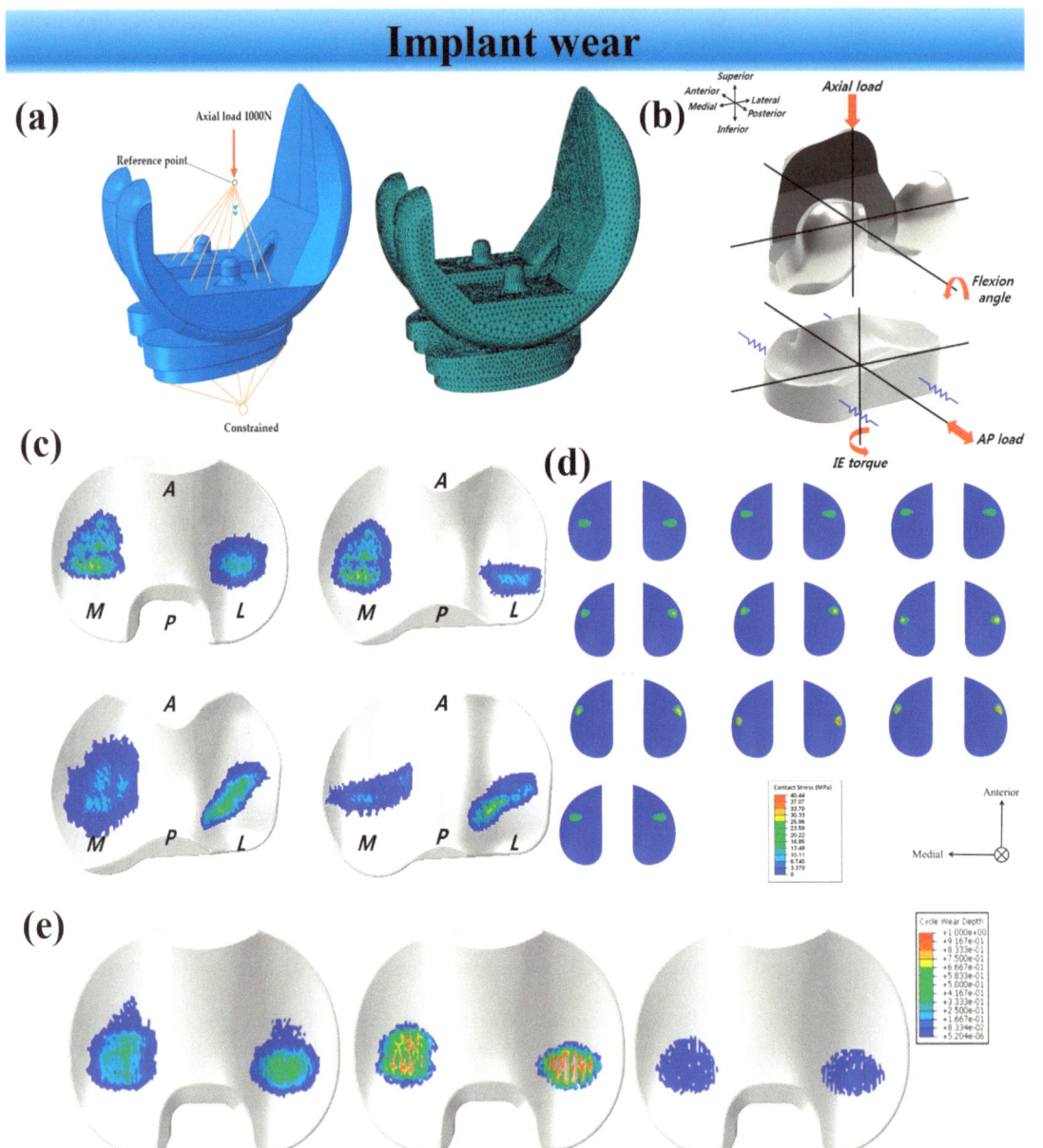

Figure 9. (**a**) FE model of the knee implant [120], (**b**) knee implant schematic [121], (**c**) predicted wear contours of the four different TKA designs: a conventional TKA, CPS-TKA, MPS-TKA, and BPS-TKA [121], (**d**) distribution of the contact stress on the UHMWPE tibial insert [120], and (**e**) predicted wear-depth contours for UHMWPE, PEEK, and CFR–PEEK in the gait simulation [122].

5.4. Other Applications

In addition to the applications of wear simulation in the POD test, gear wear, and implant wear, wear simulation is also applied in many other fields, which can be seen in Figure 10, such as the wear of seal [123,124], chain [125], tire [126], cam [127], artillery barrel [128], pump [129], metal wire [4] and so on. However, it can be observed that regardless of the field where wear simulation is applied, the main functions of wear simulation are wear profile prediction, service life prediction, and wear mechanism auxiliary analysis. Moreover, the wear simulation processes are similar across different research subjects and issues. Another striking feature can also be observed, such that Archard's wear law is widely applied in different situations: electrical contact under fretting wear [90], tribocorrosion [125], thermal-mechanical coupling wear [130], and so on. However, as mentioned in Section 4.1, in many cases, Archard's wear law needs to be combined with other models to accurately capture the wear process. The same applies to the energy dissipation wear model.

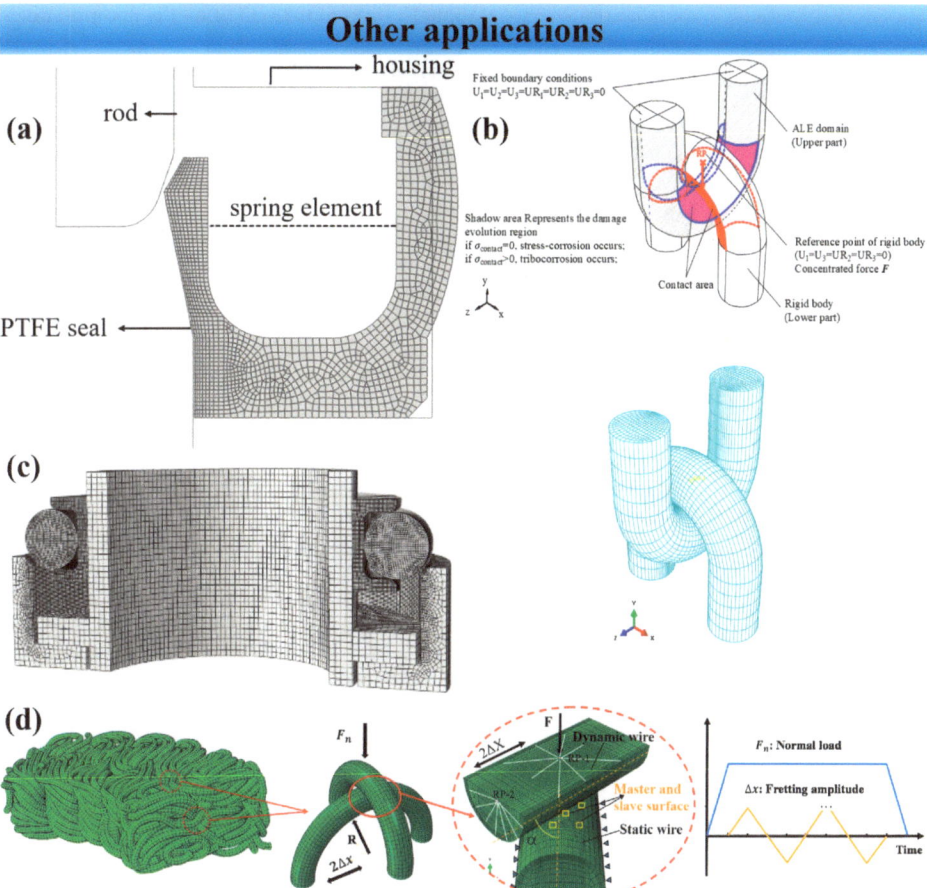

Figure 10. (**a**) An axisymmetric FE model for the spring-energized PTFE seal [123], (**b**) an FE model of the Anchor Chain [125], (**c**) an FE model of stirling engine piston rod oil-free lubrication seal [124], and (**d**) schematic diagram of the FE model of metal filament wear and simulation load loading [92].

Except for the features discussed above, an outstanding disadvantage can be observed, such that the wear processes, in practical situations, include a variety of wear mechanisms. However, this complex process is described only by Archard's theory or the energy dissipation theory. Wear simulation lacks some rationality in this aspect. Therefore, it can be concluded that the application of FEM in wear simulation is still in the research stage.

6. Summary

This paper presents the research methodology and application of ABAQUS in wear simulation, covering aspects such as FE model construction, factors influencing wear behavior, wear theory, and the practical application of wear simulation. The main conclusions of this paper are as follows:

(1) The FE model can be categorized into two types: 2D and 3D. The 2D model is suitable for cases where there is no concern about the overall wear profile, offering high computational efficiency. On the other hand, the 3D model is used for complex structures and situations with complex boundary conditions and loading, providing more accurate computational results at the cost of higher computational resources and time.

(2) To maintain contact in the model and simulate the real wear process, the model needs to be updated after each incremental step. Common methods for this include the UMESHMOTION subroutine and Python scripts. The element quality updated by the UMESHMOTION is better than what is updated by the Python script, reducing the possibility of stress concentration and convergence issues.

(3) Computational efficiency is a significant concern in finite element analysis. Two commonly used methods to address this issue are the sub-model and extrapolation method. The sub-model method is often used for large and complex structures, while the extrapolation method can be applied to general wear problems, effectively improving computational efficiency. However, it is important to note that the extrapolation method requires finding an appropriate extrapolation factor to avoid significant errors. These two optimization methods can be used in combination based on actual situations.

(4) There are many factors that influence wear behavior, which can be broadly categorized into two aspects: material properties and working conditions. These factors include hardness, roughness, lubrication, contact stress, sliding speed, and others. The effects of these factors on wear behavior are not singular, and they can also interact with each other. Therefore, it is challenging to propose a wear model that encompasses all influencing factors. Archard proposed a widely accepted theoretical model based on experiments, but it still has limitations. Many scholars have made improvements to Archard's wear law through coupled analysis. In addition to this theory, the energy dissipation theory is another wear theory that is increasingly used in current wear research. Furthermore, the results obtained from the energy dissipation model show better agreement with the experimental results compared to Archard's wear law.

(5) In practical applications, wear simulation technology can be utilized in various research fields, primarily focusing on predicting service life, wear profile, and wear mechanisms. This article provides an overview of the application of wear simulation in the POD test, gear wear, orthopedic implant wear, and other applications, leading to the following conclusions: Wear simulation technology can serve as an initial tool for product development and failure analysis, providing researchers with relatively reliable reference data.

(6) The wear simulation presented in this paper is capable of capturing micron-scale wear processes and can provide a reasonable initial estimate of material loss. However, the subtle features of a specific wear mechanism cannot be obtained.

(7) At present, the application of FEM in wear simulation is still in the research stage.

7. Perspectives

Based on the preceding discussions, it is evident that wear simulation technology holds vast potential for future applications. In this context, a future perspective on its development is provided below:

(1) Development of more precise and accurate wear models to enhance the accuracy of finite element wear analysis.
(2) Advancement of more efficient and accurate computational methods and algorithms to reduce computational costs and time.
(3) Integration of new technologies such as machine learning to optimize and improve wear models, thereby enhancing predictive capabilities.

These developments are expected to significantly enhance the effectiveness and applicability of wear simulation technology across various industrial and scientific domains. As computational capabilities continue to evolve, these advancements will play a crucial role in addressing wear-related challenges and driving innovation in engineering and materials science.

Funding: The project was supported by the Guangdong Province Natural Science Foundation (2023A1515011558), the Ministry of Education Chunhui Plan Project (HZKY20220434), the State Key Laboratory of Solid Lubrication Fund (LSL-2204), the Liaoning Province Natural Science Foundation (2022-BS-078), the Open Project of Henan Key Laboratory of Intelligent Manufacturing of Mechanical Equipment, Zhengzhou University of Light Industry (No. IM202301), the Fundamental Research Funds for the Central Universities (N2203017) and the Liaoning Province Education Department Universities Basic Scientific Research Project (LJKMZ20220345).

Data Availability Statement: No new data were created or analyzed in this study. Data sharing is not applicable to this article.

Conflicts of Interest: The authors declare no conflicts of interest.

References

1. Bose, K.K.; Ramkumar, P. Finite element method based sliding wear prediction of steel-on-steel contacts using extrapolation techniques. *Proc. Inst. Mech. Eng. Part J J. Eng. Tribol.* **2019**, *233*, 1446–1463. [CrossRef]
2. Gan, L.; Xiao, K.; Pu, W.; Tang, T.; Wang, J.X. A numerical method to investigate the effect of thermal and plastic behaviors on the evolution of sliding wear. *Meccanica* **2021**, *56*, 2339–2356. [CrossRef]
3. Xue, Y.; Chen, J.; Guo, S.; Meng, Q.; Luo, J. Finite element simulation and experimental test of the wear behavior for self-lubricating spherical plain bearings. *Friction* **2018**, *6*, 297–306. [CrossRef]
4. Llavori, I.; Zabala, A.; Mendiguren, J.; Gómez, X. A coupled 3D wear and fatigue numerical procedure: Application to fretting problems in ultra-high strength steel wires. *Int. J. Fatigue* **2021**, *143*, 106012. [CrossRef]
5. Wang, X.X.; Ping, X.C.; Zeng, X.; Wang, R.J.; Zhao, Q.; Ying, S.J.; Hu, T. Fretting fatigue experiment and simulation of WC-12Co coating taking into account the wear effects. *Surf. Coat. Technol.* **2022**, *441*, 128555. [CrossRef]
6. Xiang, D.D.; Yusheng, L.; Tianbiao, Y.; Di, W.; Xiaoxin, L.; Kaiming, W.; Lin, L.; Jie, P.; Yao, S.; Zibin, C. Review on wear resistance of laser cladding high-entropy alloy coatings. *J. Mater. Res. Technol.* **2024**, *28*, 911–934. [CrossRef]
7. Goreham-Voss, C.M.; Hyde, P.J.; Hall, R.M.; Fisher, J.; Brown, T.D. Cross-shear implementation in sliding-distance-coupled finite element analysis of wear in metal-on-polyethylene total joint arthroplasty: Intervertebral total disc replacement as an illustrative application. *J. Biomech.* **2010**, *43*, 1674–1681. [CrossRef] [PubMed]
8. Li, C.; Karimbaev, R.; Wang, S.; Amanov, A.; Wang, D.; Abdel Wahab, M. Fretting wear behavior of Inconel 718 alloy manufactured by DED and treated by UNSM. *Sci. Rep.-UK* **2023**, *13*, 1308. [CrossRef] [PubMed]
9. Bose, K.K.; Ramkumar, P. Finite Element Sliding Wear Simulation of 2D Steel-on-Steel Pin-on-Disc Tribometer. *Sae Tech. Pap.* **2018**, *28*, 11. [CrossRef]
10. Daves, W.; Kubin, W.; Scheriau, S.; Pletz, M. A finite element model to simulate the physical mechanisms of wear and crack initiation in wheel/rail contact. *Wear* **2016**, *366–367*, 78–83. [CrossRef]
11. Shen, F.; Hu, W.; Meng, Q. A damage mechanics approach to fretting fatigue life prediction with consideration of elastic-plastic damage model and wear. *Tribol. Int.* **2015**, *82*, 176–190. [CrossRef]
12. Fallahnezhad, K.; Feyzi, M.; Ghadirinejad, K.; Hashemi, R.; Taylor, M. Finite element based simulation of tribocorrosion at the head-neck junction of hip implants. *Tribol. Int.* **2022**, *165*, 107284. [CrossRef]
13. Ahmadi, A.; Sadeghi, F. A Three-Dimensional Finite Element Damage Mechanics Model to Simulate Fretting Wear of Hertzian Line and Circular Contacts in Partial Slip Regime. *J. Tribol.* **2022**, *144*, 51602. [CrossRef]
14. Bastola, A.; Stewart, D.; Dini, D. Three-dimensional finite element simulation and experimental validation of sliding wear. *Wear* **2022**, *504–505*, 204402. [CrossRef]
15. Bae, J.W.; Lee, C.Y.; Chai, Y.S. Three dimensional fretting wear analysis by finite element substructure method. *Int. J. Precis. Eng. Man.* **2009**, *10*, 63–69. [CrossRef]
16. Shu, Y.; Yang, G.; Liu, Z. Simulation research on fretting wear of train axles with interference fit based on press-fitted specimen. *Wear* **2023**, *523*, 204777. [CrossRef]

17. Bose, K.K.; Penchaliah, R. 3-D FEM Wear Prediction of Brass Sliding against Bearing Steel Using Constant Contact Pressure Approximation Technique. *Tribol. Online* **2019**, *14*, 194–207. [CrossRef]
18. Zuo, S.G.; Ni, T.X.; Wu, X.D.; Wu, K.; Yang, X.W. Prediction procedure for wear distribution of transient rolling tire. *Int. J. Automot. Technol.* **2014**, *15*, 505–515. [CrossRef]
19. Yu, H.; Lian, Z.; Lin, T.; Liu, Y.; Xu, X. Experimental and numerical study on casing wear in highly deviated drilling for oil and gas. *Adv. Mech. Eng.* **2016**, *8*, 2071834741. [CrossRef]
20. McColl, I.R.; Ding, J.; Leen, S.B. Finite element simulation and experimental validation of fretting wear. *Wear* **2004**, *256*, 1114–1127. [CrossRef]
21. Fallahnezhad, K.; Oskouei, R.H.; Taylor, M. Development of a fretting corrosion model for metallic interfaces using adaptive finite element analysis. *Finite Elem. Anal. Des.* **2018**, *148*, 38–47. [CrossRef]
22. Bevill, S.L.; Bevill, G.R.; Penmetsa, J.R.; Petrella, A.J.; Rullkoetter, P.J. Finite element simulation of early creep and wear in total hip arthroplasty. *J. Biomech.* **2005**, *38*, 2365–2374. [CrossRef] [PubMed]
23. Knight, L.A.; Pal, S.; Coleman, J.C.; Bronson, F.; Haider, H.; Levine, D.L.; Taylor, M.; Rullkoetter, P.J. Comparison of long-term numerical and experimental total knee replacement wear during simulated gait loading. *J. Biomech.* **2007**, *40*, 1550–1558. [CrossRef] [PubMed]
24. Peng, R.T.; Li, J.; Tang, X.Z.; Zhou, Z. Simulation of Tool Wear in Prestressed Cutting Superalloys. *Mater. Sci. Forum* **2016**, *836–837*, 402–407. [CrossRef]
25. Albers, A.; Reichert, S. On the influence of surface roughness on the wear behavior in the running-in phase in mixed-lubricated contacts with the finite element method. *Wear* **2017**, *376–377*, 1185–1193. [CrossRef]
26. Shankar, S.; Nithyaprakash, R.; Santhosh, B.R.; Uddin, M.S.; Pramanik, A. Finite element submodeling technique to analyze the contact pressure and wear of hard bearing couples in hip prosthesis. *Comput. Methods Biomech. Biomed. Eng.* **2020**, *23*, 422–431. [CrossRef]
27. Shankar, S.; Nithyaprakash, R.; Santhosh, B.R.; Gur, A.K.; Pramanik, A. Experimental and submodeling technique to investigate the wear of silicon nitride against Ti6Al4V alloy with bio-lubricants for various gait activities. *Tribol. Int.* **2020**, *151*, 106529. [CrossRef]
28. Nitish Prasad, K.; Ramkumar, P. FEM wear prediction of ceramic hip replacement bearings under dynamic edge loading conditions. *J. Mech. Behav. Biomed.* **2023**, *146*, 106049. [CrossRef]
29. Mukras, S.; Kim, N.H.; Sawyer, W.G.; Jackson, D.B.; Bergquist, L.W. Numerical integration schemes and parallel computation for wear prediction using finite element method. *Wear* **2009**, *266*, 822–831. [CrossRef]
30. Bortoleto, E.M.; Rovani, A.C.; Seriacopi, V.; Profito, F.J.; Zachariadis, D.C.; Machado, I.F.; Sinatora, A.; Souza, R.M. Experimental and numerical analysis of dry contact in the pin on disc test. *Wear* **2013**, *301*, 19–26. [CrossRef]
31. Schmidt, A.A.; Schmidt, T.; Grabherr, O.; Bartel, D. Transient wear simulation based on three-dimensional finite element analysis for a dry running tilted shaft-bushing bearing. *Wear* **2018**, *408–409*, 171–179. [CrossRef]
32. Arunachalam, A.P.S.; Idapalapati, S. Material removal analysis for compliant polishing tool using adaptive meshing technique and Archard wear model. *Wear* **2019**, *418–419*, 140–150. [CrossRef]
33. Joshi, V.; Ramkumar, P. Transient Wear FEA Modelling Using Extrapolation Technique for Steel-on-Steel Dry Sliding Contact. *Tribol. Online* **2022**, *17*, 162–174. [CrossRef]
34. Zhang, F.; Peng, X. Analysis on load-bearing contact characteristics of face gear tooth surface wear. *For. Chem. Rev.* **2022**, 743–754. Available online: http://forestchemicalsreview.com/index.php/JFCR/article/view/1164 (accessed on 25 December 2023).
35. Martínez-Londoño, J.C.; Martínez-Trinidad, J.; Hernández-Fernández, A.; García-León, R.A. Finite Element Analysis on AISI 316L Stainless Steel Exposed to Ball-on-Flat Dry Sliding Wear Test. *Trans. Indian Inst. Met.* **2023**, *76*, 97–106. [CrossRef]
36. Li, L.; Kang, L.; Ma, S.; Li, Z.; Ruan, X.; Cai, A. Finite element analysis of fretting wear considering variable coefficient of friction. *Proc. Inst. Mech. Eng. Part J J. Eng. Tribol.* **2019**, *233*, 758–768. [CrossRef]
37. Zhang, M.; Zeng, D.; Lu, L.; Zhang, Y.; Wang, J.; Xu, J. Finite element modelling and experimental validation of bolt loosening due to thread wear under transverse cyclic loading. *Eng. Fail. Anal.* **2019**, *104*, 341–353. [CrossRef]
38. Tandler, R.; Bohn, N.; Gabbert, U.; Woschke, E. Analytical wear model and its application for the wear simulation in automotive bush chain drive systems. *Wear* **2020**, *446–447*, 203193. [CrossRef]
39. Imran, M.; Wang, D.; Abdel Wahab, M. Three-dimensional finite element simulations of fretting wear in steel wires used in coal mine hoisting system. *Adv. Eng. Softw.* **2023**, *184*, 103499. [CrossRef]
40. Yue, T.; Abdel Wahab, M. Finite element analysis of fretting wear under variable coefficient of friction and different contact regimes. *Tribol. Int.* **2017**, *107*, 274–282. [CrossRef]
41. Hegadekatte, V.; Huber, N.; Kraft, O. Modeling and simulation of wear in a pin on disc tribometer. *Tribol. Lett.* **2006**, *24*, 51–60. [CrossRef]
42. ABAQUS Inc. *ABAQUS Analysis User's Manual*; Dassault Systèmes: Providence, RI, USA, 2017; Available online: https://help.3ds.com/HelpDS.aspx?V=2017&P=DSSIMULIA_Established&L=English&contextscope=all&F=SIMULIA_Established_FrontmatterMap/DSDocAbaqus.htm (accessed on 25 December 2023).
43. Li, H.; Ren, Z.; Su, X.; Shen, L.; Huang, J. Study on the Fretting Wear Evolution Model of Wires with Curvature Inside Metal Rubber. *Tribol. Lett.* **2023**, *71*, 22. [CrossRef]

44. Zhang, Y.; Wei, F.; Lin, S.; Sun, X.; Liu, L. Study on the Performance of Reciprocating Seals under the Coupling Effect of Elastohydrodynamic Lubrication and Rubber Wear. *Eng. Res. Express* **2024**, *6*, 015064. [CrossRef]
45. Zhang, S.; Liu, Y.; Zhou, H.; Zhang, W.; Chen, Y.; Zhu, H. Analysis of the Effect of Wear on Tire Cornering Characteristics Based on Grounding Characteristics. *World Electr. Veh. J.* **2023**, *14*, 166. [CrossRef]
46. Liu, Y.; Xiang, D.; Wang, K.; Yu, T. Corrosion of Laser Cladding High-Entropy Alloy Coatings: A Review. *Coatings* **2022**, *12*, 1669. [CrossRef]
47. Xiang, D.; Wang, D.; Zheng, T.; Chen, Y. Effects of Rare Earths on Microstructure and Wear Resistance in Metal Additive Manufacturing: A Review. *Coatings* **2024**, *14*, 139. [CrossRef]
48. Wang, K.; Liu, W.; Li, X.; Tong, Y.; Hu, Y.; Hu, H.; Chang, B.; Ju, J. Effect of hot isostatic pressing on microstructure and properties of high chromium K648 superalloy manufacturing by extreme high-speed laser metal deposition. *J. Mater. Res. Technol.* **2024**, *28*, 3951–3959. [CrossRef]
49. Li, S.; Wang, L.; Yang, G. Unified computational model of thermochemical erosion and mechanical wear in artillery barrel considering hydrodynamic friction. *Numer. Heat Transfer. Part A Appl.* **2023**, 1–21. [CrossRef]
50. Shu, L.; Hashimoto, S.; Sugita, N. Enhanced In-Silico Polyethylene Wear Simulation of Total Knee Replacements During Daily Activities. *Ann. Biomed. Eng.* **2021**, *49*, 322–333. [CrossRef]
51. Saini, V.; Maurya, U.; Thakre, G.D. Estimating the Dry-Wear Behavior of Rolling/Sliding Bearings (PB, Gunmetal, and Al6061)Tribo Materials. *J. Fail. Anal. Prev.* **2023**, *23*, 2439–2451. [CrossRef]
52. Curreli, C.; Viceconti, M.; Di Puccio, F. Submodeling in wear predictive finite element models with multipoint contacts. *Int. J. Numer. Meth. Eng.* **2021**, *122*, 3812–3823. [CrossRef]
53. Curreli, C.; Di Puccio, F.; Mattei, L. Application of the finite element submodeling technique in a single point contact and wear problem. *Int. J. Numer. Meth. Eng.* **2018**, *116*, 708–722. [CrossRef]
54. Rigney, D.A. Colmments on the sliding wear of metals. *Tribol. Int.* **1997**, *30*, 361–367. [CrossRef]
55. Lemm, J.D.; Warmuth, A.R.; Pearson, S.R.; Shipway, P.H. The influence of surface hardness on the fretting wear of steel pairs—Its role in debris retention in the contact. *Tribol. Int.* **2015**, *81*, 258–266. [CrossRef]
56. Ravikiran, A.; Jahanmir, S. Effect of contact pressure and load on wear of alumina. *Wear* **2001**, *251*, 980–984. [CrossRef]
57. Qin, W.; Wang, M.; Sun, W.; Shipway, P.; Li, X. Modeling the effectiveness of oil lubrication in reducing both friction and wear in a fretting contact. *Wear* **2019**, *426–427*, 770–777. [CrossRef]
58. Cao, G.; Zhang, J.; Guo, Y.; Liu, C.; Micheal, M.; Lv, C.; Yu, H.; Wu, H. Numerical modeling on friction and wear behaviors of all-metal progressive cavity pump. *J. Petrol. Sci. Eng.* **2022**, *213*, 110443. [CrossRef]
59. Khader, I.; Renz, A.; Kailer, A. A wear model for silicon nitride in dry sliding contact against a nickel-base alloy. *Wear* **2017**, *376–377*, 352–362. [CrossRef]
60. Arjmandi, M.; Ramezani, M.; Giordano, M.; Schmid, S. Finite element modelling of sliding wear in three-dimensional woven textiles. *Tribol. Int.* **2017**, *115*, 452–460. [CrossRef]
61. Guo, J.; Li, D.; Wang, H.; Yang, Y.; Wang, L.; Guan, D.; Qiu, Y.; He, L.; Zhang, S. Effect of contact stress on the cycle-dependent wear behavior of ceramic restoration. *J. Mech. Behav. Biomed.* **2017**, *68*, 16–25. [CrossRef] [PubMed]
62. Liu, H.; Liu, H.; Zhu, C.; Parker, R.G. Effects of lubrication on gear performance: A review. *Mech. Mach. Theory* **2020**, *145*, 103701. [CrossRef]
63. Zhao, J.; Sheng, W.; Li, Z.; Zhang, H.; Zhu, R. Effect of lubricant selection on the wear characteristics of spur gear under oil-air mixed lubrication. *Tribol. Int.* **2022**, *167*, 107382. [CrossRef]
64. Okonkwo, P.C.; Kelly, G.; Rolfe, B.F.; Pereira, M.P. The effect of sliding speed on the wear of steel-tool steel pairs. *Tribol. Int.* **2016**, *97*, 218–227. [CrossRef]
65. Chowdhury, M.A.; Khalil, M.K.; Nuruzzaman, D.M.; Rahaman, M.L. The Effect of Sliding Speed and Normal Load on Friction and Wear Property of Aluminum. *Int. J. Mech. Mechatron. Eng.* **2011**, *11*, 45–49.
66. Meng, H.C.; Ludema, K.C. Wear models and predictive equations: Their form and content. *Wear* **1995**, *181–183*, 443–457. [CrossRef]
67. Feyzi, M.; Fallahnezhad, K.; Taylor, M.; Hashemi, R. A review on the finite element simulation of fretting wear and corrosion in the taper junction of hip replacement implants. *Comput. Biol. Med.* **2021**, *130*, 104196. [CrossRef] [PubMed]
68. Archard, J.F. Contact and Rubbing of Flat Surfaces. *J. Appl. Phys.* **1953**, *24*, 981–988. [CrossRef]
69. Fouvry, S.; Arnaud, P.; Mignot, A.; Neubauer, P. Contact size, frequency and cyclic normal force effects on Ti-6Al-4V fretting wear processes: An approach combining friction power and contact oxygenation. *Tribol. Int.* **2017**, *113*, 460–473. [CrossRef]
70. Fallahnezhad, K.; Oskouei, R.H.; Badnava, H.; Taylor, M. An adaptive finite element simulation of fretting wear damage at the head-neck taper junction of total hip replacement: The role of taper angle mismatch. *J. Mech. Behav. Biomed.* **2017**, *75*, 58–67. [CrossRef] [PubMed]
71. Zhang, T.; Harrison, N.M.; McDonnell, P.F.; McHugh, P.E.; Leen, S.B. A finite element methodology for wear–fatigue analysis for modular hip implants. *Tribol. Int.* **2013**, *65*, 113–127. [CrossRef]
72. Mohd Tobi, A.L.; Shipway, P.H.; Leen, S.B. Gross slip fretting wear performance of a layered thin W-DLC coating: Damage mechanisms and life modelling. *Wear* **2011**, *271*, 1572–1584. [CrossRef]
73. Rezaei, A.; Van Paepegem, W.; De Baets, P.; Ost, W.; Degrieck, J. Adaptive finite element simulation of wear evolution in radial sliding bearings. *Wear* **2012**, *296*, 660–671. [CrossRef]

74. Shen, X.; Liu, Y.; Cao, L.; Chen, X. Numerical Simulation of Sliding Wear for Self-lubricating Spherical Plain Bearings. *J. Mater. Res. Technol.* **2012**, *1*, 8–12. [CrossRef]
75. Shu, Y.J.; Shen, F.; Ke, L.L.; Wang, Y.S. Adaptive finite element simulation and experimental verification for fretting wear of PVDF piezoelectric thin films. *Wear* **2022**, *502*, 204395. [CrossRef]
76. Zao, H.; Yumei, H.; Xingyuan, Z.; Yuanyuan, Y. A Calculation Method for Tooth Wear Depth Based on the Finite Element Method That Considers the Dynamic Mesh Force. *Machines* **2022**, *10*, 69.
77. Põdra, P.; Andersson, S. Wear simulation with the Winkler surface model. *Wear* **1997**, *207*, 79–85. [CrossRef]
78. Suh, N.P. The delamination theory of wear. *Wear* **1973**, *25*, 111–124. [CrossRef]
79. Sobis, T.; Engel, U.; Geiger, M. A theoretical study on wear simulation in metal forming processes. *J. Mater. Process. Technol.* **1992**, *34*, 233–240. [CrossRef]
80. Cheng, Q.; Zhang, H.; Zhang, T.; Li, Y.; Xu, J.; Liu, Z. Prediction method of precision deterioration of rolling guide under multi-random parameters based on frictional thermal expansion effect. *Tribol. Int.* **2023**, *189*, 108883. [CrossRef]
81. Gui, L.; Wang, X.; Fan, Z.; Zhang, F. A simulation method of thermo-mechanical and tribological coupled analysis in dry sliding systems. *Tribol. Int.* **2016**, *103*, 121–131. [CrossRef]
82. Luo, S.; Zhu, D.; Hua, L.; Qian, D.; Yan, S. Numerical analysis of die wear characteristics in hot forging of titanium alloy turbine blade. *Int. J. Mech. Sci.* **2017**, *123*, 260–270. [CrossRef]
83. Yin, J.; Lu, C.; Mo, J. Comprehensive modeling strategy for thermomechanical tribological behavior analysis of railway vehicle disc brake system. *Friction* **2024**, *12*, 74–94. [CrossRef]
84. Wang, A. A unified theory of wear for ultra-high molecular weight polyethylene in multi-directional sliding. *Wear* **2001**, *248*, 38–47. [CrossRef]
85. Kang, L.; Galvin, A.L.; Brown, T.D.; Jin, Z.; Fisher, J. Quantification of the effect of cross-shear on the wear of conventional and highly cross-linked UHMWPE. *J. Biomech.* **2008**, *41*, 340–346. [CrossRef] [PubMed]
86. Kang, L.; Galvin, A.L.; Fisher, J.; Jin, Z. Enhanced computational prediction of polyethylene wear in hip joints by incorporating cross-shear and contact pressure in additional to load and sliding distance: Effect of head diameter. *J. Biomech.* **2009**, *42*, 912–918. [CrossRef]
87. Fouvry, S.; Liskiewicz, T.; Kapsa, P.; Hannel, S.; Sauger, E. An energy description of wear mechanisms and its applications to oscillating sliding contacts. *Wear* **2003**, *255*, 287–298. [CrossRef]
88. Xie, L.; Guan, Y.; Lu, J.; Zhu, P.; Chen, R.; Lin, H. Fretting wear behavior test and numerical simulation of Inconel 690 alloy. *J. Nucl. Sci. Technol.* **2023**, *60*, 1100–1115. [CrossRef]
89. Xue, X.; Liu, J.; Jia, J.; Yang, S.; Li, Y. Simulation and Verification of Involute Spline Tooth Surface Wear before and after Carburizing Based on Energy Dissipation Method. *Machines* **2023**, *11*, 78. [CrossRef]
90. Zhang, C.; Shen, F.; Ke, L. Electrical contact resistance endurance of AgNi10 alloy under fretting wear: Experiment and numerical prediction. *Wear* **2023**, *530–531*, 205009. [CrossRef]
91. Hwang, S.; Lee, N.; Kim, N. Experiment and Numerical Study of Wear in Cross Roller Thrust Bearings. *Lubricants* **2015**, *3*, 447–458. [CrossRef]
92. Li, H.; Ren, Z.; Huang, J.; Zhong, S. Fretting wear evolution model of the metal filaments inside metal rubber. *Wear* **2022**, *506–507*, 204438. [CrossRef]
93. Zhang, Z.; Zhao, G.; Yuan, Y.; Zhang, H.; Wu, Y. Finite Element Simulation and Fretting Wear Prediction of a Tenon Connection Structure. *Lubricants* **2023**, *11*, 421. [CrossRef]
94. Tkaya, M.B.; Mezlini, S.; Mansori, M.E.; Zahouani, H. On some tribological effects of graphite nodules in wear mechanism of SG cast iron: Finite element and experimental analysis. *Wear* **2009**, *267*, 535–539. [CrossRef]
95. Banijamali, S.M.; Shariat Razavi, M.; Palizdar, Y.; Najafi, S.; Sheikhani, A.; Torkamani, H. Experimental and Simulation Study on Wear Behavior of ZK60 Alloy with 3 wt.% Yttrium Addition. *J. Mater. Eng. Perform.* **2022**, *31*, 4721–4734. [CrossRef]
96. Li, H.; Zhao, Y.; Jiang, J.; Wang, H.; He, J.; Liu, J.; Peng, J.; Zhu, M. Effect of frequency on the fatigue performance of bolted joints under axial excitation. *Tribol. Int.* **2022**, *176*, 107933. [CrossRef]
97. Li, L.; Li, G.; Wang, J.; Fan, C.; Cai, A. Fretting Wear Mechanical Analysis of Double Rough Surfaces Based on Energy Method. *Proc. Inst. Mech. Eng. Part J J. Eng. Tribol.* **2023**, *237*, 356–368. [CrossRef]
98. Zhang, T.; McHugh, P.E.; Leen, S.B. Computational study on the effect of contact geometry on fretting behaviour. *Wear* **2011**, *271*, 1462–1480. [CrossRef]
99. Cai, M.; Zhang, P.; Xiong, Q.; Cai, Z.; Luo, S.; Gu, L.; Zeng, L. Finite element simulation of fretting wear behaviors under the ball-on-flat contact configuration. *Tribol. Int.* **2023**, *177*, 107930. [CrossRef]
100. Dong, P.; Yang, Z.; Na, L.; Xinggui, W. The wear life prediction method of gear system. *J. Harbin Inst. Technol.* **2012**, *44*, 29–33, 39.
101. Janakiraman, V.; Li, S.; Kahraman, A. An Investigation of the Impacts of Contact Parameters on Wear Coefficient. *J. Tribol.* **2014**, *136*, 31602. [CrossRef]
102. Osman, T.; Velex, P. Static and dynamic simulations of mild abrasive wear in wide-faced solid spur and helical gears. *Mech. Mach. Theory* **2010**, *45*, 911–924. [CrossRef]
103. Xue, X.; Huo, Q.; Hong, L. Fretting Wear-Fatigue Life Prediction for Aero-Engine's Involute Spline Couplings Based on Abaqus. *J. Aerospace Eng.* **2019**, *32*, 4019081. [CrossRef]

104. Zhang, B.; Liu, H.; Zhu, C.; Ge, Y. Simulation of the fatigue-wear coupling mechanism of an aviation gear. *Friction* **2021**, *9*, 1616–1634. [CrossRef]
105. Feng, K.; Borghesani, P.; Smith, W.A.; Randall, R.B.; Chin, Z.Y.; Ren, J.; Peng, Z. Vibration-based updating of wear prediction for spur gears. *Wear* **2019**, *426–427*, 1410–1415. [CrossRef]
106. Liu, X.; Yang, Y.; Zhang, J. Investigation on coupling effects between surface wear and dynamics in a spur gear system. *Tribol. Int.* **2016**, *101*, 383–394. [CrossRef]
107. Changjiang, Z.; Yuying, L.; Hongbing, W.; Xu, H. Adhesive Wear Models for Helical Gears under Quasi-static and Dynamic Loads. *Chin. J. Mech. Eng.-En.* **2018**, *54*, 10–22. [CrossRef]
108. Sun, Y.; Li, Y.; Zhang, Q.; Qin, X.; Chen, K. Wear analysis and simulation of small module gear based on Archard model. *Eng. Fail. Anal.* **2023**, *144*, 106990. [CrossRef]
109. Chao, L.; Peitang, W.; Caichao, Z. Tooth contact analysis of helical beveloid gear with parallel axis. *J. Chongqing Univ. (Nat. Sci. Ed.)* **2012**, *35*, 1–6.
110. Jin, S.; Chen, H.; Li, Z.; Lai, X. A small displacement torsor model for 3D tolerance analysis of conical structures. *Proc. Inst. Mech. Eng. Part C J. Mech. Eng. Sci.* **2015**, *229*, 2514–2523. [CrossRef]
111. Xu, R.; Huang, K.; Guo, J.; Yang, L.; Qiu, M.; Ru, Y. Gear-tolerance optimization based on a response surface method. *Trans. Can. Soc. Mech. Eng.* **2018**, *42*, 309–322. [CrossRef]
112. Teoh, S.H.; Chan, W.H.; Thampuran, R. An elasto-plastic finite element model for polyethylene wear in total hip arthroplasty. *J. Biomech.* **2002**, *35*, 323–330. [CrossRef] [PubMed]
113. Uddin, M.S.; Zhang, L.C. Predicting the wear of hard-on-hard hip joint prostheses. *Wear* **2013**, *301*, 192–200. [CrossRef]
114. Xiang, D.; Cui, Y.; Wan, Z.; Wang, S.; Peng, L.; Liao, Z.; Chen, C.; Liu, W. Study on swelling, compression property and degradation stability of PVA composite hydrogels for artificial nucleus pulposus. *J. Mech. Behav. Biomed.* **2022**, *136*, 105496. [CrossRef]
115. Xiang, D.; Tan, X.; Sui, X.; He, J.; Chen, C.; Hao, J.; Liao, Z.; Liu, W. Comparative study on microstructure, bio-tribological behavior and cytocompatibility of Cr-doped amorphous carbon films for Co-Cr-Mo artificial lumbar disc. *Tribol. Int.* **2021**, *155*, 106760. [CrossRef]
116. Innocenti, B.; Labey, L.; Kamali, A.; Pascale, W.; Pianigiani, S. Development and Validation of a Wear Model to Predict Polyethylene Wear in a Total Knee Arthroplasty: A Finite Element Analysis. *Lubricants* **2014**, *2*, 193–205. [CrossRef]
117. Zhang, J.; Chen, Z.; Gao, Y.; Zhang, X.; Guo, L.; Jin, Z. Computational Wear Prediction for Impact of Kinematics Boundary Conditions on Wear of Total Knee Replacement Using Two Cross-Shear Models. *J. Tribol.* **2019**, *141*, 111201. [CrossRef]
118. Kang, K.; Son, J.; Kim, H.; Baek, C.; Kwon, O.; Koh, Y. Wear predictions for UHMWPE material with various surface properties used on the femoral component in total knee arthroplasty: A computational simulation study. *J. Mater. Sci. Mater. Med.* **2017**, *28*, 105. [CrossRef]
119. Koh, Y.; Jung, K.; Hong, H.; Kim, K.; Kang, K. Optimal Design of Patient-Specific Total Knee Arthroplasty for Improvement in Wear Performance. *J. Clin. Med.* **2019**, *8*, 2023. [CrossRef]
120. Mohd Mukhtar, N.Q.; Shuib, S.; Anuar, M.A.; Mohd Miswan, M.F.; Mohd Anuar, M.A. Design Optimisation of Bi-Cruciate Retaining Total Knee Arthroplasty (TKA) Prosthesis via Taguchi Methods. *Mathematics* **2023**, *11*, 312. [CrossRef]
121. Koh, Y.; Park, K.; Lee, H.; Park, J.; Kang, K. Prediction of wear performance in femoral and tibial conformity in patient-specific cruciate-retaining total knee arthroplasty. *J. Orthop. Surg. Res.* **2020**, *15*, 24. [CrossRef]
122. Koh, Y.; Lee, J.; Kang, K. Prediction of Wear on Tibial Inserts Made of UHMWPE, PEEK, and CFR-PEEK in Total Knee Arthroplasty Using Finite-Element Analysis. *Lubricants* **2019**, *7*, 30. [CrossRef]
123. Huang, T.C.; Tsai, J.W.; Liao, K.C. Wear and leakage assessments of canted coil Spring-Energized polytetrafluoroethylene seals under Ultra-High cycle operations. *Eng. Fail. Anal.* **2022**, *135*, 106110. [CrossRef]
124. Cao, W.; Chang, Z.; Zhou, A.; Dou, X.; Gao, G.; Gong, J. Investigation into the Influence of Parallel Offset Wear on Stirling Engine Piston Rod Oil-Free Lubrication Seal. *Machines* **2022**, *10*, 350. [CrossRef]
125. Wang, H.; Liu, T.; Zhang, Y.; Zhu, Y.; Liu, F.; Wang, T. A Fully Coupled Tribocorrosion Simulation Method for Anchor Chain Considering Mechano-Electrochemical Interaction. *Lubricants* **2022**, *10*, 330. [CrossRef]
126. Li, R.; Sun, Y.; Yu, Y.; Tian, G. Finite Element Analysis for Tread Wear of Radial Tire. In Proceedings of the 2022 5th International Conference on Mechatronics, Robotics and Automation (ICMRA), Wuhan, China, 25–27 November 2022; pp. 101–106.
127. Dai, X.; Li, J. Simulation Analysis of Cam Wear in Shedding Mechanism of Loom. *J. Physics. Conf. Ser.* **2021**, *1995*, 12020. [CrossRef]
128. Jin, Y.; Zou, L.; Huang, J.; Jiang, X.; Guo, Z.; Xie, J.; Yuan, Z. Numerical research on ablation and wear of the artillery barrel based on UMESHMOTION user-defined subroutine. *Eng. Rep.* **2023**, *5*, e12575. [CrossRef]
129. Lu, L.; Xu, Y.; Li, M.; Xue, Q.; Zhang, M.; Liu, L.; Wu, Z. Analysis of fretting wear behavior of unloading valve of gasoline direct injection high pressure pump. *J. Zhejiang Univ.-Sci. A* **2022**, *23*, 314–328. [CrossRef]
130. Li, L.; Zhang, W.; Li, G.; Wang, J.; Li, L.; Xie, M. Simulation Study of Thermal-Mechanical Coupling Fretting Wear of Ti-6Al-4V Alloy. *Appl. Sci.* **2022**, *12*, 7400. [CrossRef]

Disclaimer/Publisher's Note: The statements, opinions and data contained in all publications are solely those of the individual author(s) and contributor(s) and not of MDPI and/or the editor(s). MDPI and/or the editor(s) disclaim responsibility for any injury to people or property resulting from any ideas, methods, instructions or products referred to in the content.

MDPI AG
Grosspeteranlage 5
4052 Basel
Switzerland
Tel.: +41 61 683 77 34

Lubricants Editorial Office
E-mail: lubricants@mdpi.com
www.mdpi.com/journal/lubricants

Disclaimer/Publisher's Note: The title and front matter of this reprint are at the discretion of the Guest Editor. The publisher is not responsible for their content or any associated concerns. The statements, opinions and data contained in all individual articles are solely those of the individual Editor and contributors and not of MDPI. MDPI disclaims responsibility for any injury to people or property resulting from any ideas, methods, instructions or products referred to in the content.

www.ingramcontent.com/pod-product-compliance
Lightning Source LLC
LaVergne TN
LVHW072330090526
838202LV00019B/2385